Other Titles in This Series

168 **V. V. Kozlov, Editor,** Dynamical Systems in Classical Mechanics

167 **V. V. Lychagin, Editor,** The Interplay between Differential Geometry and Differential Equations

166 **O. A. Ladyzhenskaya, Editor,** Proceedings of the St. Petersburg Mathematical Society, Volume III

165 **Yu. Ilyashenko and S. Yakovenko, Editors,** Concerning the Hilbert 16th Problem

164 **N. N. Uraltseva, Editor,** Nonlinear Evolution Equations

163 **L. A. Bokut′, M. Hazewinkel, and Yu. G. Reshetnyak, Editors,** Third Siberian School "Algebra and Analysis"

162 **S. G. Gindikin, Editor,** Applied Problems of Radon Transform

161 **Katsumi Nomizu, Editor,** Selected Papers on Analysis, Probability, and Statistics

160 **K. Nomizu, Editor,** Selected Papers on Number Theory, Algebraic Geometry, and Differential Geometry

159 **O. A. Ladyzhenskaya, Editor,** Proceedings of the St. Petersburg Mathematical Society, Volume II

158 **A. K. Kelmans, Editor,** Selected Topics in Discrete Mathematics: Proceedings of the Moscow Discrete Mathematics Seminar, 1972–1990

157 **M. Sh. Birman, Editor,** Wave Propagation. Scattering Theory

156 **V. N. Gerasimov, N. G. Nesterenko, and A. I. Valitskas,** Three Papers on Algebras and Their Representations

155 **O. A. Ladyzhenskaya and A. M. Vershik, Editors,** Proceedings of the St. Petersburg Mathematical Society, Volume I

154 **V. A. Artamonov et al.,** Selected Papers in K-Theory

153 **S. G. Gindikin, Editor,** Singularity Theory and Some Problems of Functional Analysis

152 **H. Draškovičová et al.,** Ordered Sets and Lattices II

151 **I. A. Aleksandrov, L. A. Bokut′, and Yu. G. Reshetnyak, Editors,** Second Siberian Winter School "Algebra and Analysis"

150 **S. G. Gindikin, Editor,** Spectral Theory of Operators

149 **V. S. Afraĭmovich et al.,** Thirteen Papers in Algebra, Functional Analysis, Topology, and Probability, Translated from the Russian

148 **A. D. Aleksandrov, O. V. Belegradek, L. A. Bokut′, and Yu. L. Ershov, Editors,** First Siberian Winter School "Algebra and Analysis"

147 **I. G. Bashmakova et al.,** Nine Papers from the International Congress of Mathematicians, 1986

146 **L. A. Aĭzenberg et al.,** Fifteen Papers in Complex Analysis

145 **S. G. Dalalyan et al.,** Eight Papers Translated from the Russian

144 **S. D. Berman et al.,** Thirteen Papers Translated from the Russian

143 **V. A. Belonogov et al.,** Eight Papers Translated from the Russian

142 **M. B. Abalovich et al.,** Ten Papers Translated from the Russian

141 **H. Draškovičová et al.,** Ordered Sets and Lattices

140 **V. I. Bernik et al.,** Eleven Papers Translated from the Russian

139 **A. Ya. Aĭzenshtat et al.,** Nineteen Papers on Algebraic Semigroups

138 **I. V. Kovalishina and V. P. Potapov,** Seven Papers Translated from the Russian

137 **V. I. Arnol′d et al.,** Fourteen Papers Translated from the Russian

136 **L. A. Aksent′ev et al.,** Fourteen Papers Translated from the Russian

135 **S. N. Artemov et al.,** Six Papers in Logic

134 **A. Ya. Aĭzenshtat et al.,** Fourteen Papers Translated from the Russian

133 **R. R. Suncheleev et al.,** Thirteen Papers in Analysis

132 **I. G. Dmitriev et al.,** Thirteen Papers in Algebra

131 **V. A. Zmorovich et al.,** Ten Papers in Analysis

130 **M. M. Lavrent′ev, K. G. Reznitskaya, and V. G. Yakhno,** One-dimensional Inverse Problems of Mathematical Physics

(Continued in the back of this publication)

Dynamical Systems in Classical Mechanics

American Mathematical Society

TRANSLATIONS

Series 2 • Volume 168

Advances in the Mathematical Sciences — 25

(*Formerly Advances in Soviet Mathematics*)

Dynamical Systems in Classical Mechanics

V. V. Kozlov
Editor

American Mathematical Society
Providence, Rhode Island

Translation edited by A. A. SOSSINSKY

1991 *Mathematics Subject Classification.* Primary 34–XX, 58Fxx, 70–XX, 78–XX; Secondary 14E15, 35Qxx, 76–XX.

ABSTRACT. This is a collection of papers devoted to various mathematical problems of classical mechanics. It includes papers on the existence of invariant tori of Hamiltonian systems, dynamics of multidimensional rigid body and multidimensional generalizations of certain integrable systems, analysis of the stability of the equilibrium points, and other topics. The last paper in the collection presents some unsolved problems related to dynamical systems approach to classical mechanics.

The book is of interest to graduate students and researchers working in ordinary and partical differential equations, dynamical systems, and mechanics.

Library of Congress Card Number 91-640741
ISBN 0-8218-0427-8
ISSN 0065-9290

⊗ The paper used in this book is acid-free and falls within the guidelines
established to ensure permanence and durability.
♻ Printed on recycled paper.
This volume was printed from files prepared by the authors using
$\mathcal{A}_{\mathcal{M}}\mathcal{S}$-TeX, the American Mathematical Society's TeX macro system.

10 9 8 7 6 5 4 3 2 1 00 99 98 97 96 95

Contents

Introduction

This is a collection of papers devoted to various mathematical problems of classical dynamics. The title of the collection coincides with the topic of a seminar at the Department of Mathematics and Mechanics of Moscow Lomonosov University. Nearly all the results in this collection were first presented as talks at the seminar. Looking through this collection enables the reader to get acquainted with work done in the field of dynamical systems of classical mechanics at the Department of Mathematics and Mechanics.

Let me briefly introduce the papers of the collection. I start with the paper of V. P. Palamodov, who is well known for his work in analysis. In his paper the classical problem on the inversion of the famous Lagrange–Dirichlet theorem on the stability of equilibrium is solved in the case of analytic potential. This problem was posed by A. M. Lyapunov in 1892. The Lagrange–Dirichlet theorem guarantees stability of the equilibrium position if it is a point of strict local minimum of potential energy. The problem is: is this simple condition necessary for the stability of the equilibrium? Painlevé found an example (refined later by Wintner) where the potential energy does not have a local minimum, but the equilibrium is stable. However, in this example the potential is not analytic, though it is infinitely differentiable. Lyapunov posed the problem of inversion of the Lagrange–Dirichlet theorem in the analytic case. A lot of publications are dedicated to this problem (some of them erroneously announced a solution of the problem). I point out two of these publications. In 1977 Palamodov solved the Lyapunov problem for systems with two degrees of freedom. In 1982 Palamodov and myself studied the case when the Taylor series of the potential energy begins with a homogeneous form of order $m \geq 3$ having no minimum at the equilibrium point. The case $m = 2$ was investigated by Lyapunov. Incidentally, this result implies instability of equilibrium in a harmonic potential field. As a consequence, we have obtained Earnshow's theorem on the instability of a system of charged particles in any electrostatic field which was stated in 1839 for nondegenerate equilibria.

The contribution of S. Bolotin concerns the problem of the existence of doubly asymptotic (homoclinic) trajectories to invariant tori and invariant sets of Hamiltonian systems. The theme goes back to Poincaré, who investigated homoclinic trajectories of periodic orbits and discovered the splitting phenomenon of asymptotic surfaces, implying chaotic behavior of dynamical systems. Poincaré studied the problem of the existence of homoclinics by using perturbation methods. Bolotin studies homoclinic trajectories to hyperbolic invariant tori of perturbed Hamiltonian systems by using classical perturbation techniques, results of KAM-theory, global variational and topological methods. The idea is to perform one step of the classical Delaunay method. Then the new Hamiltonian is close to the Hamiltonian of a certain natural reversible system. This allows to reduce the problem of homoclinics for hyperbolic tori of the perturbed system to a Lagrangian variational problem with regular Lagrange function. In the case when KAM-theory fails, so that there are no hyperbolic invariant tori of the perturbed system, results of J. Mather (1991) imply the existence of certain unstable invariant sets. The existence of semihomoclinic trajectories to these sets is proved.

The field of work of Bolotin is the application of variational methods to the existence of periodic and homoclinic solutions and the integrability problem of Hamiltonian systems. For natural Hamiltonian systems, he proved the existence of periodic brake orbits with end points on the boundary of the region of possible motion (Hill region) when the boundary contains no equilibrium points (1978). H. Seifert proved this result in 1948 for an n-dimensional disk. In the same paper Bolotin obtained the existence of homoclinics to equilibrium points. One more of his results: the problem of $n \geq 3$ gravitating centers is nonintegrable (it has no analytic integrals independent of the energy integral). The cases $n = 1$ and $n = 3$ correspond to integrable problems of Kepler and Euler.

The following author is D. Treshchev. He obtained a number of results in perturbation theory of Hamiltonian systems and in dynamics of systems with impacts. I especially note his research on the mechanism of destruction of invariant resonance tori in Hamiltonian systems close to integrable ones. Our joint monograph "Billiards" was published by the American Mathematical Society.

Treshchev has contributed two papers in which he developed a new averaging method which probably will be useful in various problems concerning exponentially small effects.

In the first paper, the reducibility problem for ordinary linear differential equations with quasi-periodic coefficients is considered. The classical Floquet–Lyapunov theorem solves the reducibility problem in the periodic case. If a system has more than one frequency, the reducibility problem becomes much more complicated. In particular, in this case there are no analogs of the Floquet–Lyapunov results. Moreover, there are no general results on reducibility of linear systems close to systems with constant coefficients. Treshchev considers the case when the matrix of the system is small and shows that the system can be reduced to one with exponentially small nonconstant part.

Treshchev's second paper deals with the reducibility problem for the one-dimensional Schrödinger equation with quasi-periodic potential. It turns out that for large values of energy, the nonconstant part of the potential can be made exponentially small by a regular change of variables. The irremovable nonconstant terms in the potential are estimated in terms of its complex singularities. This result may be useful for studying the spectrum of Schrödinger operators with quasi-periodic potentials.

My joint paper with Yuriĭ Fedorov, who specializes in studying completely integrable Hamiltonian systems and systems with invariant measure, is devoted to the dynamics of a multi-dimensional rigid body. It was generally accepted that the idea appeared 20–25 years ago after it was noticed that the geodesic equations for left-invariant metrics on Lie groups have the form of the well-known Euler top problem. As a matter of fact, this observation had already been made by H. Poincaré in 1901. Moreover, even earlier, the German mathematician F. Frahm had derived the equations of motion of a multi-dimensional top and found a series of quadratic integrals for the problem. In particular, he showed that these integrals are sufficient for integrability of the equations in the four-dimensional case. Later, F. Schottky gave a scheme of a solution for this case. He also noticed a remarkable connection (in fact, an isomorphism) between the integrable cases found by Frahm and Clebsch's second integrable case of the Kirchhoff equations describing the motion of a rigid body in an ideal fluid. (Almost a century later these results were again obtained by A. Mishchenko, A. Bobenko, and others.) In addition, we present a geometrical model of the motion of an n-dimensional top, which generalizes the well-known Poinsot interpretation for $n = 3$: the inertia ellipsoid attached to the top rolls without sliding on a certain plane.

We also consider multi-dimensional generalizations of some well-known nonholonomic systems arising in classical dynamics. It turns out that these are not (at least a priori) Hamiltonian but possess an invariant measure. The latter means that the volume of an arbitrary domain in the phase space, while transferred by the phase flow of the system, is not necessarily constant, but not less then a certain fixed quantity. It is known that the qualitative behavior of many such systems in no way differs from that of integrable Hamiltonian systems. Therefore, it is among systems with an invariant measure that one should seek the first candidates to integrable ones.

In addition, we study the motion of a multi-dimensional top with left-invariant constrains. Such dynamical systems, although integrable, often do not possess an invariant measure, and their behavior is essentially different from that of Hamiltonian systems. It is worth mentioning that in these problems we may have dynamical systems on tori without a quasi-periodic motion (as in the well-known Kolmogorov paper of 1953).

A discussion of multi-dimensional generalizations of some integrable problems is continued in Fedorov's other paper. The integrable generalizations of the classical Euler and Clebsch systems found by Frahm, Manakov, and Perelomov are known to admit a Lax pair (or L–A pair) with an additional (spectral) parameter defined on the compactified complex plane. It turns out that these systems admit another Lax pair with a spectral parameter ranging over a certain unramified covering of a hyperelliptic curve ("hyperelliptic" Lax pair). Using the new L–A pair, the author discovered a remarkable connection between the multi-dimensional integrable systems mentioned above and the varieties of common tangent linear spaces of confocal quadrics in a projective space. This connection is a generalization of the nice Moser isomorphism between the well-known Neumann system and the variety of common tangent lines of confocal quadrics. It follows that several integrable systems known as essentially different have, in fact, the same origin and algebro-geometric structure.

The paper of N. Moshchevitin is devoted to integrals of quasi-periodic functions. This problem is closely connected with the small divisors problem and with Diophantine approximation theory. Everybody knows that the integral of a periodic function with zero average (as a function of the upper limit x) is also a periodic function. Poincaré noted that the integral of a quasi-periodic function with zero average may be not quasi-periodic: he found an example of a function with two frequencies and zero average for which the integral is unbounded (tends to infinity like x^α, $0 < \alpha < 1$). In the Poincaré example the function is continuous, but not differentiable. I showed that in the case of two frequencies and class of smoothness C^2 the integral in question has the return property (although it can be unbounded). Moshchevitin studied the return property for the case of $n \geq 3$ rationally incommensurable frequencies. He established the return property for certain Diophantine properties of the frequencies. These conditions are not trivial since the integral can be unbounded. Moshchevitin's paper is a survey of results on the return problem.

Recently, Moshchevitin obtained a still more interesting and surprising result: for $n \geq 3$ even in the class of analytic functions the integral of a quasi-periodic function can have no uniform (with respect to the initial phases) return property.

The paper of A. Markeev concerns orbital stability of periodic trajectories of billiard systems. The author considers the motion of a small ball between two walls (or a ray of light between two mirrors). This problem is related to short wave diffraction theory. Earlier physicists obtained stability conditions for two-gon periodic trajectories. It turns out that these conditions do not guarantee Lyapunov orbital stability of this motion. Resonances of low order between multipliers of the trajectories obstruct orbital stability. Markeev studied the stability problem in detail for two-gon periodic trajectories using normal forms and results of KAM-theory.

Finally, a few words about my two papers. The first one is devoted to the hydrodynamical description of Hamiltonian systems. From the philosophical point of view, it goes back to the Descartes vortex theory of matter. Leibniz, Huygens, J. Bernoulli, and other famous authors used Descartes' idea in mathematical investigations. The essence of our method can be illustrated by the example of the problem on geodesics of a left-invariant metric on an n-dimensional Lie group. Mathematicians who investigated this problem concentrated their attention on the Euler dynamical equations defined on the Lie algebra of this group. We approach this problem in another way. The phase flow of a right-invariant vector field gives us a set of left shifts on the group. These shifts do not change the metric. According to the Noether theorem, every such field corresponds to an integral which is linear in momenta. It is clear that there are exactly n independent Noether integrals. Fixing their constants, we obtain an n-dimensional invariant surface which projects one-to-one onto the Lie group. Restricting $2n$ Hamiltonian equations to the invariant surface and projecting these equations to the group, we obtain a dynamical system on the Lie group. This system has all characteristic properties of stationary flows of an ideal fluid.

In the concluding paper of the collection, I state some unsolved problems we discussed at the seminar on dynamical systems of classical mechanics. The scope of the problems are quite diverse. I would like to draw the attention of young mathematicians to these problems. I shall be happy if it will be possible to obtain new interesting results in these directions.

Moscow, March 1994 Valery V. Kozlov

Amer. Math. Soc. Transl.
(2) Vol. **168**, 1995

Stability of Motion and Algebraic Geometry

V. P. PALAMODOV

1. Introduction. Here we discuss the stability problem of an equilibrium for the natural mechanical system

$$(1.1)\qquad \frac{d}{dt}\left(\frac{\partial}{\partial x_1'}L(x, x')\right) - \frac{\partial}{\partial x_1}L(x, x') = 0, \qquad i = 1, \dots, n.$$

We present a constructive method to investigate the motion near an equilibrium point based on an application of algebraic geometry. In particular we get an inversion of the Lagrange–Dirichlet theorem for any critical point where the analytic potential function does not have a minimum.

The Lagrange function for the natural system has the form $L(x, x') \equiv K(x, x') - \Pi(x)$, where the kinetic energy function $K(x, x')$ is assumed to be a positive definite quadratic form in x'. J.-L. Lagrange claimed in his famous book [1] that any point x_0 which is a strict local minimum of the potential function $\Pi(x)$ is a stable equilibrium position for the system (1.1). Lagrange gave a sketch of a proof, which is in fact satisfactory if the potential is a quadratic form of $x - x_0$. A complete proof of the following result was given by G. Lejeune Dirichlet in 1846 [2].

LAGRANGE–DIRICHLET THEOREM. *Suppose that the function Π has a local strict minimum at a point x_0 of an open set $X \subset \mathbb{R}^n$ and $L \in C^1(X \times \mathbb{R}^n)$. Then for any neighborhood Y of x_0 there exists a neighborhood $Y' \subset Y$ and a number $\delta > 0$ such that any solution of (1.1) that starts from a point $x \in Y'$ with velocity x' such that $|x'| < \delta$ cannot leave Y.*

The conclusion means that the equilibrium position $x = x_0$ is stable according to the terminology of A. M. Lyapunov [3].

The system (1.1) is always conservative, which means that the total energy $E := K(x, x') + \Pi(x)$ is an integral of motion. Therefore, $\Pi(x)$ is bounded by a constant E that depends only on the initial data $(x(0), x'(0))$. On the other hand, $\Pi(x_0) \leqslant \Pi(x)$, because of the assumption of the theorem. This implies that the trajectory is captured in the preimage $\Pi^{-1}(I)$ of the interval $I = [\Pi(x_0), E]$. The set $\Pi^{-1}(I)$ tends to the equilibrium point x_0 if E tends to $\Pi(x_0)$. This completes the proof.

The *inversion problem* of Lagrange–Dirichlet is the following problem. Assume that x_0 is a critical point of Π which is not a strict minimum; is the equilibrium position $x = x_0$ unstable?

1991 *Mathematics Subject Classification.* Primary 73H10; Secondary 14E15.

The following simple example shows that this is not the case even if $n = 1$ and the potential is infinitely differentiable:

EXAMPLE 1 (Painlevé–Wintner). Let

$$K(x, x') = x'^2/2, \qquad \Pi(x) = \cos(x^{-1})\exp(-x^{-2}), \quad \Pi(0) = 0.$$

We have $\Pi \in C^\infty(\mathbb{R})$ and the graph of Π has a sequence of "pits" alternated with "hills", which concentrate to the critical point $x_0 = 0$. Any "hill" makes an obstruction for any motion with energy $E < h$, where h is the height of the "hill". Hence any motion with small energy is trapped in a small neighborhood of the origin, i.e., the equilibrium position x_0 is stable.

EXAMPLE 2 (M. Laloy [6]). This is a more sophisticated example for the case $n = 2$, where

$$K(x, x') = |x'|^2/2, \qquad \Pi = \Pi_1(x_1) + \Pi_2(x_2),$$
$$\Pi_2(x_2) = -(x_2^2 + \cos x_2^{-1})\exp(-x_2^{-1})$$

and Π_1 is the function from Example 1. Again we have $\Pi \in C^\infty(\mathbb{R}^2)$; the point $x_0 = (0, 0)$ is critical, but is not a minimum. This function makes no topological traps: the set $X_- = \{x \in \mathbb{R}^2 \mid \Pi(x) < \Pi(x_0) = 0\}$ contains two diagonals $x_1^2 = x_2^2$, therefore for any point $x \in X_-$ there is a curve $\gamma\{x' \mid \Pi(x') < 0\}$ that joins x to infinity. Nevertheless the equilibrium position x_0 is stable, since the projection of any trajectory on the x_1-axis is governed by the Lagrange equation of Example 1.

A. M. Lyapunov (1892) formulated the inversion problem in the following way: is the inverse to the Lagrange–Dirichlet theorem true for *analytic* Lagrange functions L? (cf. [3]). Note that the potential functions in the above examples are not analytic, nor does there exist a homeomorphism of a neighborhood of the critical point that makes it analytic.

In several papers a positive answer to the Lyapunov's question is given under some additional assumptions. In particular, Lyapunov [3] proved instability in the case when the second differential of Π at the critical point has no minimum. If Π has a strict maximum, instability was established by Lyapunov and Hagedorn [9]. For more details see the surveys [5, 20].

2. THEOREM 2.1 (cf. [21]). *Consider an arbitrary natural system* (1.1), *where the potential function* Π *is analytic in an open set* $X \subset \mathbb{R}^n$. *For any point* $x_0 \in X$ *there exists a neighborhood* X_0 *of* x_0 *and a positive function* T *defined on the interval* $(-\infty, U(x_0))$ *such that any motion with energy* $E < \Pi(x_0)$ *cannot spend a time lapse greater than* $T(E)$ *inside* X_0.

We give a sketch of the proof after discussing several related problems.

Since natural Lagrange systems admits diffeomorphisms of the configuration space, the theorem is true as well for any function Π analytic up to a diffeomorphism of X of class C^2.

COROLLARY 2.2. *Suppose that the potential* Π *is* C^2*-equivalent to an analytic function. If a critical point* x_0 *belongs to the closure of the set* $X_- := \{x \in X_0 \mid \Pi(x) < \Pi(x_0)\}$, *then the equilibrium position* x_0 *is unstable.*

This implies the inversion of the Lagrange–Dirichlet theorem for any analytic potential Π, except for the case when Π has a nonstrict minimum at the equilibrium position. Note that we require no extra conditions on the kinetic energy function K and, on the other hand, we get more than just instability: any motion with a subcritical energy (i.e., $E < \Pi(x_0)$) leaves X_0.

A sufficient condition for instability may be given in metric terms.

PROPOSITION 2.3. *Suppose that* $\Pi \in C^q(X)$, $q \geqslant 2$, *and the inequality*

$$|\operatorname{grad}\Pi| \geqslant C|x - x_0|^{q-2} \tag{2.1}$$

holds near a point with some positive C. *Then the conclusion of Corollary* 2.2 *still holds.*

In fact under condition (2.1) there exists a diffeomorphism $A\colon X_0 \to X_1$ between neighborhoods of the critical point x_0 of class $C^1(X_0) \cap C^2(X_0 \setminus x_0)$ such that $U(A(x))$ is a polynomial. This can be proved by the methods of [8]. Then the system (1.1) is equivalent in $X_0 \setminus x_0$ to the Lagrange system with Lagrangian $L^*(y, y') := L(A(y), (\partial A(y)/\partial y)y')$. In particular, any trajectory that belongs to the set X_- satisfies this Lagrange system, hence we can apply Corollary 2.2.

Note that an inequality like (2.1) with some positive exponent is sufficient and necessary for a C^∞-function Π to be C^∞-equivalent to an analytic function, but no analytic function with a nonisolated critical point x_0 satisfies it.

COROLLARY 2.4. *If the point* x_0 *belongs to the closure of* X_-, *then there exists an asymptotic solution of the system to the point* x_0.

An asymptotic solution is a trajectory $x = x(t)$ that tends to the critical point x_0 as $t \to \infty$ (or to $-\infty$). Such an asymptotic solution can be found by compactness arguments.

The inverse assertion is obviously true: if there exists an asymptotic solution tending to a point x_0, then this point is an unstable equilibrium. In many cases an asymptotic solution can be constructed explicitly. The case of an analytic potential was considered in [13, 14], the case $\Pi \in C^q$ treated in [15–17, 19]. The condition imposed on the potential is the absence of minimum for the first nonvanishing differential of Π at x_0 (in [15] the degenerate second differential is admitted).

3. Open problems and a conjecture.

I. *Lyapunov's problem for potential functions having a nonstrict minimum.* Only the case $n = 2$ was treated. Laloy and Peiffer [7] proved that such a critical point for an analytic potential is an unstable equilibrium. The general case is still an open problem. The condition means that $\Pi(x) \geqslant \Pi(x_0)$ in a neighborhood of x_0 and $\Pi(x) = \Pi(x_0)$ on an analytic set Z of positive dimension. Hence any point of Z is critical for Π. This is in some sense a rare case. Indeed, consider the space A of germs of real analytic functions at the point x_0 and its subset M of germs Π having a set Z of critical points of positive dimension. It is easy to show that M belongs to an algebraic subset of A of infinite codimension. This means that there exists an infinite set of algebraically independent algebraic equations of the form $p(\pi) = 0$ that hold on M. Here $\pi = \{\pi_i = D^i\Pi(x_0)\,,\ i \in N^n\}$ is the set of Taylor coefficients of Π and p is a polynomial in a finite number of π_i's.

II. *What supercritical motions are stable*? Return to the situation of Theorem 2.1 and consider an arbitrary supercritical motion (i.e., a motion with energy $E > \Pi(x_0)$) starting at a point close to $\Pi(x_0)$. Then the trajectory either leaves a fixed neighborhood X_0 of x_0 or it is constrained in X_0 forever. It is easy to show that if x_0 is not a critical point of U, there exists a neighborhood Y of x_0 such that no motion is constrained in Y. Theorem 2.1 means that if x_0 is a critical point, there are no constrained subcritical motions in a small neighborhood of x_0. In the situation of Lagrange–Dirichlet theorem, any supercritical motion close to x_0 is constrained. On the contrary, if x_0 is not a local minimum of the potential, it is plausible that in some sense almost all supercritical trajectories leave any sufficiently small neighborhood X_0 of x_0.

However, sometimes there exist constrained supercritical motions. Suppose that there is a closed smooth subvariety $W \subset X$ that is integral for (1.1). The latter means that for any point $x \in W$ and vector $x' \in T_x(W)$ we have the inclusion $x'' \in T_x(W)$, for the second derivative x'' taken from (1.1). This condition may be written in the C^∞-case in the following way: take the Hamiltonian function H corresponding to L and functions $w_1\,, \dots\,, w_k$ that generate the ideal $I(W)$ in $C^\infty(X)$. Then for any i the function $[H\,, [H\,, w_i]]$ belongs to the ideal in $C^\infty(T^*(X))$ generated by the functions $w_j\,, [H\,, w_j]$, $j = 1\,, \dots\,, k$. It follows that any motion that starts at a point $x(0) \in W$ with velocity $x'(0) \in T_x(W)$ remains in W until it leaves X.

For example, if the Lagrangian L is invariant with respect to reflections in a hyperplane W, this hyperplane is integral for the system. Let x_0 belong to an integral variety W and the function $\Pi\,|\,W$ have a strict local minimum at x_0. Then for any neighborhood Y of x_0 an arbitrary motion in W that starts at a point $x(0)$ sufficiently close to x_0 with small velocity is constrained in Y, because of the Lagrange–Dirichlet theorem.

CONJECTURE. *Suppose that $L = K - \Pi$ is an analytic function for $x \in X_0$ and $x_0 \in X_0$ is a critical point of Π. Then there exists a neighborhood Y of x_0 and an integral variety W such that $x_0 \in W$, $\Pi\,|\,W$ has a strict local minimum at x_0 and any trajectory of* (1.1) *constrained in Y will belong to W.*

To put it differently, the conjecture means that there is no reason for the existence of trajectories constrained in a small neighborhood of a critical point, except for the Lagrange–Dirichlet theorem applied to an integral variety. Theorem 2.1 would be a corollary of the conjecture, and this is the only part of the conjecture that I can prove.

III. *Stability problem for the Lagrange system with gyroscopic forces.* This is the system (1.1) with Lagrangian

$$L(x\,, x') = K(x\,, x') + \Gamma(x\,, x') - \Pi(x)\,,$$

where $\Gamma(x\,, x')$ is a function which belongs to $C^1(X \times \mathbb{R}^n)$ and is linear in the x'-variables. This term is responsible for the gyroscopic force and has the components

$$\Gamma_i := \sum_j \left[\frac{\partial^2 \Gamma}{\partial x_i \partial x_j'} - \frac{\partial^2 \Gamma}{\partial x_i' \partial x_j} \right] x_j', \qquad i = 1, \dots, n,$$

which appear at the right-hand sides of (1.1). Taking the linear combination of these equations with the coefficients x_i', we get

$$0 = \frac{d}{dt} E + \sum x_i' \Gamma_i \equiv \frac{d}{dt} E,$$

hence this system is still conservative.

There are several papers concerning the stability problem for systems with gyroscopical forces [10, 17] (see [17] for more references). In [17] an asymptotic solution to x_0 was found for the system with a gyroscopical force Γ that may be nondegenerate at x_0. We still need a better understanding of the problem.

4. A reduction. Now we describe the reduction of Theorem 2.1 to a geometric construction related to the potential function Π. We may suppose that $\Pi(x_0) = 0$. For any $x \in X$ consider the symmetric bilinear form $(y, z) \mapsto B(x; y, z)$ on $T(X)$ such that $B(x; y, y) = 2K(x, y)$. In particular $\langle y, z \rangle := B(x_0; y, z)$ is the scalar product on $T(X)$. Fix coordinates $x_1 \dots, x_n$ as in (1.1).

LEMMA 4.1. *There exists a tangent field V in a neighborhood X_0 of x_0 satisfying the following conditions:*

1. $V \in C^1(X_0)$, $V(x_0) = 0$;
2. $\langle dV(x)\xi, \xi \rangle \geqslant \langle \xi, \xi \rangle$, $\xi \in T_x(X)$, $x \in X_0$, *where $\eta = dV\xi$ is the tangent vector with coordinates $\eta(x_i) = \xi(V(x_i))$;*
3. $V(\Pi) = P\Pi$, *where P is positive and continuous in $X_0 \setminus Z$, $Z := \Pi^{-1}(0)$.*

A lemma of this kind was used by Chetaev [4].

PROOF OF THEOREM 2.1. Take the linear combination of equations (1.1) with the coefficients $V_i := V(x_i)$, $i = 1, \dots, n$:

$$\begin{aligned} 0 &= \sum V_i x_j' \frac{\partial^2 K(x, x')}{\partial x_i' \partial x_j} + \sum V_i x_j'' \frac{\partial^2 K(x, x')}{\partial x_i' \partial x_j'} \\ &\quad - \sum V_i \frac{\partial K(x, x')}{\partial x_i} + \sum V_i \frac{\partial \Pi}{\partial x_i} \\ &= \sum x_j' \frac{\partial B(x; x', V)}{\partial x_j} + B(x; x'', V) \\ &\quad - V(B(x; x', x')) + V(\Pi). \end{aligned} \tag{4.1}$$

Consider the following Lyapunov function

$$F(t) = B(x; x', V(x)) - \varepsilon B(x; x', z),$$

where the vector $z \in T_x(X)$ is defined by the equation $\langle y, z \rangle = y(\Pi)$, and ε is a small positive number. Calculate its time derivative, using (4.1):

$$\frac{d}{dt} B(x; x', V) = -V(\Pi) + V(B(x; x', x')) + B(x; x', dVx'). \tag{4.2}$$

We have

$$B(x;x',dVx') = B(0;x',dVx') + O(r)\langle x',x'\rangle$$

by Lemma 4.1.1, where $r^2 = \sum x_i^2$ and we assume that all x_i vanish at x_0. Hence the third term of the right-hand side is $\geqslant [1+O(r)]\langle x',x'\rangle$, because of Lemma 4.1.2. We have $\Pi \leqslant E < 0$ on the trajectory, since E is subcritical, and $V(\Pi) = P\Pi$ by Lemma 4.1.3. Therefore the first term of (4.2) is nonnegative. The second term is equal to $O(r)\langle x',x'\rangle$, by Lemma 4.1.1. Hence we conclude that

$$\frac{d}{dt}B(x;x',V) \geqslant [1+O(r)]\langle x',x'\rangle .$$

Then we have

$$\frac{d}{dt}B(x;x',z) = \sum x'_j \frac{\partial B(x;x',z)}{\partial x_j} + B(x;x'',z) + B(x;x',z').$$

Now we use (1.1) once more, taking the linear combination with coefficients $z_1, \dots, z_n$, where $\sum z_i\, \partial/\partial x_i = z$:

$$0 = \sum x'_j \frac{\partial B(x;x',z)}{\partial x_j} + B(x;x'',z) - z(B(x;x',x')) + \langle z,z\rangle .$$

Therefore

$$\frac{d}{dt}B(x;x',z) = z(B(x;x',x')) - \langle z,z\rangle + B(x;x',z').$$

The modulus of the first term of the right-hand side is bounded by $\mathrm{const}\langle x',x'\rangle$, since the coefficients of the form B belong to $C^1(X)$. The third term is estimated in a similar way, since the second derivatives of Π are continuous. Therefore

$$\frac{d}{dt}F \geqslant [1 - O(r) - \varepsilon C]\langle x',x'\rangle + \varepsilon\langle z,z\rangle ,$$

hence for small r and ε we finally get

$$\frac{d}{dt}F \geqslant c[\langle x',x'\rangle + \langle z,z\rangle] \tag{4.3}$$

for some positive c. The quantity $\langle z,z\rangle$ is in fact the square of the modulus of the gradient of Π calculated in the metric dual to B. It is positive on the set where $\Pi \neq 0$ and x is close to x_0, since Π is analytic. Therefore $\langle z,z\rangle \geqslant \delta > 0$, where δ depends only on E, and (4.3) implies the inequality $F(T) - F(0) \geqslant c\delta T$. Thus the theorem follows.

REMARK 1. These calculations gave a method for finding an upper estimate for the time T when the point $x(T)$ leaves X_0, in terms of the discrepancy of the initial data $(x(0), x'(0))$ from the equilibrium position (cf. [11]).

REMARK 2. The assertions of Lemma 4.1 are stable under C^1-diffeomorphisms of X_0. In fact we need such a field to be defined only in the closure of X_-.

EXAMPLES. There are several cases when such a field V can be found immediately:

1. If x_0 is not a critical point of Π, one can choose local coordinates in such a way that $\Pi(x) = x_1$. Then we can take

$$V = \sum x_i \frac{\partial}{\partial x_i}.$$

Here $V(\Pi) = \Pi$ and the operator dV is the identity.

2. Π is a homogeneous polynomial on $x_1, \dots, x_n$ of degree m. Then we can take the same field, since $V(\Pi) = m\Pi$. Note that instability was proved in this case by Chetaev (1952) (see [4]).

3. If $n = 1$, there exists an analytic coordinate y such that $y(x_0) = 0$ and $\Pi(x(y)) = cy^m + \text{const}$ for some natural m. Hence this case is contained in 2.

4. Π is a weighted-homogeneous polynomial with some positive weights $\alpha_1, \dots, \alpha_n$. Then we may set $V = \sum \alpha_i x_i\, \partial/\partial x_i$.

5. The polynomial $\Pi(x_1, x_2) := x_1^4 + cx_1^2x_2^2 + x_2^5$ is not C^1-equivalent to any weighted-homogeneous polynomial in a neighborhood of the origin 0. The set X_- is cusp-like with the origin on its edge. Let us describe a construction of a field V defined on the closure of X_-. For any small $r > 0$, we consider the arc $S_r \cap \partial X_-$ of the circle S_r of radius r centered at the origin and define V at the ends of this arc to be tangent to ∂X_- and directed to the opposite side of the origin. This condition will ensure property 4.1.3. Inside the arc the field V must "swing" between these extreme positions, but the swinging is subjected to the following condition: when we move along the arc in one direction, the field V may swing to the same direction as fast as we want; but if V swings in the opposite direction, it may do it only slowly. To be exact, the velocity V' of the opposite swinging should equal $o(1)$ as $r \to 0$. This geometrical restriction is necessary to fulfill 4.1.2. In this special case, it is easy to see that such a construction is possible.

6. It was shown in [11] for the case $n = 2$ that such a field V exists for arbitrary analytic potential. This construction is based on the same ideas as in item 5 above.

5. Basic ideas of the proof of Lemma 4.1. The main idea is to use a *resolution of singularities of the zero set* of the potential. Recall some notions from the category of real analytic manifolds and mappings.

DEFINITION. Let X be a (real analytic) manifold of dimension n, and let Z be a submanifold of codimension $m > 1$. The *monoidal transformation of X with center Z* (or blowing up of Z) is the couple (Y, π), where Y is a manifold and $\pi\colon Y \to X$ is a mapping of manifolds that can be constructed in the following way. Take some real analytic functions $x_1, \dots, x_m$ on Y, which vanish on Z and generate the ideal $I(Z)$ of all functions on Y that vanish on Z. Consider the submanifold $Y \subset X \times \mathbb{R}P^{m-1}$, defined by equations

$$\xi_i x_j = \xi_j x_i, \qquad i, j = 1, \dots, m, \tag{5.1}$$

where $\xi_1, \dots, \xi_m$ are homogeneous coordinates on $\mathbb{R}P^{m-1}$. The mapping π is the restriction of the canonical projection $X \times \mathbb{R}P^{m-1} \to X$ on Y. This mapping induces an isomorphism $\tilde{\pi}\colon Y \setminus \pi^{-1}(Z) \to X \setminus Z$; the submanifold $W := \pi^{-1}(Z)$ is a locally trivial fibration over Z with fiber $\mathbb{R}P^{m-1}$. We may think of W as a projectivized normal fibration $PN(Z)$ to Z. In fact the result of the construction does not essentially depend on the choice of the generators: taking any other set of generators $x'_1, \dots, x'_m$, we get a couple (Y', π') and an automorphism φ of $X \times \mathbb{R}P^{m-1}$ such that $\varphi(Y) = Y'$ and $\pi'\varphi = \pi$. This implies that for any point $z \in Z$ we may use any set of local generators of the ideal $I(Z)$: we implement the monoidal transformation locally and then glue the constructions together.

The manifold Y may be built in another way: take a small tubular neighborhood $T(Z)$ of Z and cut it away from X. The boundary of the remaining manifold $X \setminus T(Z)$ is a fibration over Z with fiber S^{m-1}. For any fiber we glue together any couple of opposite points of the sphere. The space we get is just Y and the image of the spheric fibration is W. This is a nonsingular hypersurface in Y with only one side. This means that if we move a normal vector v to W continuously along a closed curve $\gamma \in \pi^{-1}(Z)$, we can get the normal vector v' opposite to v at the end of γ.

The purpose of the monoidal transformation is to simplify the analytic set $Z := \Pi^{-1}(0) \cap \mathbb{R}^n$ or, to be exact, to simplify the *sheaf* $I(\Pi)$ of local ideals generated by the function Π. The simplest form it may have is that of a sheaf locally generated by the product of some powers of local coordinates. There is a very important theorem that implies that for any analytic function this sheaf may be completely simplified after a finite number of monoidal transformations. This theorem in its full generality is due to H. Hironaka [18]. According to Hironaka's theorem, there exists a sequence of monoidal transformations

$$Y := X_{r+1} \xrightarrow{\pi_r} X_r \xrightarrow{\pi_{r-1}} \cdots \longrightarrow X_j \xrightarrow{\pi_{j-1}} X_{j-1} \longrightarrow \cdots \longrightarrow X_1 \xrightarrow{\pi_0} X_0 = X, \tag{5.2}$$

which provides such a resolution. Here each π_j is a monoidal transformation whose center $Z_j \subset X_j$, $\operatorname{codim} Z_j > 1$, is an irreducible submanifold properly projecting into $\operatorname{sing} Z$. Moreover, for the first step of the resolution we may take the monoidal transformation π_0 with center $Z_0 = \{x_0\}$ and then apply [18, Basic theorem II′] to the analytic hypersurface $E_1 := \Pi_1^{-1}(0)$ and to the sheaf of ideals $I(\Pi_1)$ generated by the function $\Pi_1 := \Pi \cdot \pi_0$. This theorem implies that there exists a sequence of monoidal transformation $\pi_1, \dots, \pi_r$ with the following properties:

(5.3) for any $i > 0$ the reduced space E_i has only normal crossings with Z_i;

(5.4) E_{r+1} has only normal crossings, where E_i is defined inductively as follows: $E_{i+1} = \operatorname{red}\{\pi_i^{-1}(E_i) \cup B_i\}$, where $B_i := \pi_i^{-1}(Z_i)$.

Definition. We say that E_i has only *normal crossings with* a subspace Z, if for any point $z \in Z$ there exists a local coordinate system $x_1, \dots, x_n$ such that any irreducible component B of E_i containing z is given by the equation $x_1 = 0$ in a neighborhood of z and Z is given by the system of equations $x_1 = \cdots = x_m = 0$ with some $m \leqslant n$. If $Z = X_i$, one says that E_i has only *normal crossings*. The condition (5.4) implies that the set of roots of the function

$$\Pi_{r+1} := \Pi \cdot \pi_0 \cdot \ldots \cdot \pi_r \colon X_{r+1} \to \mathbb{R}$$

is the union of nonsingular irreducible hypersurfaces $Y_1, \dots, Y_m$ which have only normal crossings. Moreover, for any point $y \in \bigcup Y_j$ this function may be written in the following form in a neighborhood of y:

$$\Pi_{r+1} = w_1^{d_1} \cdot \ldots \cdot w_n^{d_n}, \tag{5.5}$$

where $w_1, \ldots, w_n$ are local coordinates and $d_1, \ldots, d_n$ are certain nonnegative integers. Hence for any k such that $d_k > 0$ the hypersurface $w_k = 0$ is the germ at y of a hypersurface Y_j and the number d_k depends only on j: $d_k = d(Y_j)$.

For any $j > i$, denote $\pi_i^j := \pi_i \cdot \ldots \cdot \pi_{j-1}\colon X_j \to X_i$. Note that any irreducible component E_{ij} of E_j is in fact the "proper" preimage of B_i for some $i < j$. This means that E_{ij} is the closure of $(\pi_i^j)^{-1}(B_i \setminus Z')$, where Z' is the union of all centers Z_k that are projected into B_i under (5.2).

6. Logarithmic fields. Again let (Y, π) be a monoidal transformation of a variety X with center Z and $B = \pi^{-1}(Z)$ (blowing up of Z). We call a tangent field T *logarithmic* on (Y, π), if it is defined and smooth near B and vanishes on B. It generates a linear operator in the ideal $I(B)$ in $C^\infty(Y)$ of all smooth functions that vanish on B. The ideal $I(B)^2$ is invariant for this operator, since by the Leibniz formula,

$$T(a \cdot b) = T(a) \cdot b + a \cdot T(b).$$

Therefore T induces an operator in the space $I(B)/I(B)^2$ (which is the space of sections of the conormal fiber bundle). In fact the latter coincides with the multiplication by a smooth function. We call this function *coresidue of* T and denote it $\operatorname{cres}(T, B)$.

EXAMPLES. Suppose that B is given locally by the equation $x_1 = 0$ in the coordinate space $\mathbb{R}^n$. Then the field $T_1 := x_i\,\partial/\partial x_k$ is logarithmic for any $k = 1, 2, \ldots$ and $\operatorname{cres}(T_1, B) = 1$, but $\operatorname{cres}(T_k, B) = 0$ for $k > 1$.

The benefit of logarithmic fields is demonstrated by the following

PROPOSITION 6.1. *If $X = \mathbb{R}^n$, $Z = \{0\}$, and T is a smooth logarithmic field on Y with* $\operatorname{cres}(T) = 1$, *then its image $T_* := \pi_*(T)$ is a field of class $C^i(X)$ that vanishes on Z and its differential has the asymptotics*

$$\langle dT_*(x)\xi, \xi\rangle = [1 + o(1)]\langle \xi, \xi\rangle, \tag{6.1}$$

as $x \to Z$ for an arbitrary smooth metric $\langle\cdot, \cdot\rangle$ on X.

PROOF. Choose local coordinates $x_1, \ldots, x_n$ near 0 and set $y_i := x_i \cdot \pi$. For any i, we have $T_*(x_i) \cdot \pi = T(y_i)$, by the definition of T and by the formula $T(y_i) = y_i + w_i$, where $w_i \in I(B)^2$, because of the condition on T. Therefore $T_*(x_i) = x_i + z_i$, where $z_i \cdot \pi = w_i$, hence

$$T_* = \sum x_i \frac{\partial}{\partial x_i} + \sum z_i \frac{\partial}{\partial x_i}$$

and T_* vanishes on Z. The first term defines the identity operator in the tangent space $T_0(x)$ at the origin. Let us show that the operator generated by the second term is $o(1)$. It is sufficient to check that $\xi(z_i)$ is continuous and vanishes on z for any i and $\xi \in T_0(X)$. Since $w_i \in I(B)^2$, we may assume that $w_i = u \cdot v$, where $u, v \in I(B)$, and so

$$\xi(z_i) \cdot \pi = v \cdot \xi^*(u) + u \cdot \xi^*(v), \tag{6.2}$$

where ξ^* is the lifting of the field ξ on Y. The ideal $I(B)$ is generated by the functions $y_1, \ldots, y_n$; its localization near an arbitrary point $y \in B$ is generated by the function y_1, if $y_1(y) \neq 0$. We affirm that the field $y_1\xi^*$ is smooth on Y and tangent to B. Indeed, if, for example, $\xi = \partial/\partial x_k$, then $y_1\xi = \partial/\partial p_k$, where $p_j = x_j/x_1$, $j = 2, \ldots, n$, are affine coordinates on $\mathbb{R}P^{n-1}$. Therefore, the fields $v\xi^*$ and $u\xi^*$ are smooth and tangent to B. Hence the right-hand side of (6.2) is smooth on Y and vanishes on B. This implies our assertion.

REMARK 1. Equation (6.1) can be explained geometrically in the following way: any monoidal transformation π is a *convex* mapping. The latter means that π transforms the 1-neighborhood of B (which is locally a straight strip) into a 1-neighborhood of the origin, which is a ball.

REMARK 2. This proposition admits a generalization to the case of arbitrary X and Z that has the same form; we need only to restrict ξ to the normal fibration $N_Z(X)$.

Now we start proving Lemma 4.1. Our steps follow the outline of the resolution. For the first step we try to find a logarithmic field V_1 on (X_1, B_0), where $B_0 = \pi_0^{-1}(Z_0)$, with the following properties:

i) $\operatorname{cres}(V_1, B_0) = 1$;

ii) V_1 is tangent to any component Y_j such that the mapping π_1^{r+1} is an isomorphism near Y_j. This implies that Y_j (identified with its image) has normal crossings with B_0.

This is easy to do, taking $V_1 = T_1 + f_1S_1$, where T_1 is a logarithmic field, which satisfies i), S_1 is a smooth field near B_0 tangent to B_0, and $f_1 \in I(B_0)$. Condition i) implies that any integral curve γ of the field T_1 is transversal to B_0 and this field equals $[t + o(t)]d/dt$ on γ. Therefore there exists a function f_1 that generates $I(B_0)$ and satisfies the equation

$$T_1(f_1) = f_1 . \tag{6.3}$$

Then we choose a field S_1 that satisfies ii). The corrective term f_1S_1 is a logarithmic field and $\operatorname{cres}(f_1S_1, B_0) = 0$. Therefore Proposition 6.1 can be applied to V_1. If the resolution consists of only one step π_0, then it is easy to show that the field $V + V_{1*}$ possesses all the properties we need in Lemma 4.1. This is just the case, if Π is a homogeneous polynomial with no critical points $x \neq 0$.

Otherwise we must perform more steps. For the second step, we correct the field V_1 by an extra term supported near the next center Z_1 (which belongs to some components Y_j that are not completely simplified by the transformation π_0). To do this we consider the hypersurface $B_1 := \pi_1^{-1}(Z_1)$, and the function f_2 on X_2 that generates the ideal $I(B_1)$. To be exact, we should take the double covering of a neighborhood W of B_1, where we can choose an univalent branch of f_2. Then we look for a field of the form $V_2 = V_1^* + e_1(T_2 + f_2S_2)$, where e_1 is a smooth function with compact support on X_1 which is equal to 1 in a neighborhood of Z_1 and V_1^* denotes the inverse image of the field V_1 under the mapping π_1. Note that V_1^* is a logarithmic field on B_1, since V_1 vanishes on Z_1. We subject V_2 to the following conditions:

i′) T_2 is a logarithmic field on (X_2, B_1), $\operatorname{cres}(V_1^* + T_2, B_1) = 1$;

ii′) S_2 is tangent to B_1 and V_2 is tangent to any component Y_j such that the mapping π_2^{r+1} is an isomorphism near Y_j;

iii) $T_2(f_1^*) = 0$, $f_1^* := f_1\pi_1$.

Note that the equation i′) implies that any integral curve of the field $V_1^* + T_2$ is transversal to B_1. Therefore there exists a function f_2 that generates $I(B_1)$ and satisfies the equation $(V_1^* + T_2)(f_2) = f_2$, and a field S_2 that satisfies ii′). Then, to verify 4.1.2, we find

$$dV_2 = (1 - e_1)\, dV_1 + e_1 d\,(V_1 + T_2) + e_1 d\,(f_2 S_2) + de_1(T_2 + f_2 S_2)\,. \tag{6.4}$$

According to Proposition 6.1, the first term satisfies the equation

$$\langle dV_1\xi\,,\,\xi\rangle = [1 + o(1)]\langle \xi\,,\,\xi\rangle.$$

For the second term and any point $x \in X_1$ close to Z_1, we use the following estimate from below

$$\langle d\,(V_1 + T_2 + f_2 S_2)\xi\,,\,\xi\rangle_1 \geqslant [1 + o(1)]\langle \xi\,,\,\xi\rangle_1\,, \tag{6.5}$$

where $\langle\,\cdot\,,\,\cdot\,\rangle_1$ denotes any metric on X_1 compatible with its algebraic variety structure. This follows from Proposition 6.1 applied to the variety X_1. The third term equals $o(1)\langle\xi\,,\,\xi\rangle$, since $\operatorname{cres}(f_2S_2\,,\,B_1) = 0$. For an estimate of the fourth term of (6.4), we must specify the choice of e_1.

Choose a function $e \in D(\mathbb{R})$, $\operatorname{supp} e \subset [-1/2, 1/2]$ such that $0 \leqslant e \leqslant 1$ and $e(t) = 1$ in a neighborhood of zero. Set $e_1(x) = e(g_1^\varepsilon)$, where g_1 is an arbitrary positive function that generates the ideal $I(Z_1)^2$ and $\varepsilon = \varepsilon(1)$ is a small positive parameter. We may take, for example $g_1 = \sum x_i^2$, where $x_1, \ldots, x_m$ are generators of $I(Z_1)$. We have $e_1 \in C^\infty(X_i)$ and $\operatorname{supp} e_1 \to Z_1$ as $\varepsilon(1) \to 0$.

LEMMA 6.2. *For any function $z \in I(Z_1)$ and any field v on X_1, we have*

$$ze_i \to 0\,, \tag{6.6}$$

$$zv(e_i) \to 0 \tag{6.7}$$

uniformly as $\varepsilon \to 0$.

PROOF. Equation (6.6) is a consequence of the obvious relation $te(t^\varepsilon) \to 0$ as $\varepsilon \to 0$. Then we have

$$zv(e_1) = \varepsilon zv(g_1) g_1^{\varepsilon-1} e'(g_1^\varepsilon)\,,$$

where $e' = (d/dt)e$. We have $v(g_1) = O(g_1^{1/2})$, hence $|zv(e_1)| \leqslant Cg_1$ for some positive C. Therefore

$$|zv(e_1)| \leqslant C\varepsilon g_1^\varepsilon |e'(g_1^\varepsilon)| \leqslant C\varepsilon M\,,$$

where $M := \max|te'(t)|$. This implies (6.7).

To estimate the fourth term of (6.4), we take an arbitrary smooth function y on X_0 that vanishes at x_0 and calculate the function $z = T_2(y) + f_2S_2(y)$. We have

$y = f_1 y_1$, where y_1 is a smooth function on X_1, hence $T_2(y) = f_1 T_2(y_1)$, since of iii) and $T_2(y_1) = f_2 z_1$, where z_1 is smooth on X_1, because T_2 is logarithmic on (X_2, B_1). Then we have $S_2(f_1) = w f_2$ for a smooth function w on X_2, since f_0 vanishes on B_1 and S_2 is tangent to B_1. We have $\operatorname{supp} w \subset \operatorname{supp} S_2$ and we may assume that $\operatorname{supp} w$ does not touch the closure of the set $\pi_2^{-1}(B_0) \setminus B_1$. Therefore we have $f_2 = g f_1$ with a smooth function g on $\operatorname{supp} w$ and $S_2(f_1) = w_1 f_1$, where $w_1 = gw$. Finally, we get the equation

$$z = f_1 f_2 z_2 , \qquad z_2 = z_1 + w .$$

Apply the fourth term of (6.4) to an arbitrary tangent vector ξ on X_1. We get the quantity

$$z\, de_1(\xi) = z_2 f_1 f_2 \xi(e_1) .$$

The product $f_2 \xi(e_1)$ tends to zero with $\varepsilon(1)$ according to Lemma 6.2. This gives the asymptotics

$$|\eta(y)| = o(f_1)\, \|\xi\|_1 , \tag{6.8}$$

where $\eta := de_1(\xi)(T_2 + f_2 S_2)$ and $\|\theta\|^2 = \langle \theta , \theta \rangle$. Now we specify the choice of the metric $\langle \cdot , \cdot \rangle_1$, setting, for any point $x \in X_1 \setminus B_0$,

$$\langle a , b \rangle_1 = \langle l(a) , l(b) \rangle + f_1^{-2} \langle p(a) , p(b) \rangle , \qquad a , b \in T_x(X_1) , \tag{6.9}$$

where $l(a) = a(f_1) t_1$, $t_1 := f_1^{-1} T_1$, and $p(a) = a - l(a)$. This means that $l(a)$ is the projection of a in t_1-direction and $p(a)$ is the projection on the subspace $\operatorname{Ker} df_1 \subset T(X_1)$ by (6.3). The metric (6.9) means that we stretch the component $p(a)$ of any tangent vector a with coefficient f_1^{-1} as compared with the metric $\langle \cdot , \cdot \rangle$. It follows that the metric $\langle \cdot , \cdot \rangle_1$ agrees with the smooth structure of X_1, since the mapping $d\pi_0^{-1}$ acts in the same way.

Note that (6.9) implies the inequality $\|\xi\|_1 \leqslant C f_1^{-1} \|\xi\|$. Therefore, applying (6.8) to the coordinate functions $y = x_1 , \ldots , x_n$, we get

$$\|\eta\| = o(1)\, \|\xi\| . \tag{6.10}$$

The operators l and p yield the following decomposition of the fiber bundle

$$T(X_1) = L(t_1) \oplus \operatorname{Ker} df_1 ,$$

where $L(t_1)$ is the linear span of t_1 in each fiber of $T(X_1)$. The operators dV_1 and dS_2 preserve this decomposition up to terms of size $O(f_1)$. So does the operator dT_1, because of iii). Combining inequality (6.1) with (6.5) and (6.10), we finally get

$$\langle dV_2 \xi , \xi \rangle \geqslant (1 - e_1)\langle \xi , \xi \rangle + e_1 \langle \xi , \xi \rangle + o(1) \langle \xi , \xi \rangle = [1 + o(1)] \langle \xi , \xi \rangle ,$$

which proves 4.1.2. Property 4.1.1 can be proved along the same lines.

7. The general case. If the resolution comprises more then two steps, we consider the transformation π_2, choose some fields T_3, S_3 that obey conditions like i′), ii′), iii). Then we take the function $e_2(x) = e(g_2^\varepsilon)$, where $\varepsilon = \varepsilon(2)$ is a small parameter such that $\varepsilon(2) \ll \varepsilon(1) \ll 1$ and so on. The following situation may occur: the center Z_1 belongs to several irreducible components of E_1. According to (5.3), Z_1 has only normal crossings with these components. Condition i) must be changed to

i″) $\operatorname{cres}(V_1^* + T_{i+1}, B_i) = 1/k$

for some natural k, and condition iii) to

iii″) $T_{i+1}(f_{j+1}^*) = 0$,

where j is the smallest number such that $Z_i \subset E_{ji}$ (for E_{ji} see §5). We still have the relation $V_i^*(f_{j+1}) = f_{j+1}$ similar to (6.3), since the number k is equal to the multiplicity of f_{j+1} on B_i: $f_{j+1} = g f_{i+1}^k$, where g is a smooth function that does not belong to $I(B_i)$.

Completing the induction, we get the field

$$\begin{aligned} V = {} & \pi_{1*}(T_1 + f_1 S_1) + e_1 \pi_{0*}^2 (T_2 + f_2 S_2) \\ & + e_1 e_2 \pi_{0*}^3 (T_3 + f_3 S_3) + \cdots + e_{j-1} \pi_{0*}^j (T_j + f_j S_j) + \cdots . \end{aligned}$$

It follows from ii), iii′) that the field $V_{r+1} = \pi_0^{r+1*}(V)$ is tangent to any component of $\Pi_{r+1}^{-1}(0)$. Hence $V_{r+1}(\Pi_{r+1}) = P\Pi_{r+1}$ for some bounded function P. Then we note that the field V_{r+1} is the sum of some logarithmic fields L_j on (X_{r+1}, Y_j), where Y_j, $j = 1, 2, \ldots$, are components of the root set of Π_{r+1} and the functions $c_j := \operatorname{cres}(L_j, V_j)$ are positive constants. Any Y_j is given by the equation $w_k = 0$ in any coordinate system, as in (5.5). Hence $L_j = c_j w_j \partial/\partial w_j + N_j$, where $\operatorname{cres} N_j = 0$. We deduce from (5.5) that $L_j(\Pi_{r+1}) = [c_j d(Y_j) + 0(1)]\Pi_{r+1}$, where $d(Y_j) := d(k)$ is a positive integer as in §5. Therefore, locally we have $P = c_j d(Y_j) + o(1)$, which implies that $P > 0$ in a neighborhood of the set $\Pi^{-1}(0)$. The proof of Lemma 4.1 may be completed along these lines.

8. Example. We carry the construction of §§5, 6 for the case $n = 2$,

$$\Pi(x) = (x_2 - x_1)(x_1^3 - x_2^2).$$

For the first step of the resolution, we take the point $Z_0 = (0, 0)$ as the center. Then we have

$$\Pi_1(x_1, \xi) = x_1^3(1 - \xi)(x_1 - \xi^3), \qquad \Pi_1 := \Pi\pi_0,$$

where $\xi = x_2/x_1$ is the affine coordinate on $B_0 \subset X_1$. We see that the function Π_1 has the form (5.5) everywhere except for the point $Z_1 = \{x_1 = 0, \xi = 0\}$. Blow up X_1 with center Z_1 and choose the affine coordinate $\eta = x_1/\xi$ on X_2 near B_1. We have

$$\Pi_2(\xi, \eta) = \xi^4 \eta^3 (1 - \xi)(\eta - \xi)$$

near the point $Z_2 := \{\xi = 0, \eta = 0\}$. Applying the monoidal transformation with center Z_2 and taking the affine coordinate $\theta = \eta/\xi$, we finally get

$$\Pi_3(\eta, \theta) = \theta^3 \xi^8 (1 - \xi)(\theta - 1),$$

which implies that Π is of the form (5.5). Hence the complete resolution has the form

$$X_3 \xrightarrow{\pi_2} X_2 \xrightarrow{\pi_1} X_1 \xrightarrow{\pi_0} X_0 = \mathbb{R}^2. \tag{8.1}$$

Following the lines of §6, we try to find the field $V = V_{3*}$ in the following form

$$V_3 = (T_1 + f_1 S_1) + e_1(T_2 + f_2 S_2) + e_1 e_2 (T_3 + f_3 S_3),$$

where $T_1 = x_1 \partial/\partial x_1$ in (x_1, ξ)-chart and $T_1 = x_2 \partial/\partial x_2$ in the chart $(x_2, \xi' = \xi^{-1})$. We set $S_1 = 0$, since the field T_1 itself is tangent to the component $\eta - \xi = 0$ of the set $\Pi_1^{-1}(0)$. We have $T_1^* \equiv \pi_1^*(T_1) = \eta\,\partial/\partial\eta$ for the chart (ξ, η). Hence we can take $T_2 = \xi\,\partial/\partial\xi - \eta\,\partial/\partial\eta$ and since $T_2(f_1) = 0$, we can set $f_1 = x_1$, obtaining

$$\operatorname{cres}(T_1^* + T_2, B_1) = 1.$$

We may set $S_2 = 0$, since no component of $\Pi_2^{-1}(0)$ transversal to B_1 appears. The field $T_1^{**} + T_2^* \equiv \pi_2^*(T_1^* + T_2)$ equals $\xi\,\partial/\partial\xi - \theta\,\partial/\partial\theta$ in the chart (ξ, θ) and equals $\theta'\,\partial/\partial\theta'$ in the chart $(\eta, \theta' = \theta^{-1})$. We must choose the field T_3 to satisfy the condition i$''$) and the condition

iii$''$) $T_3(f_1) = 0$

instead of iii), since the center Z_2 is contained in the intersection of two components $B_1 = E_{12}$ and E_{02} of E_2. We have $f_1 = x_1 = \eta\xi = \theta\xi^2$, therefore we can take $T_3 = \theta\,\partial/\partial\theta - (1/2)\xi\,\partial/\partial\xi$. Then we have $T_1^{**} + T_2^* + T_3 = (1/2)\xi\,\partial/\partial\xi$ in the (ξ, θ)-chart and $T_1^{**} + T_2^* + T_3 = (1/2)\eta\,\partial/\partial\eta$ in the (η, θ)-chart. Hence

$$\operatorname{cres}(T_1^{**} + T_2^* + T_3, B_2) + 1/2$$

which agrees with the multiplicity of f_1 on B_0: $f_1 = \theta f_3^2$, where $f_3 = \xi$. The field $T_1^{**} + T_2^* + T_3$ is tangent to the component $Y_2 = \{\theta - 1 = 0\}$ of $\Pi_3^{-1}(0)$. Hence we can set $S_3 = 0$ and the construction is completed by the field

$$\begin{aligned} V &= \pi_{1*}(T_1) + \pi_{0*}^2(T_2) + \pi_{0*}^3(T_3) \\ &= x_1 \frac{\partial}{\partial x_1} + x_2 \frac{\partial}{\partial x_2} + e_1(\xi)\pi_{0*}^2\left(\xi \frac{\partial}{\partial \xi} - \eta \frac{\partial}{\partial \eta}\right) \\ &\quad + e_1(\xi) e_2(\eta) \pi_{0*}^3\left(\theta \frac{\partial}{\partial \theta} - \frac{1}{2}\xi \frac{\partial}{\partial \xi}\right) \end{aligned}$$

defined in a neighborhood of E_3. We calculate

$$\pi_{0*}^2\left(\xi \frac{\partial}{\partial \xi} - \eta \frac{\partial}{\partial \eta}\right) = x_2 \frac{\partial}{\partial x_2}, \qquad \pi_{0*}^3\left(\theta \frac{\partial}{\partial \theta} - \frac{1}{2}\xi \frac{\partial}{\partial \xi}\right) = -\frac{1}{2} x_2 \frac{\partial}{\partial x_2},$$

which finally gives

$$V = x_1 \frac{\partial}{\partial x_1} + \left[1 + e_1\left(\frac{x_2}{x_1}\right)\left(1 - \frac{1}{2} e_2\left(\frac{x_1^2}{x_2}\right)\right)\right] x_2 \frac{\partial}{\partial x_2}$$

in a neighborhood of the origin.

Let us check that the field cV satisfies the conditions of Lemma 4.1 provided that $c > 1$ and the parameters $\varepsilon(1)$ and $\varepsilon(2)$ are small enough. First we calculate the matrix dV:

$$\frac{\partial}{\partial x_1} V(x_1) = 1, \qquad \frac{\partial}{\partial x_2} V(x_1) = 0,$$

$$\frac{\partial}{\partial x_1} V(x_2) = x_2 \frac{\partial}{\partial x_1} e(x), \qquad \frac{\partial}{\partial x_2} V(x_2) = 1 + e(x) + x_2 \frac{\partial}{\partial x_2} e(x),$$

where

$$e(x) := \left[e_2\left(\frac{x_2}{x_1}\right)\left(1 - \frac{1}{2} e_2\left(\frac{x_1}{x_2}\right)\right)\right] \geqslant 0.$$

We have

$$x_2 \frac{\partial}{\partial x_1} e = o(1), \qquad x_2 \frac{\partial}{\partial x_2} e = o(1), \tag{8.2}$$

hence dV is bounded and $dV \geqslant E + o(1)$, which proves 4.1.1 and 4.1.2. To verify (8.2) we find, for example, the derivative $x_2(\partial/\partial x_1) e_2(x_2/x_1) = -\xi e'(\xi)$. This function is $o(1)$ as $\varepsilon(2) \to 0$ by (6.6). The other terms in (8.2) are estimated in the same way.

For small parameters $\varepsilon(1)$, $\varepsilon(2)$, the supports of the functions e_1 and e_2 are close to Z_1 and Z_2 respectively. Therefore $e(x) = 1$ and

$$V = x_1 \frac{\partial}{\partial x_1} + x_2 \frac{d}{dx_2} \quad \text{for } \left|\frac{x_2}{x_1}\right| \geqslant 1 \text{ and small } x_1^2 + x_2^2,$$

if $\varepsilon(1)$ is small enough. In particular, V is tangent to the line $x_1 - x_2 = 0$. If the point x belongs the curve $x_1^3 - x_2^2 = 0$ and tends to the origin, we have $x_2/x_1 \to 0$, $x_1^2/x_2 \to 0$. Hence $e(x) = 3/2$ and

$$V = x_1 \frac{\partial}{dx_1} + \frac{3}{2} x_2 \frac{\partial}{\partial x_2}.$$

Therefore $V(\Pi)$ vanishes when $\Pi = 0$, which implies that $V(\Pi) = P\Pi$.

References

1. J. L. Lagrange, *Mècanique analytique*, Veuve Desaint, Paris, 1788.
2. Lejeune G. Dirichlet, *Über die Stabilitat des Gleichgewichts*, J. Reine Angew. Math. **32** (1846), 85–88.
3. A. M. Lyapunov, *General problem of the stability of motion*, Kharkov Math. Soc., Kharkov, 1892; French transl., Ann. of Math. Studies, vol. 17, Princeton Univ. Press, Princeton, NJ, 1947.
4. N. G. Chetaev, *The stability of motion*, Pergamon Press, New York, 1961.
5. N. Rouche, P. Habets, and M. Laloy, *Stability theory by Lyapunov's direct method*, Springer-Verlag, New York, 1977.
6. M. Laloy, *On equilibrium instability for conservative and partially dissipative systems*, Internat. J. Non Linear Mech. **11** (1976), 295–301.
7. M. Laloy and K. Peiffer, *On the instability of equilibrium when the potential function has a non-strict local minimum*, Arch. Rational Mech. Anal. **78** (1982), 213–222.
8. J. N. Mather, *Stability of C^∞-mappings,* III *Finitely determined map-germs*, Inst. Hautes Études Sci. Publ. Math. **35** (1968).

9. P. Hagedorn, *Die Umkehrung der Stabilitatssatze von Lagrange-Dirichlet und Routh*, Arch. Rational Mech. Anal. **42** (1971), 281–316.
10. P. Hagedorn, *Uber die Instabilitat konservativer Systeme mit gyrockopischen Kraften*, Arch. Rational Mech. Anal. **58** (1975), 1–9.
11. V. P. Palamodov, *Equilibrium stability in a potential field*, Funktsional. Anal. i Prilozhen. **11** (1977), no. 4, 42–55; English transl., Functional Anal. Appl. **11** (1977), 277–289.
12. S. D. Taliaferro, *An inversion of the Lagrange–Dirichlet stability theorem*, Arch. Rational Mech. Anal. **73** (1980), 183–190.
13. V. V. Kozlov, *Asymptotic solutions of equations of classical mechanics*, J. Appl. Math. Mech. **46** (1982), 454–457.
14. V. V. Kozlov and V. P. Palamodov, *On asymptotic solutions of the equations of classical mechanics*, Dokl. Akad. Nauk SSSR **263** (1982), no. 2, 285–289; English transl., Soviet Math. Dokl. **25** (1982), 335–339.
15. V. V. Kozlov, *Asymptotic motions and the inversion of the Lagrange–Dirichlet theorem*, J. Appl. Math. Mech. **50** (1987), 719–725.
16. A. N. Kuznetsov, *On existence of asymptotic solutions to a singular point of an autonomic system, possessing a formal solution,*, Funktsional. Anal. i Prilozhen. **23** (1989), no. 4, 63–74; English transl. in Functional Anal. Appl. **23** (1989).
17. V. Moauro and P. Negrini, *On the inversion of Lagrange–Dirichlet theorem*, Differential Integral Equations **2** (1989), no. 4, 471–478.
18. H. Hironaka, *Resolution of singularities of an algebraic variety over a field of characteristic zero*, Ann. of Math. (2) **79** (1964), 107–274.
19. S. D. Taliaferro, *Instability of an equilibrium in a potential field*, Arch. Rational Mech. Anal. **109** (1990), 183–194.
20. L. Salvadori, *Stability problems for holonomic mechanical systems*, Atti Acad. Sci. Torino Cl. Sci Fis. Mat. Natur. **126** (1992), suppl. 2, 151–168.
21. V. P. Palamodov, *On stability of motion with several degrees of freedom*, Conference "Nonlinear Hamiltonian Mechanics" (1992), Trento, June 1–5.

Translated by THE AUTHOR

DEPARTMENT OF MATHEMATICS, MOSCOW STATE UNIVERSITY, MOSCOW 119899, RUSSIA

Amer. Math. Soc. Transl.
(2) Vol. **168**, 1995

Homoclinic Orbits to Invariant Tori of Hamiltonian Systems

SERGEY V. BOLOTIN

§0. Introduction

Let $N = \mathbb{R}/\mathbb{Z}\{x\} \times \mathbb{R}\{y\}$ be the infinite cylinder and $f\colon N \to N$ a smooth integrable twist map:

$$f(x, y) = (x + \omega(y), y), \qquad \partial\omega/\partial y > 0.$$

Let $f_\varepsilon\colon N \to N$ be a smooth perturbation of f. This means that f_ε smoothly depends on the parameter $\varepsilon \geqslant 0$ and $f_0 = f$. Suppose that f_ε is area-preserving and exact: for any curve γ which winds round the cylinder once

$$\oint_\gamma y'\,dx' - y\,dx = 0, \qquad f_\varepsilon(x, y) = (x', y').$$

If $\omega = \omega(y_0)$ is a Diophantine number and $\varepsilon > 0$ is sufficiently small, then by KAM-theory the perturbed map has an invariant curve with rotation number ω near the circle $M = \mathbb{R}/\mathbb{Z} \times \{y_0\} \subset N$. If $\omega = m/n$ is a rational number, then M consists of periodic points for f, i.e., $f^n|_M$ is the identity map. Poincaré [36] proved that in general for $\varepsilon > 0$ such an invariant circle M disintegrates, but the perturbed map has at least one hyperbolic and one elliptic n-periodic orbit near M. Using the area-preserving property, he proved that the separatrices of the hyperbolic orbit intersect: in an $O(\sqrt{\varepsilon})$-neighborhood of M there exist at least two homoclinic orbits. Poincaré also noticed that this implies the existence of an infinite number of homoclinics. Near M, the phase portrait of the iterated map f_ε^n looks like the phase portrait of a periodically perturbed pendulum. In the analytic case, the angle of intersection of separatrices is exponentially small as $\varepsilon \to 0$ [36], and in the smooth case it is smaller than any power of ε. Hence the existence of a homoclinic orbit cannot be established by classical perturbations theory.

In the present paper, the Poincaré result is generalized to the multidimensional case. Suppose that the unperturbed symplectic diffeomorphism f of a symplectic manifold N has a compact invariant Lagrangian manifold M foliated by nonresonant invariant tori. The main example is a resonant invariant torus of a completely integrable symplectic diffeomorphism. We assume that f has a positive definite

1991 *Mathematics Subject Classification*. Primary 58F05, 58F30.

normal torsion near M and the rotation vector satisfies the Diophantine condition. For the perturbed exact symplectic diffeomorphism f_ε, almost all invariant tori disintegrate, but generically at least one of them survives and is hyperbolic for $\varepsilon > 0$. This can be proved by using KAM-type results on the conservation of hyperbolic invariant tori [21, 40, 43]. If a certain topological condition is satisfied, we prove the existence of homoclinic orbits to the hyperbolic invariant torus of the perturbed map f_ε and provide certain estimates for their number (Theorem 1.1).

The main results of the first part of the paper are presented in §1. To prove the existence of a homoclinic orbit, instead of using the conservation of area as Poincaré did in the two-dimensional case, we reduce the map f_ε to the Poincaré map of a time-periodic positive definite Lagrangian system. This is done in §§2–3 by means of classical perturbation theory (the Delaunay method [36]). Then we apply the variational method suggested in [10–12]. It is described in the second part of the paper (§§4–8). For the case when no invariant tori exist (the perturbation is only C^1-small or the rotation vector does not satisfy the Diophantine condition), in §9 we prove the existence of invariant sets in an $o(\sqrt{\varepsilon})$-neighborhood of M and the existence of semihomoclinic orbits (Theorem 1.1′). These sets are minimal in the sense of Mather [32]. We also obtain analogous results in the case of perturbations of invariant manifolds of autonomous and time-periodic Hamiltonian systems (Theorems 1.2–1.3 and 1.2′–1.3′).

The existence of a homoclinic to an invariant torus is usually the first step in proving nonintegrability and complicated behavior of a dynamical system (see, for example, [4, 26, 28, 45]). Possibly, it can also help in proving Arnold diffusion [3, 16, 20] for a generic perturbation of a completely integrable symplectic map or Hamiltonian system.

Poincaré's results can be generalized to the nonperturbative situation under the assumption that the area-preserving exact twist map of the cylinder is monotone, i.e., $\partial x'/\partial x > 0$ and $x' - x \to \pm\infty$ as $x \to \pm\infty$ on the covering plane. In this case, according to Aubry–Mather theory [5, 30], for each ω the map has a compact invariant set with rotation number ω. Homoclinic and heteroclinic orbits to these sets were studied in [31].

The proper generalization of monotone twist maps to the multidimensional case are Poincaré maps of time-periodic positive definite Lagrangian systems: the Lagrangian is strictly convex in velocity. Such symplectic diffeomorphisms are called positive definite (or optical). For area-preserving maps of the cylinder, Moser proved [33] that every monotone exact twist map is positive definite. The multidimensional generalization of Aubry–Mather theory was obtained by Mather [32].

In the second part of the paper (§§4–8), we consider invariant tori of positive definite Lagrangian systems (or, equivalently, of positive definite symplectic diffeomorphisms) that are supports of action-minimizing invariant measures [32]. By using Lagrangian variational methods, we prove the existence of semiasymptotic (in Birkhoff's sense [8]) and semihomoclinic orbits to these invariant tori. Under certain additional assumptions, these orbits are homoclinic. The existence of homoclinics to perturbed invariant tori (Theorems 1.1–1.3) follows from this. The results are stated in §4 and the main Theorem 4.2 is proved in §§6–7. In §5 and §8 arbitrary minimal invariant sets are considered. In §5 the existence of semiasymptotic orbits is established. In §8, under certain topological assumptions, we prove the existence of semihomoclinic orbits to minimal invariant sets and obtain an estimate for the number of homoclinics (Theorem 4.3). In §9 these results are used in the proof of Theorems 1.1′–1.3′.

Main results of the present paper were announced in [13].

§1. Perturbations of invariant manifolds and existence of homoclinic orbits

Let N be a $2m$-dimensional symplectic manifold with symplectic structure Ω and $f: N \to N$ a symplectic diffeomorphism.

DEFINITION. An f-invariant set $\Gamma \subset N$ is called an *invariant torus* with rotation vector $\omega \in \mathbb{T}^n$ if Γ is the image of an embedding $g: \mathbb{T}^n \to N$ of the standard n-dimensional torus $\mathbb{T}^n = \mathbb{R}^n/\mathbb{Z}^n$ and f acts on Γ as a translation: for any $\theta \in \mathbb{T}^n$, we have $f(g(\theta)) = g(\theta + \omega)$.

We shall assume that ω is nonresonant: $\langle \omega, k\rangle \notin \mathbb{Z}$ for all nonzero integer vectors $k \in \mathbb{Z}^n \setminus \{0\}$. In this case the restriction $f|_\Gamma$ is ergodic and Γ is isotropic: $\Omega|_\Gamma = 0$ [25].

DEFINITION. An invariant torus Γ is called *hyperbolic* if the sets of points $x \in N$ such that $g^k x \to \Gamma$ as $k \to +\infty$ and $k \to -\infty$, are smooth m-dimensional submanifolds W^s and W^u in N (stable and unstable manifolds), intersecting along Γ, and the following nondegeneracy condition is satisfied.

For each point $x \in \Gamma$ the tangent spaces $T_x W^s$ and $T_x W^u$ are direct sums: $T_x W^s = T_x\Gamma \oplus E_x^s$ and $T_x W^u = T_x\Gamma \oplus E_x^u$, where the vector bundles E^s and E^u are f-invariant:

$$df(x)v \in E^s_{f(x)}, \quad \text{for all } v \in E^s_x,$$
$$df(x)v \in E^u_{f(x)}, \quad \text{for all } v \in E^u_x.$$

There exists a Riemannian metric $\|\cdot\|$ on M and a constant λ, $0 < \lambda < 1$, such that for any $x \in \Gamma$

$$\|df(x)v\| \leqslant \lambda\|v\|, \quad \text{for all } v \in E^s_x,$$
$$\|df^{-1}(x)v\| \leqslant \lambda\|v\|, \quad \text{for all } v \in E^u_x.$$

The definition of a hyperbolic invariant torus for a Hamiltonian system is similar. Arnold [3] called such tori whiskered. See also [16, 23, 43, and 45].

DEFINITION. The orbit $f^k x$, $k \in \mathbb{Z}$, of the diffeomorphism f is called *homoclinic* to Γ, if $x \notin \Gamma$ and $f^k x$ tends to Γ as k tends to $\pm\infty$. In other words, $x \in W^s \cap W^u \setminus \Gamma$ is a homoclinic point.

The goal of this section is to state some existence results for homoclinic orbits to hyperbolic invariant tori of perturbations of conditionally integrable (integrable on an invariant submanifold) symplectic diffeomorphisms. The proofs are given in subsequent sections.

Let $M^m \subset N$ be a compact connected Lagrangian [2, 44] invariant manifold of a symplectic diffeomorphism f.

DEFINITION. Let $\omega \in \mathbb{T}^n$. The diffeomorphism f is called *quasiperiodic* on M, if M is foliated by n-dimensional invariant tori and the restriction of f to each of them is conjugate to the translation of $\mathbb{T}^n$ by ω. The vector ω is called the *rotation vector* of f on M.

This means that there exists a smooth free group action $g\colon \mathbb{T}^n \times M \to M, \theta \in \mathbb{T}^n$, $x \in M \to g_\theta x \in M$, on M such that $f|_M = g_\omega$. Without loss of generality, we can assume that ω is nonresonant, so that f is ergodic on each of the fibers of M.

If ω is resonant, i.e., $\langle k , \omega\rangle \in \mathbb{Z}$ for some nonzero $k \in \mathbb{Z}^n$, we define the lower dimensional torus $\mathbb{T}^l$, $l < n$, as the subgroup

$$\mathbb{T}^l = \{\theta \in \mathbb{T}^n : \langle k , \theta\rangle \in \mathbb{Z} \Leftrightarrow \langle k , \omega\rangle \in \mathbb{Z}\} \subset \mathbb{T}^n.$$

We obtain a group action of the torus $\mathbb{T}^l$ on M. Thus M is foliated by l-dimensional f-invariant tori and f is ergodic on each of them.

Let f_ε be a perturbation of the symplectic diffeomorphism f, i.e., a smooth family of symplectic diffeomorphisms of N, defined for small $\varepsilon \geqslant 0$, such that $f_0 = f$. Then under suitable conditions at least one f-invariant torus among the fibers of M survives the perturbation: there exists a family Γ_ε, $\varepsilon > 0$, of hyperbolic f_ε-invariant tori with the same rotation vector ω, such that $\Gamma_0 = g_{\mathbb{T}^n} x_0 \subset M$, $x_0 \in M$, is an invariant torus for f. We shall prove that there exists a homoclinic orbit to Γ_ε.

Now let us state assumptions on the diffeomorphism f and on the perturbation f_ε that ensure the existence of the torus Γ_ε. The following condition is usual in KAM-theory (see, for example, [4]).

DIOPHANTINE CONDITION. Suppose that there exist C and $\nu > 0$ such that the rotation vector ω satisfies the inequality

$$|\langle k , \omega\rangle + j| \geqslant C(|k| + |j|)^{-\nu} \tag{1.1}$$

for any integer vector $k \in \mathbb{Z}^n \setminus \{0\}$ and $j \in \mathbb{Z}$. Here we have fixed some vector $\omega \in \mathbb{R}^n$ corresponding to $\omega \in \mathbb{T}^n = \mathbb{R}^n/\mathbb{Z}^n$.

Since M is Lagrangian, there exists a 1-form λ on some neighborhood U of M such that $\Omega = d\lambda$ and $\lambda(x) = 0$ for all $x \in M$. Moreover, there exists a symplectic diffeomorphism of the neighborhood U onto a neighborhood of M in T^*M transforming λ into the standard Liouville 1-form on T^*M [44]. We shall use this identification later, but now we prefer more invariant conditions.

Define the generating function $S\colon U \to \mathbb{R}$ of the diffeomorphism $f|_U$ by the equation $dS = f^*\lambda - \lambda$. If U is small enough, the generating function exists and is unique up to a constant. The invariant manifold M consists of critical points for S. It follows that for each $x \in M$ the tangent space T_xM is contained in the kernel of the bilinear form $d^2S(x)$ and thus $d^2S(x)$ defines a quadratic form K on T_xN/T_xM. Since M is Lagrangian, $T_xN/T_xM \cong T_x^*M$ [44]. We obtain a function K on T^*M which is a quadratic form on T_x^*M for all $x \in M$.

We shall denote a point in T^*M by (x , p), where $x \in M$ and $p \in T_x^*M$. For the given group action g of the torus $\mathbb{T}^n$ on M define the group action $g^*\colon \mathbb{T}^n \times T^*M \to$

T^*M by the covariant law: for any $\theta \in \mathbb{T}^n$ the map $g_\theta^*: T^*M \to T^*M$ is given by the formula

$$g_\theta^*(x, p) = (g_\theta(x), ((dg_\theta(x))^*)^{-1}p), \qquad x \in M, \ p \in T_x^*M. \tag{1.2}$$

Let $\overline{K}: T^*M \to \mathbb{R}$ be the average of the function K under this action:

$$\overline{K} = \int_{\mathbb{T}^n} K \circ g_\theta^* \, d\theta. \tag{1.3}$$

Then $\overline{K}$ is quadratic in the momentum $p \in T_x^*M$ and invariant under the group action g^*. We shall prove later that this function does not depend on the choice of the 1-form λ and thus it is invariantly defined.

DEFINITION. We call the invariant manifold M of the symplectic diffeomorphism f *positive definite* if for each $x \in M$ the function $\overline{K}$ is a positive definite quadratic form on T_x^*M. In other words, $\overline{K}$ is a Riemannian metric on M.

This is a generalization of the definition of an invariant Lagrangian torus with positive definite normal torsion [25].

EXAMPLE 1.1. COMPLETELY INTEGRABLE SYMPLECTIC MAP. Suppose that $N = \mathbb{T}^m\{x\} \times D\{y\}$, $D \subset \mathbb{R}^m$, with the standard symplectic structure $dy \wedge dx$ and let f be a completely integrable symplectic diffeomorphism:

$$f(x, y) = (x + \omega(y), y), \qquad \omega = \partial h(y)/\partial y.$$

Let $M = \mathbb{T}^m \times \{y_0\}$ be a resonant invariant torus with multiplicity $m - n$: the rotation vector $\omega = \omega(y_0)$ satisfies $m - n$ independent resonance conditions, i.e., the rank of the resonance group $K = \{k \in \mathbb{Z}^m : \langle \omega, k \rangle \in \mathbb{Z}\}$ equals $m - n$. Define the torus $\mathbb{T}^n \subset \mathbb{T}^m$ by the formula $\mathbb{T}^n = \{\theta \in \mathbb{T}^m : \langle \theta, k \rangle = 0, \text{ for all } k \in K\}$ and the group action of $\mathbb{T}^n$ on M by translations. Thus M is foliated by ergodic invariant tori and the symplectic diffeomorphism f is quasiperiodic on M. It is easy to see that the generating function of f is $S = \langle y, \omega(y) \rangle - h(y)$. Hence, the invariant manifold M is positive definite if and only if the symmetric matrix $\partial\omega(y_0)/\partial y = \partial^2 h(y_0)/\partial y^2$ is positive definite. This is the usual positive definiteness condition (see, for example, [25]).

Now consider a smooth perturbation f_ε of the diffeomorphism f. Suppose that the family f_ε of symplectic diffeomorphisms is exact in a neighborhood U of M. This means that for small $\varepsilon \geqslant 0$ there exists the generating function S_ε of the diffeomorphism $f_\varepsilon|_U$, ${f_\varepsilon}^*\lambda - \lambda = dS_\varepsilon$. The necessary and sufficient condition is that the restriction $({f_\varepsilon}^*\lambda - \lambda)|_M$ is exact. Of course, up to a constant, $S_\varepsilon = S + \varepsilon S_1 + O(\varepsilon^2)$, where S is the generating function for f. Define the function V on M by the formula

$$V = -S_1|_M. \tag{1.4}$$

Note that the family f_ε is exact if and only if the locally Hamiltonian vector field $\xi_\varepsilon = \partial f_\varepsilon/\partial\varepsilon \circ f_\varepsilon^{-1}$ is Hamiltonian in some neighborhood U of M, i.e., the closed

1-form $i_{\xi_\varepsilon}\Omega|_M$ is exact: $i_{\xi_\varepsilon}\Omega|_M = -dH_\varepsilon$, where H_ε is a smooth function on M. It is easy to see that $V = H_0$. This proves that V does not depend on the choice of the 1-form λ.

Let $\overline{V}$ be the average of V under the $\mathbb{T}^n$-action on M:

$$\overline{V} = \int_{\mathbb{T}^n} V \circ g_\theta \, d\theta \,. \tag{1.5}$$

The function $\overline{V}$ is $\mathbb{T}^n$-invariant and thus it defines a smooth function on the quotient manifold $\widetilde{M} = M/\mathbb{T}^n$.

For the main part of the next theorem we need the following

TOPOLOGICAL CONDITION. Let Γ_0 be a fiber of M and let $i\colon \Gamma_0 \to M$ be the inclusion. Suppose that the induced homomorphism of the homology groups

$$i_*\colon H_1(\Gamma_0\,, \mathbb{R}) \to H_1(M\,, \mathbb{R})$$

is injective.

For example, this condition is satisfied if $H_2(M\,, \Gamma_0\,, \mathbb{R}) = 0$, or if the fiber bundle $M \to \widetilde{M}$ is trivial [42]. Thus it is satisfied in the case of a completely integrable symplectic diffeomorphism.

THEOREM 1.1. *Let a symplectic diffeomorphism f be quasiperiodic on a compact invariant Lagrangian manifold M. Suppose that the rotation vector $\omega \in \mathbb{R}^n$ satisfies the Diophantine condition and the manifold M is positive definite. Let f_ε, $\varepsilon \geqslant 0$, be an exact perturbation of the diffeomorphism f in a neighborhood U of M. Suppose that the maximum of the function $\overline{V}$ on $\widetilde{M}$ is strict and nondegenerate. Then for all sufficiently small $\varepsilon > 0$, the following results hold.*

(1) *The perturbed diffeomorphism f_ε has a smooth n-dimensional invariant torus $\Gamma_\varepsilon \subset U$ with rotation vector ω. The torus Γ_ε smoothly depends on $\varepsilon \geqslant 0$ and $\Gamma_0 \subset M$ is the set of maximum points of the function $\overline{V}$.*

This means that there exists a smooth family of embeddings $h_\varepsilon\colon \mathbb{T}^n \to N$, $\varepsilon \geqslant 0$, such that $\Gamma_\varepsilon = h_\varepsilon(\mathbb{T}^n)$, $f_\varepsilon(h_\varepsilon(\theta)) \equiv h_\varepsilon(\theta + \omega)$, and $h_0(\theta) = g_\theta(x_0)$, $x_0 \in \Gamma_0$.

(2) *For $\varepsilon > 0$ the torus Γ_ε is hyperbolic. The stable and unstable manifolds $W^s = W^s(\Gamma_\varepsilon)$ and $W^u = W^u(\Gamma_\varepsilon)$ are smooth m-dimensional Lagrangian manifolds in N and they smoothly depend on $\sqrt{\varepsilon}$. The intersection angle of W^s and W^u along Γ_ε is of order $\sqrt{\varepsilon}$.*

This means that for any $x \in \Gamma_\varepsilon$ the angle in $T_xM/T_x\Gamma_\varepsilon$ between any vector in $T_xW^s/T_x\Gamma_\varepsilon$ and any vector in $T_xW^u/T_x\Gamma_\varepsilon$ is of the order $\sqrt{\varepsilon}$. As ε tends to zero, bounded neighborhoods of Γ_ε in the stable and unstable manifolds tend to $M \subset N$.

Suppose further that the topological condition is satisfied. Then:

(3) *Let $\pi\colon U \to M$ be the projection obtained by the identification of the neighborhood $U \subset N$ with a neighborhood of M in T^*M. Then*

$$\pi(W^s) = \pi(W^u) = M\,.$$

(4) *There exists an orbit of f_ε which is homoclinic to Γ_ε and contained in an $O(\sqrt{\varepsilon})$-neighborhood W of M: $W^s \cap W^u \setminus \Gamma_\varepsilon \neq \varnothing$. More precisely, the homoclinic orbit is contained in the neighborhood*

$$W = W_\varepsilon = \Big\{x \in U : d(x\,, M) \leqslant \Big(2\varepsilon \max_M (\overline{V}|_{\Gamma_0} - \overline{V}) + o(\varepsilon)\Big)^{1/2}\Big\}.$$

The distance is defined by the imbedding $U \subset T^*M$ *and the Riemannian metric* $\overline{K}$ *on* T^*M.

(5) *If the homology group* $H_1(M, \Gamma_0, \mathbb{R})$ *is nonzero, the number of homoclinic orbits to* Γ_ε *in* W *is not less than* $2 \dim H_1(M, \Gamma_0, \mathbb{R})$.

COROLLARY 1.1. *If* M *is a resonant invariant torus with multiplicity* $m - n$ *of a completely integrable positive definite symplectic diffeomorphism (Example* 1.1*), then for small* $\varepsilon > 0$ *there exist at least* $2(m-n)$ *homoclinic orbits to the perturbed* n*-dimensional hyperbolic torus.*

REMARKS. 1. For the case $n = 0$ (invariant tori are fixed points of the diffeomorphism), fixed points of the perturbed map f_ε exist if we replace the positive definiteness assumption by the weaker condition that for each $x \in M$ the quadratic form $\overline{K}$ is nondegenerate on T_x^*M. Counting critical points of the generating function, one can prove that in this case for small ε there exist at least cat(M)[1] fixed points of the perturbed diffeomorphism, and to each nondegenerate critical point of $\overline{V}$ there corresponds a fixed point [2]. However, without the positive definiteness assumption, it can happen that none of these points are hyperbolic. Therefore our results do not hold in this case: it is possible that no homoclinic orbits exist.

2. If $n > 0$, it is also possible to prove the existence of an f_ε-invariant hyperbolic torus replacing the positive definiteness assumption by the nondegeneracy assumption (see [43], where a similar situation of perturbation of the resonant invariant torus of a completely integrable analytic Hamiltonian system is considered). However, in this case for the existence of an invariant torus of the perturbed map, further conditions on critical points of $\overline{V}$ are needed, and generically they are not necessarily satisfied. Examples of Herman show that it is possible that there exist no homoclinic orbits to perturbed invariant tori. It would be interesting to find sufficient conditions for the existence of homoclinics in the nonpositive definite case.

3. Homoclinic orbits constructed in Theorem 1.1 are "short" in a certain sense (see Lemma 3.5). If one of these homoclinics is isolated, it is possible to prove the existence of an infinite number of "long" homoclinics. Therefore, in any case the number of homoclinics is infinite. In the transversal case, this was known to Poincaré and Birkhoff (see, for example, [45]). In the nontransversal case the proof can be obtained by using the variational methods developed in [18, 19, 22, 38, 41] for the case of homoclinics to a position of equilibrium. It is not given here.

4. Statements (3)–(5) do not hold without the topological condition. We shall give an example later for the case of Hamiltonian systems (Example 1.2).

5. If the perturbed diffeomorphism f_ε is only C^1-close to f, or if the rotation vector ω does not satisfy the Diophantine condition, then KAM-theory fails and in general the hyperbolic invariant torus Γ_ε does not exist. In this case, using the methods of Mather (see [32], where the case of a C^1-small perturbation of a positive definite Diophantine Lagrangian invariant torus is considered), it is possible to prove the existence of a compact invariant set in a small neighborhood of Γ_0. This set is hyperbolic in a certain sense and a partial generalization of Theorem 1.1 holds.

For simplicity, we shall consider the case when $f_\varepsilon \in C^2(N \times \mathbb{R}\{\varepsilon\}, N)$. Then for each ε the diffeomorphism f_ε is C^2, but the vector field $\partial f_\varepsilon / \partial \varepsilon$ is only C^1 on

[1] For the definition of cat(M) see the second paper of V. Kozlov in this volume.

$N \times \mathbb{R}\{\varepsilon\}$. Thus f_ε is C^1-close to f. Hence, the generating function S_ε belongs to $C^2(U \times \mathbb{R}\{\varepsilon\})$. Though $\partial S_\varepsilon/\partial\varepsilon$ is in general only of class $C^1(U \times \mathbb{R}\{\varepsilon\})$, it can be proved that the function V defined in (1.4) is of the class $C^2(M)$. This follows from the alternative definition of V mentioned after (1.4). Thus, all the definitions given before Theorem 1.1 make sense in the present case.

Let $i : \Gamma_0 \to M$ be the inclusion and

$$i^* : H^1(M, \mathbb{R}) \to H^1(\Gamma_0, \mathbb{R}) = H^1(\mathbb{T}^n, \mathbb{R}) = \mathbb{R}^n$$

the induced map of cohomology groups. Denote $E = i^*(H^1(M, \mathbb{R})) \subset \mathbb{R}^n$. If the topological condition is satisfied, then i^* is surjective and thus $E = \mathbb{R}^n$.

THEOREM 1.1′. *Suppose that the perturbed diffeomorphism f_ε is C^1-close to f and the rotation vector ω is nonresonant. Suppose that the invariant manifold M is positive definite and the maximum of $\overline{V}$ on $\widetilde{M}$ is strict. Let $\Gamma_0 = \{\overline{V} = \max\}$. For any $C > 0$ and all sufficiently small $\varepsilon > 0$, there exists a $\delta = \delta(\varepsilon)$ such that $\delta \to 0$ as $\varepsilon \to 0$ and*:

(1) *For any vector $I \in E \subset \mathbb{R}^n$ such that $|I| \leqslant C\varepsilon$, the diffeomorphism f_ε has a compact invariant set $\Gamma = \Gamma(I)$ in the neighborhood*

$$D = D_\varepsilon = \left\{x \in U : \pi(x) \in U_\delta,\ d(x, M) \leqslant \left(2\varepsilon \max_{U_\delta}(\overline{V}|_{\Gamma_0} - \overline{V}) + o(\varepsilon\delta^2)\right)^{1/2}\right\},$$

of the torus Γ_0. Here $U_\delta = \{y \in M : d(y, \Gamma_0) \leqslant \delta\}$. Thus, as $\varepsilon \to 0$, the set Γ tends to Γ_0 and is $o(\sqrt{\varepsilon})$-close to M. The projection π of Γ into U_δ is injective.

(2) *For any $x \in M$, there exists a positive half-orbit $f_\varepsilon^k(y) \in W$, $k \geqslant 0$, of the map f_ε starting in $y \in \pi^{-1}(x) \cap W$ and with the ω-limit set Ω_+ contained in D and intersecting Γ (the neighborhood $W = W_\varepsilon$ is defined in Theorem* 1.1). *Moreover, any minimal invariant subset of Ω_+ is contained in Γ. The same holds for negative half-orbits, if the ω-limit set is replaced by the α-limit set Ω_-.*

(3) *If $H_1(M, \Gamma_0, \mathbb{R}) \neq 0$, then f_ε has in W at least $2\dim H_1(M, \Gamma_0, \mathbb{R})$ orbits that are not contained in D and such that their limit sets $\Omega_\pm \subset D$ satisfy the conditions of part* (2).

Orbits in (2) and (3) are semiasymptotic to Γ in Birkhoff's sense [8] (see §4). Theorem 1.1′ is stated in [13]. The proof is contained in §9.

REMARKS. 1. We sketch the definition of the invariant set $\Gamma = \Gamma(I)$. It is based on the construction of Mather [32], which is described in §4. By means of this construction and the results of §§2–3, to every $O(\varepsilon)$-small cohomology class $c \in H^1(M, \mathbb{R})$ there corresponds an invariant set $\Lambda_c \subset D = D_\varepsilon$. It can be proved (see §9) that for all $O(\varepsilon)$-small $c, c' \in H^1(M, \mathbb{R})$ such that $i^*c = i^*c' = I \in \mathbb{R}^n$, we have $\Lambda_c = \Lambda_{c'}$. Hence, the set Λ_c depends only on $I = i^*c$. Thus, for any $I \in E$, $|I| \leqslant C\varepsilon$, we get an f_ε-invariant set $\Gamma(I) \subset D_\varepsilon$. If the topological condition is satisfied, the set Γ is defined for any $I \in \mathbb{R}^n$, $|I| \leqslant C\varepsilon$.

2. The vector $I \in \mathbb{R}^n$ is a sort of an "action variable" for the set Γ. If Γ is an invariant torus close to Γ_0, then I is exactly the action variable for Γ:

$$I_i = \int_{\gamma_i} \lambda, \qquad i = 1, \dots, n,$$

where γ_i are the basis cycles of Γ (see Lemma 4.1). If the assumptions of Theorem 1.1 are satisfied, then for some $I = I(\varepsilon)$ the set $\Gamma(I(\varepsilon))$ is the invariant torus Γ_ε. However, if $n > 0$, generically even in this case for a dense set of I the sets $\Gamma(I)$ are not invariant tori.

3. The invariant sets Γ consist of supports of f_ε-invariant probability measures μ on U. To each measure μ there corresponds a homology class $\rho(\mu) \in H_1(M, \mathbb{R})$ [32] (the correspondence is not canonical). If $\operatorname{supp}(\mu) \subset D$, then $\rho = i_* \nu$, $\nu \in H_1(\Gamma_0, \mathbb{R}) = \mathbb{R}^n$. If the topological condition is satisfied, then $\nu(\mu) \in \mathbb{R}^n$ is correctly defined and close to ω. The vector $\nu \in \mathbb{T}^n$ can be considered as the "rotation vector" for μ. If μ is the measure supported by an ergodic invariant torus close to Γ_0, then ν is its rotation vector (see Lemma 4.1). However, it seems impossible to prove the existence of an ergodic measure with a given rotation vector ν.

4. The set $\Gamma(I)$ is the union of supports of invariant measures μ, $\operatorname{supp}(\mu) \subset D$, that minimize the Mather action functional

$$A_I(\mu) = \int S_\varepsilon \, d\mu - \langle I, \nu(\mu) \rangle .$$

5. Only if the set $\Gamma(I)$ is ergodic, then its existence has dynamical consequences. It can be proved that for any C and small ε there exist at least $n + 1$ points I_k, $|I_k| \leqslant C\varepsilon$, such that the sets $\Gamma(I_k)$ are ergodic.

6. It is possible to obtain an estimate for $\delta = \delta(\varepsilon)$. For example, if the assumptions of Theorem 1.1 are satisfied, then our proof yields $\delta = o(\varepsilon^{1/4})$. Probably in this case $\delta = o(\varepsilon^{1/2})$.

COROLLARY 1.2. *If M is an arbitrary resonant Lagrangian invariant torus with multiplicity $m - n$ of a completely integrable positive definite symplectic diffeomorphism, and the perturbation f_ε is C^1-small, then to any $O(\varepsilon)$-small $I \in \mathbb{R}^n$ there corresponds an unstable invariant set in an $o(\sqrt{\varepsilon})$-neighborhood of the M and at least $2(m - n)$ semihomoclinic orbits.*

This corollary partly fills the gap between the results of Bernstein and Katok [7], where periodic orbits of the perturbed diffeomorphism were considered for the case $n = 0$, and of Mather [32], where $n = m$ and the torus M is Diophantine. However, in these works the smallness assumptions on the perturbation are milder. In [7] they do not depend on the resonant torus.

As usual, Theorem 1.1, dealing with symplectic maps, can be reformulated for Hamiltonian systems. There are three different but similar versions: time-periodic, autonomous, and autonomous isoenergetic. We shall give all the details only for the second case, which is the simplest [12, 43]. Incidentally, the two other and the case of diffeomorphisms can be reduced to it by the well-known Poincaré trick [36].

Consider a smooth family $H = H_0 + \varepsilon H_1 + O(\varepsilon^2)$ of functions on N. Denote by $g_\varepsilon^t \colon N \to N$ the corresponding Hamiltonian phase flow. Suppose that the unperturbed Hamiltonian system with the Hamiltonian H_0 has a compact connected Lagrangian invariant manifold M and the flow on M is quasiperiodic. This means that $g_0^t|_M = g_{\omega t}$, where g is a free group action of the torus $\mathbb{T}^n$ on M and $\omega \in \mathbb{R}^n$ is the frequency vector. In other words, the unperturbed Hamiltonian vector field v is constant on each invariant torus in M. In the case of autonomous Hamiltonian systems, ω is nonresonant if $\langle \omega, k \rangle \neq 0$ for all $k \in \mathbb{Z} \setminus \{0\}$. The Diophantine condition (1.1) is replaced by the following

DIOPHANTINE CONDITION. There exist $C, \nu > 0$ such that

$$|\langle \omega, k \rangle| \geqslant C|k|^{-\nu} \tag{1.6}$$

for all $k \in \mathbb{Z}^n \setminus \{0\}$.

It is easy to see that the invariant manifold M consists of critical points of the Lagrange function $L = \langle v, \lambda \rangle - H_0$, and thus, as before, the second differential $-d^2L(x)$, $x \in M$, defines a quadratic function K on T^*M. Let $\overline{K}$ be the averaged function (1.3) on T^*M. We call the invariant manifold M *positive definite*, if, as in the case of symplectic diffeomorphisms, $\overline{K}$ is a Riemannian metric on T^*M. Denote $V = H_1|_M$ and let $\overline{V}$ be the average (1.5).

The topological condition is the same as in the case of diffeomorphisms. A sketch of the proof of the next theorem was given in [12].

THEOREM 1.2. *Suppose that the maximum of $\overline{V}$ on $\widetilde{M} = M/\mathbb{T}^n$ is strict and nondegenerate. Then for small $\varepsilon > 0$ the perturbed Hamiltonian flow g_ε^t satisfies statements* (1)–(5) *of Theorem* 1.1 *if we replace everywhere the diffeomorphism f_ε by the flow g_ε^t. In short, the perturbed Hamiltonian system has a hyperbolic invariant torus Γ_ε and there exist homoclinic orbits to it.*

Suppose that the frequency vector is nonresonant but does not satisfy the Diophantine condition, and the perturbation is only C^2-small: $H \in C^2(N \times \mathbb{R}\{\varepsilon\})$ and, moreover, the differential dH (calculated for fixed ε) is of class C^2 on $N \times \mathbb{R}\{\varepsilon\}$. Then $H_1 \in C^2(N)$. In this case, an analog of Theorem 1.1′ holds.

THEOREM 1.2′. *Suppose that the maximum of $\overline{V}$ on $\widetilde{M}$ is strict. Then for small $\varepsilon > 0$ the perturbed Hamiltonian flow g_ε^t satisfies statements* (1)–(3) *of Theorem* 1.1′, *if we replace everywhere the diffeomorphism f_ε by the flow g_ε^t.*

Let us show that the topological condition in Theorems 1.1 and 1.2 is necessary. We shall provide a counterexample to Theorem 1.2. A counterexample to Theorem 1.1 can be obtained by a small modification.

EXAMPLE 1.2. *Heavy rigid body with a fixed point.* Let $G = SO(3)$ be the orthogonal group and let $N = T^*G$ with the standard symplectic structure. As usual, for any $X \in G$ the tangent space T_XG is identified with the angular velocity space $\mathbb{R}^3\{\Omega\}$ by using the equation

$$X^{-1}\dot{X}a = \Omega \times a, \qquad \text{for all } a \in \mathbb{R}^3.$$

The cotangent space T_X^*G is identified with the angular momentum space $\mathbb{R}^3\{M\}$ (see [2, 4]). Let $e \in \mathbb{R}^3$ be the vertical unit vector and $\gamma = X^{-1}e \in S^2$ the Poisson vector [2, 4]. Consider the Hamiltonian system with Hamiltonian function

$$H = \langle M, I^{-1}M \rangle/2 + \varepsilon P\langle \gamma, \rho \rangle + (\omega - \varepsilon)\langle M, \gamma \rangle$$

on the phase space N. This is the Hamiltonian of a heavy rigid body with a fixed point [2, 4], considered in the rotating frame of reference. Here I is the inertia tensor, εP the product of the weight and the distance from the fixed point to the center of mass, and ρ the unit radius-vector of the mass center.

For $\varepsilon = 0$ the unperturbed Hamiltonian system has an invariant Lagrangian manifold $\{M = 0\} \cong SO(3)$. Since for $M = 0$ we have $\Omega = \partial H/\partial M = \omega\gamma$, every orbit in $\{M = 0\}$ corresponds to a rotation of the rigid body with the constant vertical angular velocity ω. Hence, this manifold is fibered by $2\pi/\omega$-periodic orbits of the unperturbed system and the corresponding fiber bundle is the standard Hopf fibration $SO(3) \to S^2\{\gamma\}$ [42]. If ω is a Diophantine number, this Hamiltonian system satisfies all assumptions of Theorem 1.2 except the topological condition. Thus statements (1) and (2) of Theorem 1.2 hold true. Hence for small $\varepsilon > 0$ the perturbed system has a hyperbolic periodic orbit Γ_ε. This orbit is given by the equations $\gamma = \rho$, $\Omega = \omega\gamma$, $M = \varepsilon I\rho$. Let us show, for example, that statement (3) of Theorem 1.2 is not satisfied.

For simplicity, we shall consider the case when all momenta of inertia are equal: I is the unit matrix. Then the Hamiltonian equations have three integrals: H, $J = \langle M, \gamma\rangle$, and $K = \langle M, \rho\rangle$. On Γ_ε, and, consequently, on any orbit asymptotic to Γ_ε, we have $K = J = \varepsilon$ and $H = \varepsilon(P + \omega) - \varepsilon^2/2$. Let $\pi\colon T^*G \to G$ be the projection. It follows that for any $\gamma \in S^2$ corresponding to a point in the set $\pi(W^s(\Gamma_\varepsilon))$ or $\pi(W^u(\Gamma_\varepsilon))$, we have

$$U(\gamma) = \min\{H(M, \gamma) : K(M, \gamma) = J(M, \gamma) = \varepsilon\} \leqslant \varepsilon(P + \omega) - \varepsilon^2/2. \tag{1.7}$$

The function $U(\gamma)$ is called the *reduced* or *effective potential* [2, 4]. A calculation yields

$$U(\gamma) = \varepsilon^2/(1 + \langle \rho, \gamma\rangle) + \varepsilon(P\langle \rho, \gamma\rangle + \omega - \varepsilon).$$

Substituting $U(\gamma)$ into (1.7), we see that the sets $\pi(W^{s,u}(\Gamma_\varepsilon))$ are contained in the region of possible motion

$$\{X \in G : \langle \rho, \gamma\rangle \leqslant 1 - \varepsilon/P\} \neq G.$$

Thus, $\pi(W^{s,u}(\Gamma_\varepsilon)) \neq G$ and therefore statement (3) of Theorem 1.2 is not satisfied.

REMARKS. 1. In the time-periodic case (the Hamiltonian $H \in C^\infty(N \times \mathbb{T})$ is 1-periodic in $t \in \mathbb{T} = \mathbb{R}/\mathbb{Z}$), we require that the restriction of the Hamiltonian flow of the unperturbed system on the Lagrangian manifold $M \subset N$ be autonomous and quasiperiodic. Thus, it equals $g_{\omega t}\colon M \to M$, where g is a free $\mathbb{T}^n$-action on M. The Diophantine condition (1.6) is replaced by (1.1). The functions K on $T^*M \times \mathbb{T}$ and V on $M \times \mathbb{T}$ are defined as in the autonomous case. The functions $\overline{K}$ on T^*M and $\overline{V}$ on M are obtained by additional averaging in time: for example

$$\overline{K}(x, p) = \int_{\mathbb{T}^n}\int_0^1 K(g_\theta^*(x, p), t)\, d\theta\, dt. \tag{1.8}$$

We shall show that the functions $\overline{K}$ and $\overline{V}$ coincide with the functions defined by (1.3) and (1.5), where the symplectic diffeomorphism f_ε is the time-1 Poincaré map of the Hamiltonian system.

The statements of the results for the time-periodic case coincide with the statements of Theorems 1.1 *and* 1.1′. For further references, we call these results Theorem 1.3 and Theorem 1.3′.

2. In the isoenergetic case [12], the positive definiteness condition is replaced by the quasiconvexity assumption $\overline{K}(x, p) > 0$ for $\langle p, v(x)\rangle = 0$, $p \neq 0$, where v is the unperturbed Hamiltonian vector field on M, and the Diophantine condition by the usual isoenergetic KAM condition on the ratios of frequencies. In this case the frequency vector of the hyperbolic torus of the perturbed system is only proportional to the frequency vector of the unperturbed system, and the energy is fixed. The rest is the same as in the autonomous case.

In fact both cases can be reduced to Theorem 1.2 (or 1.2′). Let us recall how to do this for the time-periodic case, since we shall deduce Theorem 1.1 from Theorem 1.3. The first to use this trick was Poincaré, who applied it to the three-body problem [36].

Identify a neighborhood U of M in N with a neighborhood of M in T^*M. Since the unperturbed vector field v on M is quasiperiodic, the perturbed Hamiltonian H on $U \times \mathbb{T}$ is

$$H(x, p, t) = \langle v(x), p\rangle + K(x, p, t) + O(\|p\|^3) + O(\varepsilon),$$

where the function K is quadratic in p. Define the new configuration space $\widehat{M} = M \times \mathbb{T}\{t\}$ and denote by E the momentum dual to t. Let the new Hamiltonian function on the phase space $T^*\widehat{M}$ be

$$\begin{aligned}\widehat{H}(x, p, t, E) &= \exp(H + E)\\ &= \langle v(x), p\rangle + E + \widehat{K}(x, p, t, E) + O(\|p\|^3 + |E|^3) + O(\varepsilon),\end{aligned}$$

where $\widehat{K} = K(x, p, t) + \langle v(x), p\rangle^2/2 + E^2/2$. The unperturbed Hamiltonian system is quasiperiodic on $\widehat{M}$ and the frequency vector is $(\omega, 1) \in \mathbb{R}^{n+1}$. If the positive definiteness condition is satisfied, then the average $\overline{K}$ is a positive definite quadratic form in p. Then the obtained autonomous Hamiltonian system on $T^*\widehat{M}$ satisfies the autonomous positive definiteness condition: the average of $\widehat{K}$ is positive definite in p and E. The Diophantine condition (1.1) becomes (1.6). The projections of the trajectories of the obtained system into $T^*M \times \mathbb{T}$ coincide with the trajectories of the initial time-periodic Hamiltonian system. Thus, *the time-periodic case is a corollary of the autonomous case* (Theorem 1.3 and 1.3′ follow from Theorems 1.2 and 1.2′). A similar trick can be applied in the isoenergetic case.

EXAMPLE 1.3 [43]. Consider a smooth perturbation

$$H = H_0(y) + \varepsilon H_1(x, y) + O(\varepsilon^2), \qquad x \in \mathbb{T}^m, \ y \in D \subset \mathbb{R}^m, \tag{1.9}$$

of a completely integrable Hamiltonian system. Let $M = \mathbb{T}^m \times \{y_0\}$ be a resonant invariant torus of the unperturbed system: the frequency vector $\omega = \partial H_0(y_0)/\partial y$ satisfies $m - n$ independent resonance conditions $\langle \omega, k\rangle = 0$, $k \in \mathbb{Z}^m$. Then M is foliated by n-dimensional invariant tori and M is positive definite if and only if the Hessian of H_0 at y_0 is positive definite. In this case, the function $\overline{V}$ on M is the sum of resonance harmonics

$$\overline{V} = \sum_{k \in \mathbb{Z}^m, \ \langle \omega, k\rangle = 0} h_k \exp(i\langle k, x\rangle)$$

of the function $H_1|_M$. If the maximum of $\overline{V}$ on $\widetilde{M} \cong \mathbb{T}^{m-n}$ is nondegenerate, Theorem 1.2 ensures the existence of at least $2(m-n)$ "short" homoclinic orbits to an n-dimensional hyperbolic invariant torus of the perturbed system.

The first to study this situation was Poincaré [36]. He considered the nondegenerate case: $\det(\partial^2 H_0(y_0)/\partial y^2) \neq 0$. For $n = 1$ (M is foliated by periodic orbits of the unperturbed system), Poincaré proved that to each nondegenerate critical point of the averaged function $\overline{V}$ on $\mathbb{T}^{m-1}$ there corresponds a periodic orbit of the perturbed system. He calculated the characteristic exponents of this orbit and so obtained statements (1) and (2) of Theorem 1.2 for $n = 1$. This result is based on the implicit functions theorem. When $n > 1$, KAM-theory is needed, and in general for the perturbed system no hyperbolic invariant tori with frequency ω exist unless additional assumptions are made. Poincaré only obtained formal series representing the torus and its stable and unstable invariant manifolds, but was unable to prove convergence.

For $n > 1$, the existence of a hyperbolic invariant torus of the perturbed system (1.9) was proved by Treshchev [43] under certain nondegeneracy assumptions. Treshchev studied the analytic case and used Graff's results [23] on the conservation of hyperbolic invariant tori. By using the methods of [46] or [40], Treshchev's theorem can be extended to the smooth case (in [46] hyperbolic tori of symplectic maps were also studied). We shall not do this in the present paper, but only reduce the KAM-part of Theorems 1.1–1.3 (statements (1) and (2)) to a small generalization of Treshchev's results. This is done in §2.

The main result of Theorems 1.1–1.3 is (4). Statements (3) and (5) are relatively easy. Statements (3)–(5) follow from general existence results on homoclinic orbits of positive definite Lagrangian systems. They are proved by global variational methods introduced in [10–12]. Different Lagrangian variational methods for homoclinic orbits were used later by Rabinowitz, Giannoni, Tanaka [21, 22, 37–39] and others (see references in these papers and [1, 29]). Recently, the existence problem for homoclinic orbits to equilibria of Hamiltonian systems was studied by Hamiltonian variational methods which are quite different from the methods used in the present paper (see, for example, [18, 19, 41]). Probably these methods can be applied to generalize Theorems 1.1–1.3 to the nonpositive definite case.

Theorems 1.1′–1.3′ are based only on classical perturbations theory and existence theorems of the variational calculus. They are treated in the §9.

§2. The Delaunay method

In this section we reduce the case of symplectic diffeomorphisms (Theorem 1.1) to the case of time-periodic Hamiltonian systems (Theorem 1.3). Recall that Theorem 1.2 implies Theorem 1.3. The former theorem is reduced to the case of nearly natural Hamiltonian systems. We shall also show how statements (1) and (2) of Theorem 1.2 follow from a small generalization of Treshchev's result [43]. All this is done by a global version of a classical method described by Poincaré [36] as the Delaunay method.

Suppose that the assumptions given before Theorem 1.1 are satisfied. First we shall prove that the Riemannian metric $\overline{K}$ on M is well defined, i.e., $\overline{K}$ is a symplectic invariant.

LEMMA 2.1. *The function $\overline{K}$ on T^*M does not depend on the choice of the Liouville 1-form λ.*

PROOF. Replace λ by $\widetilde{\lambda} = \lambda + dh$, where $dh(x) = 0$ for all $x \in M$. If $\widetilde{S}$ is the corresponding generating function for the map f, then

$$d\widetilde{S} = f^*\widetilde{\lambda} - \widetilde{\lambda} = dS + d\,(h \circ f - h)\,.$$

Hence $\widetilde{S} = S + h \circ f - h$. It follows that for $x \in M$

$$d^2\widetilde{S}(x) = d^2S(x) + d^2h(f(x)) \circ df(x) - d^2h(x)\,.$$

Since M is Lagrangian and f-invariant, after the identification $T_xN/T_xM = T_x^*M$, $v \to i_v\Omega|_{T_xM}$, the differential $df(x)$ is replaced by $((df(x)|_{T_xM})^*)^{-1}$. Denote by K, $\widetilde{K}$, and W the quadratic forms on T^*M defined by second differentials of S, $\widetilde{S}$, and h respectively. The latter equation implies that

$$\widetilde{K} = K + W \circ (f|_M)^* - W = K + W \circ g_\omega^* - W\,.$$

Taking the average (1.3) with respect to the $\mathbb{T}^n$-action (1.2) on M, we get

$$\overline{\widetilde{K}} = \overline{K} + \overline{W} \circ g_\omega^* - \overline{W} = \overline{K}\,,$$

since $\overline{W}$ is g_ω^*-invariant. Lemma 2.1 is proved. The assumptions of Theorems 1.2 and 1.3 are correct by the same reason.

By the Weinstein theorem [44], a neighborhood U of M in N is symplectically diffeomorphic to a neighborhood of M in T^*M with the standard symplectic structure. Since the statement of Theorem 1.1 is local, from now on we can assume that $N = T^*M$ and $f\colon U \subset T^*M \to T^*M$. Thus, f_ε is a smooth family of exact symplectic diffeomorphisms of U into T^*M.

For technical reasons, it is convenient not to deal with symplectic diffeomorphisms, but with time-periodic Hamiltonian systems. Choose an arbitrary vector $\omega \in \mathbb{R}^n$ corresponding to the rotation vector $\omega \in \mathbb{T}^n$ and define the vector field v on M by the formula

$$v(x) = \left.\frac{\partial}{\partial t}\right|_{t=0} g_{\omega t}(x) = \sum_{i=1}^{n} \omega_i v_i\,, \qquad v_i(x) = \left.\frac{\partial}{\partial \theta_i}\right|_{\theta=0} g_\theta(x)\,. \tag{2.1}$$

Then the flow $g_{\omega t}^*$ on T^*M is generated by the Hamiltonian system with the Hamiltonian function

$$h_0 = \langle v(x)\,,\, p\rangle = \langle \omega\,,\, J\rangle\,, \qquad J_i = \langle p\,,\, v_i\rangle\,.$$

LEMMA 2.2. *Let the neighborhood U be small enough. Then for sufficiently small $\varepsilon \geqslant 0$, there exists a smooth family H_ε of functions on $T^*M \times \mathbb{T}$:*

$$H_\varepsilon(x, p, t) = \langle v(x), p\rangle + O(\|p\|^2) + O(\varepsilon), \qquad t \in \mathbb{T}, \tag{2.2}$$

such that the diffeomorphism $f_\varepsilon|_U$ is the time-1 Poincaré map of the time-periodic Hamiltonian system with the Hamiltonian H_ε.

PROOF. Let $F_\varepsilon = g^*_{-\omega} \circ f_\varepsilon \colon U \to T^*M$. Then for $\varepsilon = 0$, we have $F_0|_M = \mathrm{id}_M$. Since F_0 is symplectic, it is easy to see that if U is small, then F_0 is homotopic to id_U in the class of (nonsymplectic) diffeomorphisms $U \to T^*M$. For small ε, this holds for F_ε. Thus, there exists a smooth homotopy $F^t_\varepsilon \colon U \to T^*M$, $0 \leqslant t \leqslant 1$, such that $F^0_\varepsilon = \mathrm{id}$, $F^1_\varepsilon = F_\varepsilon$, and $F^t_0|_M = \mathrm{id}_M$.

Let $f^t_\varepsilon = g^*_{\omega t} \circ F^t_\varepsilon \colon U \to T^*M$. We can assume that $f^t_\varepsilon \equiv \mathrm{id}$ for $0 \leqslant t \leqslant \delta$ and $f^t_\varepsilon = f_\varepsilon$ for $1 - \delta \leqslant t \leqslant 1$. Let $V \subset U$ be a small neighborhood of M. Since the family f_ε is exact, by the Moser method (see, for example, [17]), we see that there exists a smooth family of diffeomorphisms $h^t_\varepsilon \colon U \to T^*M$ such that $h^t_\varepsilon \equiv \mathrm{id}$ for $0 \leqslant t \leqslant \delta$ and $1 - \delta \leqslant t \leqslant 1$, $h^t_0|_M = \mathrm{id}_M$ and the diffeomorphism $g^t_\varepsilon = h^t_\varepsilon \circ f^t_\varepsilon \circ (h^t_\varepsilon)^{-1} \colon V \to T^*M$ is symplectic. Then $g^t_\varepsilon \equiv \mathrm{id}$ for $0 \leqslant t \leqslant \delta$ and $g^t_\varepsilon = f_\varepsilon|_V$ for $1 - \delta \leqslant t \leqslant 1$.

Let $H^t_\varepsilon \colon g^t_\varepsilon(V) \subset T^*M \to \mathbb{R}$, $0 \leqslant t \leqslant 1$, be the Hamiltonian function corresponding to the flow g^t_ε. Then $H^t_\varepsilon \equiv 0$ for $0 \leqslant t \leqslant 1$ and $1 - \delta \leqslant t \leqslant 1$. Thus H^t_ε can be extended to a smooth 1-periodic in t function $H_\varepsilon \colon T^*M \times \mathbb{T}\{t\} \to \mathbb{R}$. Since $g^t_0|_M = g_{\omega t}$, (2.1) implies $H_0 = \langle v(x), p\rangle + O(\|p\|^2)$. This proves (2.2). □

From now on the case of symplectic diffeomorphisms (Theorem 1.1) and the case of time-periodic Hamiltonian systems (Theorem 1.3) are treated simultaneously. From the Taylor expansion of (2.2), we get

$$H_\varepsilon(x, p, t) = \langle v(x), p\rangle + G(x, p, t, \varepsilon) + O(\|p\|^3) + \varepsilon O(\|p\|) + O(\varepsilon^2), \tag{2.3}$$

where

$$G = \langle A(x, t)p, p\rangle/2 + \varepsilon U(x, t). \tag{2.4}$$

The idea of the Delaunay method is as follows. Let $\mu = \sqrt{\varepsilon}$. If we perform a nonsymplectic transformation of variables $(x, p) \to (x, p/\mu)$, the Hamiltonian flow with Hamiltonian function (2.3) is replaced by the Hamiltonian flow with the Hamiltonian

$$\begin{aligned} h_\mu(x, p, t) &= H_\varepsilon(x, \mu p, t)/\mu \\ &= h_0(x, p) + \mu h_1(x, p, t) + \mu^2 O(\|p\|) + O(\mu^3), \end{aligned} \tag{2.5}$$

where

$$h_1 = \langle A(x, t)p, p\rangle/2 + U(x, t). \tag{2.6}$$

This is the standard form of a Hamiltonian function for the averaging method, since the flow $g^*_{\omega t}$ corresponding to the unperturbed Hamiltonian h_0 is generated by an ergodic flow on the torus $\mathbb{T}^{n+1} = \mathbb{T}^n \times \mathbb{T}\{t\}$. Thus, there exists a time-periodic canonical transformation, smoothly depending on μ and identical for $\mu = 0$, such that after this transformation the new Hamiltonian function takes the form

$$(2.7) \qquad h(x, p, t, \mu) = h_0(x, p) + \mu \overline{h}_1(x, p) + \mu^2 O(\|p\|) + O(\mu^3).$$

Here, as in (1.8), $\overline{h}_1$ denotes the average of the function (2.6) with respect to the $\mathbb{T}^{n+1}$-action on $T^*M \times \mathbb{T}$ corresponding to the $\mathbb{T}^n$-action (1.2). The first two terms in (2.7) are $\mathbb{T}^n$-invariant and hence they commute.

Reduction of the Hamiltonian function to the form (2.7) is the first step of the Delaunay method [36]. It is convenient to rewrite (2.7) in the canonical variables (recall that (2.7) was obtained from (2.3) by a nonsymplectic transformation).

LEMMA 2.3. *There exists a smooth family φ_ε^t of time-periodic symplectic diffeomorphisms of T^*M such that $\varphi_\varepsilon^0 = \mathrm{id}$, $\varphi_0^t|_M = \mathrm{id}$, that transforms the Hamiltonian flow with the Hamiltonian function* (2.3) *into the flow with the Hamiltonian*

$$(2.8) \quad F(x, p, t, \varepsilon) = \langle v(x), p\rangle + \overline{G}(x, p, \varepsilon) + O(\|p\|^3) + \varepsilon O(\|p\|) + O(\varepsilon^2).$$

PROOF. It is standard (see, for example, [4]). The symplectic diffeomorphism φ_ε^t is defined as the time-1 map of the Hamiltonian flow corresponding to a Hamiltonian function W on $T^*M \times \mathbb{T}$. Note that the "time" here is not t, but some additional parameter. The function W is a t-periodic solution of the equation

$$(2.9) \qquad W_t + \{W, h_0\} = \overline{G} - G, \qquad \{W, h_0\} = \sum_{s=1}^{n} \omega_s \{W, J_s\}.$$

Here $\{\cdot, \cdot\}$ is the Poisson bracket. Each smooth function on $T^*M \times \mathbb{T}$ can be represented as a Fourier series (a sum of Fourier harmonics). By definition, every Fourier harmonic f is a function satisfying the equations

$$f|_{t+\tau} = e^{ij\tau} f|_t, \quad f \circ g_\theta^* = e^{i\langle k, \theta\rangle} f, \qquad i = \sqrt{-1},$$

for some $k \in \mathbb{Z}^n$ and $j \in \mathbb{Z}$. Thus

$$\frac{\partial}{\partial t} f = ijf, \qquad \{f, J_s\} = \frac{\partial}{\partial \theta_s}\Big|_{\theta=0} f \circ g_\theta^* = ik_s f, \qquad s = 1, \dots, n.$$

Hence, for any harmonic f, the left-hand side of (2.9) equals $i(\langle k, \omega\rangle + j)f$. Therefore, a smooth solution of (2.9) exists because of the Diophantine condition (1.1) [4]. □

REMARKS. 1. Using a sequence of normalizing symplectic diffeomorphisms, it is possible to make the Hamiltonian (2.7) autonomous and $\mathbb{T}^n$-invariant up to an arbitrary order μ^k [4].

2. If f_ε is only C^1-close to f (see assumptions of Theorem 1.1′), it is possible to prove that dH_ε is of class C^2 on $T^*M \times \mathbb{R}\{\varepsilon\}$. Thus equations (2.3) and (2.4) hold, but the functions A and U in (2.4) are only C^1. Due to the small denominators phenomenon [4], it is possible that the solution of (2.9) does not exist. The same happens if the frequency vector ω does not satisfy the Diophantine condition. If the assumptions of Theorem 1.1′ or Theorem 1.3′ are satisfied, then the following weaker version of Lemma 2.3 holds.

LEMMA 2.3$'$. *For any $\alpha > 0$ there exists a smooth family φ_ε^t, $\varepsilon \geqslant 0$, of t-periodic symplectic transformations such that $\varphi_\varepsilon^0 = \mathrm{id}$, $\varphi_0^t|_M = \mathrm{id}$, and the new Hamiltonian takes the form*

$$F(x, p, t, \varepsilon) = \langle v(x), p\rangle + \overline{G}(x, p, \varepsilon) + \langle w(x, t)p, p\rangle/2 + \varepsilon u(x, t) + O(\|p\|^3) + \varepsilon O(\|p\|) + O(\varepsilon^2), \tag{2.8$'$}$$

where $\|w\|_{C^1} \leqslant \alpha$ and $\|u\|_{C^1} \leqslant \alpha$. The C^1-norms are defined by the Riemannian metric on M.

PROOF. Let us approximate the functions A and U by C^∞ functions on $M \times \mathbb{T}$ that are C^1-close to A and U respectively. Expanding these smooth functions into Fourier series, we obtain Fourier polynomials $\widetilde{A}(x, t)$ and $\widetilde{U}(x, t)$ such that $w = A - \widetilde{A}$ and $u = U - \widetilde{U}$ satisfy the conditions of Lemma 2.3$'$ for arbitrary $\alpha > 0$. If we replace in (2.4) A by $\widetilde{A}$ and U by $\widetilde{U}$, the corresponding equation (2.9) has a Fourier polynomial solution $\widetilde{W}$. After the symplectic transformation generated by $\widetilde{W}$, the Hamiltonian takes the form (2.8$'$). □

We shall use (2.8$'$) in the proof of Theorems 1.1$'$–1.3$'$.

LEMMA 2.4. *Let $\overline{K}$ and $\overline{V}$ be the functions defined in (1.3) and (1.5). Then $\overline{G} = \overline{K} + \varepsilon\overline{V}$.*

PROOF. By the classical formula for the variation of the action function (see, for example, [2]), the generating function of the time-1 mapping f_ε of the flow g_ε^t on T^*M defined by the Hamiltonian (2.8) is

$$S = \int_0^1 L \circ g_\varepsilon^t \, dt.$$

Here L is the Lagrangian function on $T^*M \times \mathbb{T}$:

$$\begin{aligned} L(x, p, t, \varepsilon) &= \langle F_p, p\rangle - F \\ &= \langle \overline{A}(x)p, p\rangle/2 - \varepsilon\overline{U}(x) + O(\|p\|^3) + \varepsilon O(\|p\|) + O(\varepsilon^2). \end{aligned}$$

Since for small ε and $\|p\|$ the flow g_ε^t is close to $g_{\omega t}^*$, and the first two terms in L are $\mathbb{T}^n$-invariant, we obtain

$$S = \langle \overline{A}(x)p, p\rangle/2 - \varepsilon\overline{U}(x) + O(\|p\|^3) + \varepsilon O(\|p\|) + O(\varepsilon^2).$$

Now Lemma 2.4 follows from the definitions of K and V. Note that we used here the fact that the functions $\overline{K}$ and $\overline{V}$ are symplectic invariants (Lemma 2.1), and hence after all canonical transformations are performed they are the same as in (1.3) and (1.5). □

Lemma 2.4 implies that the definitions of the functions $\overline{K}$ and $\overline{V}$ given for the case of diffeomorphisms coincide with the definitions given for the case of time-periodic Hamiltonian systems. Thus, Theorem 1.1 and Theorem 1.3 are equivalent. We already showed in §1 that the time-periodic case follows from the autonomous one.

Therefore, Theorems 1.1 and 1.3 are corollaries of Theorem 1.2. Theorems 1.1′ and 1.3′ are also reduced to the case of autonomous Hamiltonian systems (Theorem 1.2′).

Let us sketch the idea of the proof of Theorem 1.2. At first we shall drop the last three terms in (2.8) which are arbitrarily small for small ε and $\|p\|$. Then the Hamiltonian $F = h_0 + \varepsilon \overline{G}$ is autonomous and invariant under the action of the torus $\mathbb{T}^n$ on T^*M. The Hamiltonian equations have the integral of momentum $J \colon T^*M \to \mathbb{R}^n$.

Let us consider the time-dependent symplectic transformation $g^*_{\omega t}$ of T^*M. It is easy to see that after this transformation the Hamiltonian function F is replaced by $\overline{G}$. By (2.4),

$$\overline{G} = \|p\|^2/2 + \varepsilon \overline{V}(x), \tag{2.10}$$

where we denoted $\overline{K} = \|p\|^2/2$.

We obtained a classical autonomous reversible Hamiltonian system with configuration space M and $\mathbb{T}^n$-invariant kinetic and potential energy. The kinetic energy is positive definite in momentum and defines a Riemannian metric on M. If the assumptions of Theorem 1.2 are satisfied, the set $\Gamma_0 \subset M$ of maximum points of the potential energy $\overline{V}$ is a nondegenerate critical manifold for $\overline{V}$. Hence, Γ_0 consists of unstable equilibria of our classical Hamiltonian system, and $2(m-n)$ characteristic exponents are real and nonzero. This proves statements (1) and (2) of Theorem 1.1. The existence of a homoclinic orbit to a nondegenerate critical manifold Γ_0 of maximum points of the potential energy was proved in [9] by classical variational methods based on the Maupertuis principle: homoclinics are critical points of the suitably regularized Maupertuis functional

$$S(\gamma) = \int_\gamma 2\sqrt{\varepsilon\,(\overline{V}|_{\Gamma_0} - \overline{V}(\gamma(t)))}\,\|\dot{\gamma}(t)\|\,dt$$

on the space of curves with endpoints in Γ_0. Similar methods were used in [6].

The sets Λ_I in Theorem 1.1′ can be obtained as the sets of critical points of the Hamiltonian (2.10) on $\{J = I\} \subset T^*M$.

To make this rigorous, we should remember about the dropped terms in (2.8). If we perform the canonical transformation $g^*_{\omega t}$, we obtain a Hamiltonian system which is a small t-quasiperiodic perturbation of the system (2.10). If the normalization procedure of the averaging method is convergent, we can make the Hamiltonian autonomous and $\mathbb{T}^n$-invariant. Then the existence of a hyperbolic invariant torus of the perturbed system is a corollary of the implicit functions theorem. However, in general the normal form is divergent (even for $n = 0$).

The system (2.8) is a perturbation of a system with a hyperbolic invariant torus. To prove (1) and (2) in the general case, we can use the results of [23, 43, 46] on perturbations of hyperbolic tori. They are based on KAM-theory. Unfortunately none of them can be directly applied, although Treshchev's theorem is very close to what is needed here. The difference is that he studied only the analytic case. The analyticity condition can be dropped if we modify Treshchev's proof using the methods of [46] or [40] instead of [23]. Treshchev proved that the invariant torus smoothly depends on $\mu = \sqrt{\varepsilon}$ for $\varepsilon \geqslant 0$, and the statement of Theorem 1.1 is that it smoothly depends on ε. This can be obtained by taking into account the Birkhoff normal form up to an arbitrary order. Although in general the normal form is divergent, it shows that the Taylor series in μ parametrizing the hyperbolic torus contain only even powers of μ (see also §3). This completes the sketch of the proof of statements (1) and (2) of Theorems 1.1–1.3.

The proofs of the remaining statements of Theorems 1.1–1.3 are based on Lagrangian variational methods.

§3. Reduction to a positive definite Lagrangian system

The next step in the proofs of Theorems 1.1–1.3 is to reduce the Hamiltonian system to a positive definite Lagrangian system, so that Lagrangian variational methods can be applied. Without loss of generality, we may consider only the case of autonomous Hamiltonian systems. Lemmas 2.3–2.4 imply that after performing a symplectic transformation in a neighborhood U of M in T^*M, the Hamiltonian function (2.8) takes the form

$$F = \langle v(x), p\rangle + \langle A(x)p, p\rangle/2 + \varepsilon V(x) + O(\|p\|^3) + \varepsilon O(\|p\|) + O(\varepsilon^2), \tag{3.1}$$

where the Riemannian metric $\langle A(x)p, p\rangle/2$ on T^*M and the function V on M are $\mathbb{T}^n$-invariant and coincide with the functions $\overline{K}$ and $\overline{V}$ defined in §1 (for simplicity we have dropped the bar). From now on we shall denote by $\|\cdot\|$ the Riemannian metric $\|p\|^2 = \langle A(x)p, p\rangle$. By (3.1), there exist sufficiently small $\delta > 0$ and positive constants $a < 1 < b$, arbitrarily close to 1, such that for small $\varepsilon \geqslant 0$ and all $(x, p, t) \in T^*M \times \mathbb{T}$, $\|p\|^2 \leqslant \delta$, we have

$$a\|u\|^2 \leqslant \langle F_{pp}(x, p, \varepsilon)u, u\rangle \leqslant b\|u\|^2, \quad \text{for all } u \in T_xM. \tag{3.2}$$

Let us modify the function (3.1) outside the region $\|p\| \leqslant \delta$ in such a way that:

(1) F is a smooth function on T^*M, smoothly depending on ε;
(2) F satisfies inequality (3.2) everywhere in T^*M, maybe with different $a < 1 < b$ close to 1;
(3) F is a polynomial in p of degree 2 for sufficiently large $\|p\|$.

It is easy to see that such a function exists. We already know that for small $\varepsilon > 0$, the Hamiltonian system with Hamiltonian (3.1) has the hyperbolic invariant torus Γ_ε smoothly depending on ε and coinciding with the torus $\Gamma_0 = \{x \in M : V(x) = \max V\} = g_{\mathbb{T}^n}x_0 \subset T^*M$ for $\varepsilon = 0$. Since Γ_ε is ergodic, $F|_{\Gamma_\varepsilon}$ is a constant depending only on ε. We shall assume that it is equal to zero. In particular, $V|_{\Gamma_0} = 0$.

Suppose that the topological condition of §1 is satisfied. Our goal is to prove the following

LEMMA 3.1. *Let $C > 2\max_M(-V)$. If $\varepsilon > 0$ is sufficiently small, then*

(1) *there exists an orbit of the Hamiltonian system with Hamiltonian F which is homoclinic to the invariant torus Γ_ε and contained in the region $W_\varepsilon = \{(x, p) \in T^*M : \|p\|^2 \leqslant C\varepsilon\}$;*

(2) *if the homology group $G = H_1(M, \Gamma_0, \mathbb{R})$ is nonzero, then there exist* $2\dim G$ *homoclinic orbits with the same property;*

(3) *for each $x \in M$ there exists an orbit $t \to (x(t), p(t))$, $x(0) = x$, which is asymptotic to Γ_ε as $t \to \infty$ and contained in W_ε for all $t \geqslant 0$.*

If $\varepsilon < \delta/C$, statements (4) and (5) of Theorem 1.2 follow from statements (1) and (2) of Lemma 3.1, since the corresponding homoclinic orbits are contained in the region where the Hamiltonian function F was not modified. Thus to prove (4) and (5), it is sufficient to prove Lemma 3.1. Statement (3) of Theorem 1.2 follows from (3) of Lemma 3.1.

LEMMA 3.2. *There exists a symplectic transformation of T^*M, smoothly depending on ε for small $\varepsilon \geqslant 0$ and identical for $\varepsilon = 0$, such that after this transformation the new Hamiltonian function $F(x, p, \varepsilon)$ has all the stated properties and*

(1) *the torus Γ_ε does not depend on ε and coincides with the torus $\Gamma_0 \subset M \subset T^*M$;*

(2) *the function $u = F(x, 0, \varepsilon)$ on M attains its strict maximum 0 on Γ_0 and Γ_0 is a nondegenerate critical manifold for u.*

PROOF. Let $\pi\colon T^*M \to M$ be the projection. The invariant torus Γ_ε is the image of a smooth embedding $h_\varepsilon\colon \mathbb{T}^n \to T^*M$, $\Gamma_\varepsilon = h_\varepsilon(\mathbb{T}^n)$, and the flow on Γ_ε is given by the formula $g_\varepsilon^t(h_\varepsilon(\theta)) = h_\varepsilon(\theta + \omega t)$. For $\varepsilon = 0$, we have $\Gamma_\varepsilon|_{\varepsilon=0} = \Gamma_0$ and $h_0(\theta) \equiv g_\theta(x_0)$. Extend the mapping from $\pi(\Gamma_\varepsilon)$ onto Γ_0 defined by the formula $\pi(h_\varepsilon(\theta)) \to h_0(\theta)$, $\theta \in \mathbb{T}^n$, to an arbitrary diffeomorphism $\varphi_\varepsilon\colon M \to M$ smoothly depending on the parameter ε. If we perform a symplectic transformation

$$(x, p) \to (\varphi_\varepsilon(x), (d\varphi_\varepsilon(x)^*)^{-1}p), \qquad (x, t) \in T^*M,$$

then in the new variables $\pi(\Gamma_\varepsilon) = \Gamma_0$ does not depend on ε and $\pi(h_\varepsilon(\theta)) \equiv h_0(\theta) = g_\theta(x_0)$.

By (2) of Theorem 1.2, the invariant torus Γ_ε is hyperbolic and its stable and unstable manifolds $W^{s,u} \subset T^*M$ smoothly depend on $\mu = \sqrt{\varepsilon}$. The manifolds $W^{s,u}$ are Lagrangian and in some neighborhood of Γ_ε their projections from T^*M into M are diffeomorphisms. Thus, near Γ_ε they are graphs of closed 1-forms $\alpha^\pm$ defined in a small tubular neighborhood U of Γ_0 in M:

$$W^{s,u} \supset \{(x, p) \in T^*M : x \in U,\ p = \alpha^\pm(x)\}.$$

Here the value of a 1-form α at a point $x \in M$ is identified with a covector in T_x^*M. Since the manifolds $W^{s,u}$ intersect along Γ_ε, for $x \in \Gamma_0$ we obtain $\alpha^+(x) = \alpha^-(x)$. There exists a closed 1-form α on U, smoothly depending on μ, such that $\alpha^+(x) = \alpha^-(x) = \alpha(x)$ for $x \in \Gamma_0$. Then the 1-forms $\alpha^\pm - \alpha$ on U are exact: $\alpha^\pm - \alpha = dS^\pm$. The functions $S^\pm$ smoothly depend on μ. We can extend them to smooth functions on M (also denoted by $S^\pm$) smoothly depending on μ.

By the topological assumption in §1, the closed 1-form α on U is a restriction of some closed 1-form β on M: $\alpha = \beta|_U = i^*\beta$, where $i\colon U \to M$ is the inclusion. We can choose β smoothly depending on μ. Since neighborhoods of Γ_ε in $W^{s,u}$ tend to M as $\varepsilon \to 0$, we can assume that for $\varepsilon = 0$ the functions $S^\pm$ and the 1-form β are zero. Now perform the canonical transformation $(x, p) \to (x, p - \beta(x))$ of T^*M. Then in the new variables, the manifolds $W^{s,u}$ are given by the equations $p = dS^\pm(x)$. From now on we can assume that $\alpha = 0$ and the forms $\alpha^\pm$ are exact: $\alpha^\pm = dS^\pm$, where

$$S^\pm = \mu S_1^\pm + \mu^2 S_2^\pm + O(\mu^3). \tag{3.3}$$

Since the manifolds $W^{s,u}$ are invariant and, consequently, contained in the zero energy hypersurface, everywhere in U the Hamilton–Jacobi equation is satisfied:

(3.4) $$F(x, \alpha^{\pm}(x), \varepsilon) = \langle v(x), dS^{\pm}(x)\rangle + \|dS^{\pm}(x)\|^2/2 + \varepsilon V(x) + O(\mu^3) = 0.$$

Substituting (3.3) into (3.4) and taking together terms of order μ and μ^2, we get

(3.5) $$v \cdot S_1^{\pm} = 0, \qquad v \cdot S_2^{\pm} + \|dS_1^{\pm}\|^2/2 + V = 0.$$

Here for any function h on M, $v \cdot h$ is the derivative along a trajectory $x \to g_{\omega t}x$ of the vector field v. Since the trajectory ωt is dense in $\mathbb{T}^n$, (3.5) implies that the functions $S_1^{\pm}$ are $\mathbb{T}^n$-invariant [4]. Since V and K are $\mathbb{T}^n$-invariant, the second equation (3.5) implies that the functions $S_2^{\pm}$ are $\mathbb{T}^n$-invariant and

(3.6) $$\|dS_1^{\pm}\|^2/2 + V = 0.$$

It is easy to see that (3.6) implies $S_1^+ = -S_1^-$ (the Hamiltonian system with the Hamiltonian $\overline{G} = \|p\|^2/2 + V$ is reversible and thus the unstable manifold is mapped onto the stable manifold by the reflection $p \to -p$). Let $S = (S^+ + S^-)/2$. Then

(3.7) $$S = \varepsilon S_2 + O(\mu^3), \qquad S_2 = (S_2^+ + S_2^-)/2,$$

where the function S_2 on U is $\mathbb{T}^n$-invariant. For $x \in \Gamma_0$, we have $dS^+(x) = dS^-(x)$. Hence, Γ_ε is given by the formula

$$\Gamma_\varepsilon = \{(x, p) : x \in \Gamma_0, \ p = dS(x)\}.$$

Incidentally, (3.7) implies that Γ_ε is a C^1-function of ε for $\varepsilon \geqslant 0$. In a similar way it can be proved by induction that Γ_ε smoothly depends on ε (this is a part of the statement of Theorem 1.2).

It is possible to extend the function $S\colon U \to \mathbb{R}$ in such a way that $S \in C^\infty(M \times \mathbb{R}\{\mu\})$ and S is given by (3.7), where the function $S_2\colon M \to \mathbb{R}$ is smooth and $\mathbb{T}^n$-invariant.

LEMMA 3.3. *Let $\varepsilon > 0$ be small enough. Then the function $u(x, \varepsilon) = F(x, dS(x), \varepsilon)$ on M attains its maximum* 0 *when $x \in \Gamma_0$, and Γ_0 is a nondegenerate critical manifold for u.*

PROOF. First consider the case $x \in U$. Since the function $F(x, p, \varepsilon)$ is uniformly convex in p, we obtain

(3.8) $$\begin{aligned} u(x, \varepsilon) &= F(x, dS^+(x)/2 + dS^-(x)/2, \varepsilon) \\ &\leqslant F(x, dS^+(x), \varepsilon)/2 + F(x, dS^-(x), \varepsilon)/2 = 0, \end{aligned}$$

where the Hamilton–Jacobi equation (3.4) was used. Here $u(x, \varepsilon) = 0$ if and only if $dS^+(x) = dS^-(x)$, i.e., if $x \in \Gamma_0$. The intersection of the vector spaces T_xW^s and T_xW^u is $T_x\Gamma_\varepsilon$. From (3.2) and (3.8) it follows that Γ_0 is a nondegenerate critical manifold for u.

Now consider the case $x \in M \setminus U$. Since Γ_0 is the set of maximum points for the function V, there exists $c > 0$ such that $V(x) \leqslant -c$ for all $x \in M \setminus U$. By (3.5) and (3.7), for small $\varepsilon > 0$ and $x \in M \setminus U$, we have

$$u(x, \varepsilon) = v \cdot S + \|dS\|^2/2 + \varepsilon V(x) + O(\varepsilon^2) = \varepsilon V(x) + O(\varepsilon^2) \leqslant -\varepsilon c + O(\varepsilon^2) < 0.$$

Here we used the fact that $v \cdot S_2 = 0$. □

Let us perform a symplectic transformation $(x, p) \to (x, \hat{p})$, $p = \hat{p} + dS(x)$. Since $\langle v(x), dS_2(x)\rangle = 0$, in the new variables the Hamiltonian function $\widehat{F}$ is of same type (3.1), but the torus Γ_ε is given by the equation $\Gamma_\varepsilon = \{(x, 0) \in T^*M : x \in \Gamma_0\}$ and hence coincides with Γ_0. Lemma 3.3 implies that the function $\widehat{F}(x, 0, \varepsilon)$ has a nondegenerate maximum 0 for $x \in \Gamma_0$. □

We shall continue to denote the canonical variables on T^*M by (x, p), although they are different from the variables (x, p) in (3.1). Since the Hamiltonian function F is convex in momentum, if we carry out the Legendre transformation

$$(x, p) \in T^*M \to (x, \dot{x}) \in TM, \qquad \dot{x} = F_p,$$

then the Hamiltonian system transforms to a Lagrangian system with the Lagrangian function

$$L(x, \dot{x}, \varepsilon) = \max_p (\langle \dot{x}, p\rangle - F(x, p, \varepsilon)). \tag{3.9}$$

From (3.1) it follows that

$$\begin{aligned} L(x, \dot{x}, \varepsilon) &= \|\dot{x} - v(x)\|^2/2 - \varepsilon V(x) + O(\varepsilon^2) \\ &\quad + \varepsilon O(\|\dot{x} - v(x)\|) + O(\|\dot{x} - v(x)\|^3), \end{aligned} \tag{3.10}$$

where we denoted by the same symbol $\|\cdot\|$ the Riemannian metric on TM dual to the Riemannian metric $\|\cdot\|$. It is given by the formula $\|\dot{x}\|^2 = \langle \dot{x}, A(x)^{-1}\dot{x}\rangle$, where the operator A is defined in (3.1). By (3.1) and (3.10), we obtain the expressions for the Legendre transform

$$\begin{aligned} \dot{x} &= v(x) + A(x)p + O(\varepsilon) + O(\|p\|^2), \\ p &= A^{-1}(x)(\dot{x} - v(x)) + O(\varepsilon) + O(\|\dot{x} - v(x)\|^2). \end{aligned} \tag{3.11}$$

As a consequence of (3.2), for all $(x, \dot{x}) \in TM$ we have

$$\|\xi\|^2/b \leqslant \langle \xi, L_{\dot{x}\dot{x}}(x, \dot{x}, \varepsilon)\xi\rangle \leqslant \|\xi\|^2/a, \qquad \text{for all } \xi \in T_xM. \tag{3.12}$$

For small $\varepsilon \geqslant 0$, (3.10) and (3.12) imply that

$$\|\dot{x} - v\|^2/(2b) - \varepsilon V - K\varepsilon^2 \leqslant L(x, \dot{x}, \varepsilon) \leqslant \|\dot{x} - v\|^2/(2a) - \varepsilon V + K\varepsilon^2, \tag{3.13}$$

where the constant $K > 0$ does not depend on ε (this requires a small change of $a < 1 < b$). Moreover, the Lagrangian is quadratic in velocity for large $\|\dot{x}\|$.

LEMMA 3.4. *In the Lagrangian variables, the hyperbolic invariant torus Γ_ε is given by the equation*

$$\Gamma = \Gamma_\varepsilon = \{(x, \dot{x}) \in TM : x \in \Gamma_0,\ \dot{x} = v(x)\} \subset TM .$$

For small $\varepsilon > 0$, the Lagrangian L attains its minimum 0 on Γ and Γ is a nondegenerate critical manifold for L.

By (3.9), we have $\min_{\dot{x}} L(x, \dot{x}, \varepsilon) = -F(x, 0, \varepsilon)$. Thus, Lemma 3.4 follows from Lemma 3.2. □

Now everything is ready for applying Lagrangian variational methods. We shall use them to prove the following

LEMMA 3.5. *There exists a constant $c > 0$ such that for sufficiently small $\varepsilon > 0$:*

(1) *there is a trajectory $x: \mathbb{R} \to M$ of the Lagrangian system which is homoclinic to the invariant torus Γ and such that*

$$\int_{-\infty}^{+\infty} L(x(t), \dot{x}(t), \varepsilon)\, dt < c\sqrt{\varepsilon}; \tag{3.14}$$

(2) *if the homology group G is nonzero, then there exist* $2 \dim G$ *homoclinic orbits satisfying* (3.14);

(3) *for any $x_0 \in M$ there exists an asymptotic orbit $x: [0, \infty) \to M$ such that $x(0) = x_0$ and $\int_0^\infty L\, dt < c\sqrt{\varepsilon}$.*

We shall show in the addendum to Corollary 6.3 that the constant c in (3.14) depends only on M, Γ_0, the foliation of M defined by the group action $g: \mathbb{T}^n \times M \to M$, the Riemannian metric $\|\cdot\|$, and $\min_M V$.

PROOF OF LEMMA 3.1. Denote by $(x(t), p(t)) \in T^*M$ the homoclinic solution of the Hamiltonian equations which is constructed in Lemma 3.5. In order to deduce (1) and (2) of Lemma 3.1 from Lemma 3.5, we must only show that for any $\delta > 0$ and small $\varepsilon > 0$, we have $\|p(t)\|^2 \leqslant C\varepsilon$ for all t, where $C = 2(\max_M(-V) + \delta)$.

Fix arbitrary small $\lambda > 0$ and let Λ be the set of $t \in \mathbb{R}$ such that $L(x(t), \dot{x}(t), \varepsilon) \leqslant \lambda\varepsilon$. By (3.13), for any $t \in \Lambda$ we have

$$\|\dot{x}(t) - v(x(t))\|^2 \leqslant 2b\varepsilon\,(\lambda + K\varepsilon^2).$$

From (3.11) it follows that for small $\varepsilon > 0$ and $\lambda < \delta/(2b)$:

$$\|p(t)\|^2 \leqslant \delta\varepsilon, \qquad \text{for all } t \in \Lambda. \tag{3.15}$$

Differentiating the function $E = \|p(t)\|^2/2 + \varepsilon V(x(t))$ and taking into account the Hamiltonian equations with Hamiltonian (3.1), for small ε and $\|p\|$ we get

$$\begin{aligned} \dot{E} &= \{E, F\} = \{E, \langle v, p\rangle + E + O(\varepsilon^2) + \varepsilon O(\|p\|) + O(\|p\|^3)\} \\ &= \{E, O(\varepsilon^2) + \varepsilon O(\|p\|) + O(\|p\|^3)\} = O(\varepsilon^2) + \varepsilon O(\|p\|^2) + O(\|p\|^4). \end{aligned}$$

We used here the fact that the "energy" E is $\mathbb{T}^n$-invariant, and thus commutes with $\langle v, p\rangle$. Hence, there exists a positive constant B such that if $\|p\|^2 \leqslant C\varepsilon$ and ε is small enough, then

$$\dot{E} \leqslant B\varepsilon^2. \tag{3.16}$$

Let $\tau \in \Lambda$. Since $V \leqslant 0$, (3.15) implies $E(\tau) \leqslant \varepsilon\delta/2$. Denote

$$T = \sup\{s \in [0, c\varepsilon^{-1/2}\lambda^{-1}] : \|p(t)\|^2 \leqslant C\varepsilon, \text{ for all } t \in [\tau, \tau + s]\}.$$

Integrating the inequality (3.16) from $t = \tau$ to $t = \tau + T$, we obtain

$$E(t) \leqslant E(\tau) + B\varepsilon^2 T \leqslant \delta\varepsilon, \quad \text{for all } t \in [\tau, \tau + T]$$

if $\varepsilon > 0$ is sufficiently small. Therefore

$$\|p(t)\|^2 = 2(E - \varepsilon V) \leqslant 2(\delta + \max(-V))\varepsilon = C\varepsilon.$$

Hence $T = c\varepsilon^{-1/2}\lambda^{-1}$. By (3.14), the measure of the set $\mathbb{R}\setminus\Lambda$ is less than $c\varepsilon^{-1/2}\lambda^{-1}$. Since $\tau \in \Lambda$ is arbitrary, the latter inequality holds for all $t \in \mathbb{R}$. Statements (1) and (2) of Lemma 3.1 are proved. The proof of (3) is the same. □

To complete the proof of Theorems 1.1–1.3, we need now only to prove Lemma 3.5. It is deduced from the results of the next section.

Suppose now that the perturbation of the Hamiltonian is only C^2-small (in the case of diffeomorphisms, the perturbed map f_ε is only C^1-close to f), and the Diophantine condition is replaced by the nonresonance assumption. If the assumptions before Theorem 1.2 are satisfied, then by Lemma 2.3′, (3.1) is replaced by

$$\begin{aligned} F = \langle v(x), p\rangle + \langle A(x)p, p\rangle/2 + \varepsilon V(x) + \langle w(x)p, p\rangle/2 \\ + \varepsilon u(x) + O(\|p\|^3) + \varepsilon O(\|p\|) + O(\varepsilon^2). \end{aligned} \tag{3.1′}$$

It is easy to see that for each sufficiently small α there exists $\delta > 0$ such that (3.2) holds, where the constants $a < 1 < b$ are arbitrarily close to 1 and do not depend on α. Thus, it is possible to modify F for $\|p\| > \delta$ in such a way that conditions (1)–(3) are satisfied. An appropriate value of α is chosen in §9.

In the present case, KAM-theory fails and thus all results of this section that used the existence of the hyperbolic invariant torus Γ_ε, also fail (the torus is replaced by an invariant set). In order to prove Theorems 1.1′–1.3′, we need to save at least a part of Lemma 3.4. The definition (3.9) holds, but (3.10) is replaced by

$$\begin{aligned} L(x, \dot{x}, \varepsilon) = \|\dot{x} - v(x)\|^2/2 - \varepsilon V(x) + \langle w'(x)(\dot{x} - v(x)), \dot{x} - v(x)\rangle/2 \\ - \varepsilon u(x) + O(\varepsilon^2) + \varepsilon O(\|\dot{x} - v(x)\|) + O(\|\dot{x} - v(x)\|^3), \end{aligned} \tag{3.10′}$$

where $w' = (A + w)^{-1} - A^{-1}$. By Lemma 2.3′, for any $\alpha > 0$ we can assume

$$\|w\| \leqslant \alpha, \quad \|\nabla w\| \leqslant \alpha, \quad |u| \leqslant \alpha, \quad \|\nabla u\| \leqslant \alpha \tag{3.17}$$

and the $\mathbb{T}^n$-average of u is zero. The function (3.10′) satisfies the inequalities (3.12), and (3.13) is replaced by

$$\|\dot{x} - v\|^2/(2b) - \varepsilon V - \alpha\varepsilon - K\varepsilon^2 \leqslant L \leqslant \|\dot{x} - v\|^2/(2a) - \varepsilon V + \alpha\varepsilon + K\varepsilon^2, \tag{3.13′}$$

where $K > 0$ does not depend on ε but may depend on α, and $a < 1 < b$ are arbitrarily close to 1. These properties are sufficient to prove Theorems 1.1′–1.3′. We shall do this in §9.

§4. Homoclinic orbits of positive definite Lagrangian systems

Consider a time-periodic Lagrangian system with a compact configuration space M^m and Lagrangian function $L \in C^\infty(P)$ on the extended phase space $P = TM \times \mathbb{T}$. The following conditions on L were suggested by Mather [32].

Positive definiteness. For all $x \in M$ and $t \in \mathbb{T}$ the Lagrangian function $L(x, v, t)$ is strictly convex in velocity $v \in T_xM$: the Hessian $L_{vv}(x, v, t)$ is positive definite.

Superlinearity. For all $x \in M$ and $t \in \mathbb{T}$, we have $L(x, v, t)/\|v\| \to \infty$ as $\|v\| \to \infty$.

Completeness. All maximal solutions of the Lagrangian equations are defined on the entire real line $\mathbb{R}$.

If L is positive definite and quadratic in velocity for large $\|v\|$, then the two latter conditions are satisfied. These conditions can be modified (for example, the completeness condition can be replaced by certain assumptions on the behavior of derivatives of L as $\|v\| \to \infty$), but the given form seems the simplest and most natural. It follows that the Legendre transformation is a diffeomorphism $P \to T^*M \times \mathbb{T}$ and thus we can rewrite the equations of motion in Hamiltonian form. The completeness condition implies that the corresponding Poincaré map $g: T^*M \to T^*M$ is everywhere defined. A symplectic diffeomorphism of T^*M is called *positive definite* (or *optical*) if it is obtained from a positive definite Lagrangian system in such a way. Positive definite symplectic diffeomorphisms provide a generalization for finite compositions of exact monotone area preserving twist maps of the cylinder [6, 30, 33]. However, the results of this section are formulated more naturally in the language of Lagrangian systems. Let us translate some definitions of §1 to this case.

Denote by $g^t: P \to P$ the phase flow of the Lagrangian system.

DEFINITION. A set $\Gamma \subset P$ is called an *invariant torus* with *frequency vector* $\omega \in \mathbb{R}^n$ if Γ is the image of a smooth embedding $F: \mathbb{T}^{n+1} \to P$ of the form: $F(\theta, \tau) = (F_\tau(\theta), \tau)$, $(\theta, \tau) \in \mathbb{T}^{n+1}$, where F_τ is a mapping from $\mathbb{T}^n$ into TM and

$$g^t F(\theta, \tau) = F(\theta + \omega t, \tau + t) \tag{4.1}$$

for all $\theta \in \mathbb{T}^n$ and $\tau, t \in \mathbb{R}$.

The torus $\Gamma_0 = F_0(\mathbb{T}^n) \subset TM$ is an invariant torus of the Poincaré map $TM \to TM$. Although this is not necessary everywhere, we assume that the frequency vector ω is nonresonant, i.e., $\langle \omega, k \rangle \notin \mathbb{Z}$ for all nonzero $k \in \mathbb{Z}^n$. Then the flow g^t on Γ is ergodic.

Let p be the projection from TM to M. Denote $f_\tau = p \circ F_\tau: \mathbb{T}^n \to M$ and $f(\theta, \tau) = f_\tau(\theta)$. Since the Lagrangian equations are of the second order, differentiating (4.1), we get

$$F_\tau(\theta) = (f_\tau(\theta), Df_\tau(\theta)\omega + (\partial/\partial\tau) f_\tau(\theta)), \tag{4.2}$$

where $Df_\tau : \mathbb{R}^n \to TM$ is the derivative of the map f_τ. Let $V_\tau = f_\tau(\mathbb{T}^n) \subset M$.

DEFINITION. An $n+1$-dimensional ($n < m$) invariant torus Γ is called *hyperbolic* (or *whiskered*) if the unions of orbits that are asymptotic to Γ as $t \to \pm\infty$ are $m+1$-dimensional submanifolds W^s and W^u in P and the following condition is satisfied. For any point $x \in \Gamma$ the tangent spaces to W^s and W^u are direct sums $T_x W^s = T_x\Gamma \oplus E_x^s$ and $T_x W^u = T_x\Gamma \oplus E_x^u$, where the vector bundles E^s and E^u are g^t-invariant: if $v \in E_x^s$, then $dg^t(x)v \in E_{g^t(x)}^s$ and the same for E_x^u. There exist a Riemannian metric $\|\cdot\|$ on P and a constant $\lambda > 0$ such that for all $x \in \Gamma$ and $t \geqslant 0$ we have

$$\|dg^t(x)v\| \leqslant e^{-\lambda t}\|v\|, \quad \text{for all } v \in E_x^s,$$
$$\|dg^t(x)v\| \geqslant e^{\lambda t}\|v\|, \quad \text{for all } v \in E_x^u.$$

Thus Γ_0 is a hyperbolic invariant torus of the Poincaré map.

THEOREM 4.1. *Suppose that the hyperbolic torus Γ is the set of minimum points of the Lagrangian L. Then*

(1) $\pi(W^{s,u}(\Gamma)) = M \times \mathbb{T}$, *where* $\pi\colon P \to M \times \mathbb{T}$ *is the projection*;

(2) *there exists a homoclinic orbit to* Γ: $W^s(\Gamma) \cap W^u(\Gamma) \setminus \Gamma \neq \varnothing$;

(3) *if the homology group* $G = H_1(M, V_0, \mathbb{R})$ *is nonzero, then there exist at least* $2\dim G$ *homoclinic orbits*;

(4) *if the Lagrangian satisfies the inequality* $L(x, v, t) \leqslant A\|v\|^2 + B$ *for all* $(x, v, t) \in P$, *and* $L|_\Gamma = 0$, *then the Hamiltonian actions of all these homoclinic orbits are bounded by a constant depending only on* A, B, M, Γ, *and the Riemannian metric* $\|\cdot\|$.

Lemma 3.5 is a corollary of Theorem 4.1 and the apriori bound for the Hamiltonian actions of constructed orbits which is obtained in the proof. In order to prove Theorems 1.1–1.3, it is now sufficient to prove Theorem 4.1. We shall prove more general results. First let us state some properties of minimizing invariant measures of Lagrangian systems, due mainly to Mather [32].

Let $\mathfrak{M}$ be the set of Borel probability measures μ in P with compact support that are invariant with respect to the phase flow g^t of the Lagrangian system.

DEFINITION [32]. Let $\mu \in \mathfrak{M}$. Define the *action* $A(\mu)$ of the measure μ as

$$A(\mu) = \int L\, d\mu \tag{4.3}$$

and the *homology class* $\rho(\mu) \in H_1(M, \mathbb{R})$ by the equation

$$\langle \rho(\mu), [\lambda] \rangle = \int \lambda\, d\mu \tag{4.4}$$

for any closed 1-form λ on M. In (4.4) the 1-form λ is regarded as a function on $P = TM \times \mathbb{T}$ (it is independent of the last variable), and $[\lambda] \in H^1(M, \mathbb{R})$ is the de Rham cohomology class of the form λ.

Note that

$$A(\mu) = \int S\, dv,$$

where S is the generating function for the Poincaré map $g: T^*M \to T^*M$, and ν the corresponding g-invariant measure on T^*M.

By (4.1), any invariant torus Γ defines an invariant measure $\mu_\Gamma \in \mathfrak{M}$ by the formula

$$\int h\, d\mu_\Gamma = \int_{\mathbb{T}^{n+1}} h \circ F\, d\theta\, d\tau \tag{4.5}$$

for any function $h \in C^0(P)$. If $\mu = \mu_\Gamma$, then (4.2) and (4.3) imply that

$$A(\mu_\Gamma) = I(f, \omega) = \int L\left(f_\tau(\theta), Df_\tau(\theta)\omega + \frac{\partial}{\partial \tau} f_\tau(\theta), \tau\right) d\theta\, d\tau . \tag{4.6}$$

Note that $I(f, \omega)$, $f \in C^\infty(\mathbb{T}^{n+1}, M)$, is the value of the Percival functional [35] defined on the space of tori $C^\infty(\mathbb{T}^{n+1}, M)$. Critical points of I are invariant tori with frequency vector ω. Usually for $n \geqslant 1$ the functional $I(f, \omega)$ does not assume its minimal value on $C^\infty(\mathbb{T}^{n+1}, M)$.

Mather proved that for each $c \in H^1(M, \mathbb{R})$ the functional $A_c(\mu) = A(\mu) - \langle \rho(\mu), c\rangle$ assumes its minimal value on $\mathfrak{M}$ and the union of supports of minimizing measures is a compact invariant subset $\Lambda_c \subset P$. The set Λ_c is the graph of a Lipschitz section of the fiber bundle $\pi: P \to M \times \mathbb{T}$; this section is defined on a closed subset $\pi(\Lambda_c)$ of $M \times \mathbb{T}$. For each $\rho \in H_1(M, \mathbb{R})$ the functional A assumes its minimum on $\{\mu \in \mathfrak{M} : \rho(\mu) = \rho\}$. We call the minimizing measure *ρ-minimal.*

DEFINITION. An invariant set $\Gamma \subset P$ is called *minimal,* or *c-minimal,* if $\Gamma = \Lambda_c$ for some $c \in H^1(M, \mathbb{R})$.

The support of every ρ-minimal invariant measure is contained in a c-minimal set for some c. If a c-minimal set Γ is ergodic, then it is the support of some ρ-minimal measure. If Γ is an invariant torus, we call it a *c-minimal torus.* If the frequency vector ω is nonresonant, then Γ is c-minimal if μ_Γ is a point of strict minimum for the functional A_c on $\mathfrak{M}$. Then the corresponding map $f \in C^\infty(\mathbb{T}^{n+1}, M)$ is the minimum point of the Percival functional (4.6) on the homotopy class of f. However, it is possible that a minimum point of the Mather functional is not a minimal torus. An example (for the case of a periodic orbit) is due to Hedlund [24].

LEMMA 4.1. *Let $f_{0*}: \mathbb{R}^n = H_1(\mathbb{T}^n, \mathbb{R}) \to H_1(M, \mathbb{R})$ be the homomorphism induced by the map $f_0: \mathbb{T}^n \to M$. Then:*

(1) $\rho(\mu_\Gamma) = f_{0*}(\omega)$;

(2) *if the torus is c-minimal, then $I = f_0^* c \in H^1(\mathbb{T}^n, \mathbb{R}) = \mathbb{R}^n$ is its action variable defined by the formula*

$$I = \int_{\mathbb{T}^n} (Df_0(\theta))^* p\, d\theta , \qquad I_i = \oint_{\gamma_i} \langle p, dx\rangle , \quad i = 1, \dots, n,$$

where $p = L_v(F(\theta, 0)) \in T^*_{f_0(\theta)}M$ is the momentum, and γ_i are the basis cycles for Γ_0.

Statement (1) follows from (4.3) and (4.4). We omit the proof, since it is nearly the same as in the case of a periodic orbit [32]. Statement (2) can be obtained from (4.9). □

Without loss of generality, we can assume that $c = 0$, because if we replace the Lagrangian L by $L - \lambda$, where $[\lambda] = c$, then the Lagrangian equations remain the same, but according to (4.3) and (4.4), the function $A(\mu)$ is replaced by $A(\mu) - \langle \rho(\mu), c\rangle$. The definition of a minimal torus can be reformulated in a different way [11].

DEFINITION. An invariant torus Γ is *minimal* (we assume that $c = 0$), if for any recurrent in Birkhoff's sense [8] trajectory $t \to \gamma(t) \in P$ that is not contained in Γ, we have

$$\liminf_{T\to\infty} \frac{1}{T}\int_0^T L(\gamma(t))\,dt < A(\mu_\Gamma). \tag{4.7}$$

This definition of a minimal torus was suggested in [11] for the case of periodic orbits. The equivalence of the definitions follows from the Birkhoff theorem [8]: in the ω-limit set of any stable trajectory in Lagrange's sense there exists a recurrent trajectory.

As a consequence of the Mather theorem, we see that for any minimal invariant torus $\Gamma \subset P$ the projection $\pi\colon \Gamma \to M \times \mathbb{T}$ is a diffeomorphism onto its image $V \subset M \times \mathbb{T}$. In particular, for all $t \in \mathbb{T}$ the map $f_t\colon \mathbb{T}^n \to M$ is an embedding and $f_t(\mathbb{T}^n) = V_t \subset M$ is a submanifold in M diffeomorphic to the n-dimensional torus.

Let us give the simplest sufficient condition for minimality and hyperbolicity of a torus Γ.

PROPOSITION 4.1. *Suppose that there exist a smooth closed* 1*-form* λ *and a smooth function* W *on* $M \times \mathbb{T}$ *such that the minimum of the function* $L + \dot W + \lambda$ *on* P *is attained on an invariant torus* Γ, *and* Γ *is a nondegenerate critical manifold for* $L + \dot W + \lambda$. *Then* Γ *is* $[\lambda]$*-minimal and hyperbolic.*

The minimality of Γ is evident. We omit the proof of the hyperbolicity. □

The next proposition provides a partial inversion.

PROPOSITION 4.2. *Let* Γ *be a* c*-minimal hyperbolic invariant torus of the Lagrangian system. Then there exist a smooth function* W *on* $M \times \mathbb{T}$ *and a closed* 1*-form* λ *on* M, $[\lambda] = c$, *such that*

(1) *the function* $\widehat{L} = L + \dot W + \lambda$ *on* P *is constant on* Γ;

(2) *there exists a small neighborhood* U *of* $V \subset M \times \mathbb{T}$ *such that* $\widehat{L} > \widehat{L}|_\Gamma$ *on* $\pi^{-1}(U) \setminus \Gamma$;

(3) Γ *is a nondegenerate critical manifold for* $\widehat{L}$.

The proof is given in the end of the section.

All minimal invariant tori of dimension less than $m + 1$ are unstable and in a certain sense hyperbolic. Moreover, the same holds for any minimal invariant set.

DEFINITION (Birkhoff [8]). An orbit $\gamma\colon \mathbb{R} \to P$, which is stable in Lagrange's sense, is called *semiasymptotic* to the invariant set Γ as $t \to \infty$, if every invariant subset of the ω-limit set of γ that is minimal in Birkhoff's sense is contained in Γ. The definitions of an orbit that is semiasymptotic to Γ as $t \to -\infty$ and of a semihomoclinic orbit are similar.

PROPOSITION 4.3. *Let Γ be a minimal invariant set and let $W^{\pm}$ be the unions of orbits that are semiasymptotic to Γ as $t \to \pm\infty$ respectively. Then*

(1) $\pi(W^{+}) = \pi(W^{-}) = M \times \mathbb{T}$;

(2) *if the conditions of Proposition* 4.1 *are satisfied, then* $\pi(W^{s}) = \pi(W^{u}) = M \times \mathbb{T}$.

In the next theorem it is essential that Γ is a hyperbolic torus.

THEOREM 4.2. *Let Γ be a minimal hyperbolic invariant torus of the Lagrangian system. Then*

(1) *there exists a semihomoclinic orbit to Γ: the sets $W^{u} \cap W^{+} \setminus \Gamma$ and $W^{-} \cap W^{s} \setminus \Gamma$ are nonempty. Thus, $W^{s} \cap \overline{W^{u}} \setminus \Gamma \neq \varnothing$ and $W^{u} \cap \overline{W^{s}} \setminus \Gamma \neq \varnothing$;*

(2) *if the assumptions of Proposition* 4.1 *are satisfied, then there exists a homoclinic orbit.*

Statement (1) of Theorem 4.1 follows from Proposition 4.3, and (2) from Theorem 4.2. For Γ a periodic orbit (one-dimensional torus), (2) of Theorem 4.2 was obtained in [11] by Lagrangian variational methods (the proof was given only for simply-connected M). Similar results were proved in [6]. The case of invariant tori can be studied in a same way (see §6). This was done in [12], where a sketch of the existence proof for a homoclinic orbit to an invariant torus of a Hamiltonian system that is a small perturbation of a system with an invariant Lagrangian manifold was given. Theorem 4.2 can be generalized to the noncompact case, if some completeness assumptions at infinity are satisfied and the topological space M/V_0 is not homotopically equivalent to a point (if M is compact, this condition is always satisfied). This was done by Giannoni [21] for the case when Γ is an equilibrium point and an assumption similar to that of Theorem 4.2 is satisfied.

THEOREM 4.3. *Let Γ be a minimal invariant set such that the homology group $G = H_1(M \times \mathbb{T}, V, \mathbb{R})$ is nonzero, $V = \pi(\Gamma)$. Then*

(1) *there exists a semihomoclinic orbit to Γ;*

(2) *if $c = 0$ and there exists a C^1 function S on $M \times \mathbb{T}$ such that*

$$S_t + H(x, S_x(x, t), t) + A(\mu_\Gamma) \leqslant 0,$$

then there exist at least $2 \dim G$ *action-minimizing homoclinic orbits.*

For reversible systems, this was obtained in [10]. The assumption in (2) implies that the invariant set Γ is isotropic. Since in general Γ is not a manifold, the usual definition of an isotropic set must be modified.

By Mather's theorem [32], the set Γ is a *Lipschitz graph*, i.e., the inverse map $(\pi|_\Gamma)^{-1}\colon V = \pi(\Gamma) \to \Gamma$ is Lipschitz:

$$\Gamma = \{(x, v, t) \in P = TM \times \mathbb{T} : v = u(x, t) \in T_xM,\ (x, t) \in V\},$$

where u is a Lipschitz section of the fiber bundle $\pi\colon P \to M \times \mathbb{T}$. Denote by $y(x, t) = L_v(x, u(x, t), t) \in T_x^*M$, $(x, t) \in V$, the corresponding section of the fiber bundle $P^* \to M \times \mathbb{T}$.

DEFINITION. The invariant set Γ is called *isotropic* if for a small neighborhood N of any point of V and some function $S \in C^{1+Lip}(N)$, we have $y(x, t) = S_x$ and $-H(x, y(x, t), t) = S_t$ for $(x, t) \in N \cap V$. Thus, the Poincaré–Cartan 1-form

$$\alpha = \langle y(x, t), dx\rangle - H(x, y(x, t), t)\, dt \tag{4.8}$$

is "locally exact" on V.

It seems that every minimal invariant set is isotropic, but in general this is not proved (for a special case, the proof is given in [10]).

If Γ is topologically complicated (for example, Γ is a Cantor set), then the group $H_1(M \times \mathbb{T}, V, \mathbb{R})$ is the Čech homology group [42], defined as the inverse limit $\lim_{V\subset U} H_1(M \times \mathbb{T}, U, \mathbb{R})$, where U is an open neighborhood of V in $M \times \mathbb{T}$. If Γ is a minimal invariant torus, then $H_1(M \times \mathbb{T}, V, \mathbb{R}) = H_1(M/V_0, \mathbb{R})$.

Let us show how to associate an element of the group G to any homoclinic orbit to Γ. Let U be an open neighborhood of V in $M \times \mathbb{T}$ and $\gamma\colon \mathbb{R} \to M$ be the solution corresponding to the homoclinic orbit. The dual vector space to $E = H_1(M \times \mathbb{T}, U, \mathbb{R})$ is $E^* = H^1(M \times \mathbb{T}, U, \mathbb{R})$. Any $c \in E^*$ is defined by a closed 1-form λ on $M \times \mathbb{T}$ such that $\lambda|_U = 0$. Define the homology class $[\gamma] \in E$ of the homoclinic orbit γ by the formula

$$\langle[\gamma], c\rangle = \int_\gamma \lambda, \qquad \text{for all } c \in E^*,$$

where the integral is taken along the curve $t \to (\gamma(t), t) \in M \times \mathbb{T}$. Since the restriction of λ on U is zero, the integral is convergent and does not depend on the choice of the 1-form λ. Thus, for any open neighborhood U we get an element in $H_1(M\times\mathbb{T}, U, \mathbb{R})$. Hence, we obtain an element of the group G. If the assumptions of (2) of Theorem 4.3 are satisfied, then the homology classes of homoclinic orbits generate G over $\mathbb{R}^+$ (see §8).

Theorem 4.1 follows from Proposition 4.3 and Theorems 4.2–4.3. Proposition 4.3 is proved in the next section, Theorem 4.2 in §§6–7, and Theorem 4.3 in §8. In §9 a modification of the proof of this theorem yields Theorems 1.1′–1.3′.

Reversible case. Suppose that the system is reversible:

$$L(x, -v, -t) = L(x, v, t), \quad \text{for all } x, v, t.$$

Let I be the action functional on the set of curves $\Omega = C^\infty([0, 1/2], M)$:

$$I(\gamma) = \int_0^{1/2} L(\gamma(t), \dot\gamma(t), t)\, dt, \qquad \gamma \in \Omega.$$

If γ is a critical point for I, then γ is a solution of the Lagrangian equations such that its continuation is a reversible 1-periodic orbit: $\gamma(t) \equiv \gamma(-t) \equiv \gamma(t+1)$. Let $\widehat{\gamma}$ be the corresponding closed curve in P.

PROPOSITION 4.4. *The minimal invariant set in P corresponding to the zero cohomology class in $H^1(M, \mathbb{R})$ is the union of curves $\widehat{\gamma} \subset P$ such that $\gamma \in \Omega$ is a point of minimum for I.*

To prove this, we must only note that for any curve $x\colon [0, 1] \to M$, by reversibility $I(x) = I(x|_{[0,1/2]}) + I(x|_{[1/2,1]}) \geqslant 2I(\gamma)$. □

For the same reason, in this case all constructed semihomoclinic orbits are homoclinic.

COROLLARY 4.1 [10, 11]. *If the minimum of the functional I is strict and nondegenerate, then the corresponding periodic orbit is hyperbolic and there exists a homoclinic orbit to it. Without the nondegeneracy assumption, there exist at least $2\dim H_1(M, \mathbb{R})$ homoclinic orbits.* □

PROOF OF PROPOSITION 4.2. Let Γ be a minimal hyperbolic torus of the Lagrangian system. Without loss of generality, we can assume that $c = 0$ and $A(\mu_\Gamma) = 0$ (add an arbitrary constant to the Lagrangian). It is convenient to use the Hamiltonian formalism. Let $P^* = T^*M \times \mathbb{T}$ be the extended phase space and

$$H(x, p, t) = \max_v (\langle p, v\rangle - L(x, v, t))$$

the Hamiltonian function on P^*. We shall identify P and P^* by the Legendre transformation and use the same notation for the images in P^* of the invariant torus Γ and invariant manifolds $W^{s,u}$.

LEMMA 4.2. *The invariant torus Γ is isotropic and the manifolds $W^{s,u}$ are Lagrangian.*

SKETCH OF THE PROOF. Since the torus Γ is ergodic, by the result of Herman [25] it is isotropic. This means that the restriction on Γ of the Poincaré–Cartan 1-form $\lambda = \langle p, dx\rangle - H\,dt$ is closed. Let us show that the manifold W^s is Lagrangian. By the definition of a hyperbolic torus, Γ is a normally hyperbolic attractor for $g^t|_{W^s}$. Hence for any closed contractible curve $\gamma \subset W^s$ and some sequence $t_k \to \infty$, the curve $g^{t_k}\gamma$ tends to a closed contractible curve in Γ (faster than its length increases). Since Γ is isotropic, the Poincaré–Cartan theorem implies that

$$\oint_\gamma \lambda = \oint_{g^{t_k}\gamma} \lambda \to 0 \quad \text{as } k \to \infty. \quad \square$$

By the Mather theorem, the projection of the torus Γ onto its image $V \subset M \times \mathbb{T}$ is a diffeomorphism. Hence

$$\Gamma = \{(x, p, t) \in P^* : (x, t) \in V,\ p = y(x, t) \in T_x^*M\},$$

where y is smooth. Since Γ is isotropic, the 1-form (4.8) on V is closed.

LEMMA 4.3. *The 1-form $\alpha|_V$ is exact: $\alpha = dW$, where W is a smooth function on V.*

PROOF. Let $f : \mathbb{T}^{n+1} \to M$ be the map defining the invariant torus Γ and ω the frequency vector. It is easy to prove that if Γ is a minimal torus, then the pair (f, ω) is a critical point of the Percival functional (4.6) on the space $C^\infty(\mathbb{T}^{n+1}, M)\{f\} \times \mathbb{R}^n\{\omega\}$. Hence, the derivative of $I(f, \omega)$ by ω is zero:

(4.9)
$$\begin{aligned}\frac{\partial I(f, \omega)}{\partial \omega} &= \frac{\partial}{\partial \omega}\int L\Big(f_\tau(\theta), Df_\tau(\theta)\omega + \frac{\partial}{\partial \tau} f_\tau(\theta), \tau\Big)\,d\theta\,d\tau \\ &= \int (Df_\tau(\theta))^* L_v\Big(f_\tau(\theta), Df_\tau(\theta)\omega + \frac{\partial}{\partial \tau} f_\tau(\theta), \tau\Big)\,d\theta\,d\tau \\ &= \int (Df_\tau(\theta))^* y(f_\tau(\theta), \tau)\,d\theta\,d\tau = 0.\end{aligned}$$

Here the derivative $\partial I(f, \omega)/\partial\omega$ is a vector in $\mathbb{R}^n$ and the integrand is a function of $(\theta, \tau) \in \mathbb{T}^{n+1}$ with values in $\mathbb{R}^n$.

Since the torus V is the image of the embedding f of the standard torus $\mathbb{T}^{n+1}$ into $M \times \mathbb{T}$, $f(\theta, t) = (f_t(\theta), t)$, it is sufficient to prove that the 1-form $f^*\alpha$ on $\mathbb{T}^{n+1}$ is exact. We obtain

$$\begin{aligned} f^*\alpha &= \langle y(f(\theta, t)), d(f_t(\theta))\rangle - H(f_t(\theta), y(f(\theta, t)), t)\,dt \\ &= \langle y(f(\theta, t)), Df_t(\theta)\,d\theta + (\partial/\partial t)f_t(\theta)\,dt\rangle \\ &\quad - \big(\langle y(f(\theta, t)), Df_t(\theta)\omega + (\partial/\partial t)f_t(\theta)\rangle \\ &\qquad - L(f_t(\theta), Df_t(\theta)\omega + (\partial/\partial t)f_t(\theta), t)\,dt\big) \\ &= \langle y(f(\theta, t)), Df_t(\theta)(d\theta - \omega\,dt)\rangle \\ &\qquad + L(f_t(\theta), Df_t(\theta)\omega + (\partial/\partial t)f_t(\theta), t)\,dt. \end{aligned}$$

A closed 1-form on the torus $\mathbb{T}^{n+1}$ is exact if and only if the average over $\mathbb{T}^{n+1}$ of the coefficients at independent differentials vanish. The average of the coefficient at $d\theta - \omega\,dt$ is equal to the integral (4.9), and the average of the coefficient at dt is equal to the value $I(f, \omega) = A(\mu_\Gamma)$ of the Percival functional. Since the latter was assumed to be zero, Lemma 4.3 is proved.

Let us extend the function W in Lemma 4.3 to a smooth function on $M \times \mathbb{T}$ and replace the Lagrangian L by $L - \dot{W}$. This means that we performed the symplectic transformation $p \to p - W_x$, $H \to H + W_t$ (see (4.11)). Then the torus Γ remains c-minimal with $c = 0$, but $L|_\Gamma \equiv 0$ and in the new Hamiltonian variables the 1-form α on V is zero.

LEMMA 4.4. *The projection $\pi: P \to M \times \mathbb{T}$ is a diffeomorphism of small neighborhoods of Γ in $W^{s,u}$ onto their image $U \subset M \times \mathbb{T}$.*

PROOF. It is similar to the proof of the nonexistence of conjugate points on an action-minimizing curve. Let Ω be the set of all piecewise smooth curves $\gamma: \mathbb{R} \to M$ such that for some $\theta_\pm \in \mathbb{T}^n$ we have $\gamma(t) \to \gamma_\pm(t) = f_t(\theta_\pm + \omega t)$ and $\dot\gamma(t) \to \dot\gamma_\pm(t)$ as $t \to \pm\infty$. Moreover, the convergence is sufficiently fast (for example, exponential). A manifold structure on Ω is not needed here. Since $L|_\Gamma \equiv 0$, the action functional

$$S: \Omega \to \mathbb{R}, \qquad S(\gamma) = \int_{-\infty}^{\infty} L(\gamma(t), \dot\gamma(t), t)\,dt$$

is well defined. We claim that $S \geqslant 0$.

Suppose that $S(\gamma) < 0$, $\gamma \in \Omega$. Since the torus Γ is ergodic, there exist arbitrary large integers $k_\pm \to \pm\infty$ such that $d(\theta_- + \omega k_-, \theta_+ + \omega k_+) \to 0$. The definition of Ω implies that $S(\gamma|_{[k_-, k_+]}) \to S(\gamma) < 0$. Since the curve $\gamma|_{[k_-, k_+]}$ is nearly closed, perturbing it slightly, we get a closed curve with negative action. Hence, there exists a closed orbit carrying an invariant measure μ such that $A(\mu) < 0$. This contradicts the assumption $\min_{\mathfrak{M}} A = 0$. The claim is proved.

Let $\gamma \in \Omega$, $\gamma(t) = f_t(\theta + \omega t)$, be the curve corresponding to an orbit in Γ. The tangent space $T_\gamma\Omega$ is the set of piecewise smooth vector fields ξ along γ such that

$\xi(t) \to X_\pm(t) = Df_t(\theta + \omega t)u_\pm$, $u_\pm \in \mathbb{R}^n$, and $\dot\xi(t) \to \dot X_\pm(t)$ exponentially as $t \to \pm\infty$. For any $\xi \in T_\gamma\Omega$, the second variation $\delta^2 S(\gamma)(\xi) \geqslant 0$.

We already know that $\pi|_\Gamma\colon \Gamma \to V$ is a diffeomorphism. Suppose that the projection of W^s into $M \times \mathbb{T}$ is degenerate at some point $x \in \Gamma$. Then $d\pi(x)w = 0$ for nonzero $w \in T_x W^s$. Let $w = u + v$, where $u \in T_x\Gamma$ and $v \in E^s_x$, $v \neq 0$. Then $d\pi(x)v = -d\pi(x)u \in T_x V$. Let $x(t) \in \Gamma$, $x(\tau) = x$, be the trajectory passing through x, and let $\pi(x(t)) = (\gamma(t), t) \in V$, $\gamma(t) \in V_t$. Then $\gamma(t) = f_t(\theta + \omega t)$ for some $\theta \in \mathbb{T}^n$. Denote by $v(t) = dg^{t-\tau}(x)v \in T_{x(t)}P$ the solution of the variational equation with the initial condition $v(\tau) = v$. Since $v \in E^s_x$, it follows that $v(t) \to 0$ exponentially as $t \to \infty$. Let $d\pi(x(t))v(t) = (\xi(t), 0)$, where $\xi(t) \in T_{\gamma(t)}M$ is a Jacobi field. By the assumption, $\xi(\tau) \in V_\tau$ and thus $\xi(\tau) = X(\tau)$, where $X(t) = Df_t(\theta + \omega t)u$, $u \in \mathbb{R}^n$. Let

$$\eta = (d/dt)|_{t=\tau}(\xi(t) - X(t)) \in T_{\gamma(\tau)}M.$$

The derivative belongs not to $T(TM)$ but to TM, since for $t = \tau$ the vector field $\xi - X$ is zero. The assumption $v \notin T_x\Gamma$ implies that $\eta \neq 0$. Let $\eta(t)$ be a smooth vector field along $\gamma(t)$ such that $\eta(\tau) = \eta$ and $\eta(t) \equiv 0$ for sufficiently large $|t|$. Denote

$$\zeta(t) = \begin{cases} \xi(t), & t \geqslant \tau, \\ X(t), & t \leqslant \tau, \end{cases}$$

and $\zeta_\varepsilon(t) = \zeta(t) + \varepsilon\eta(t)$, $\varepsilon > 0$. Then $\zeta_\varepsilon \in T_\gamma\Omega$ and a calculation yields

$$\delta^2 S(\gamma)(\zeta_\varepsilon) = \delta^2 S(\gamma)(\zeta) + 2\varepsilon\delta^2 S(\gamma)(\zeta, \eta) + \varepsilon^2\delta^2 S(\gamma)(\eta) = -2\varepsilon\langle L_{vv}\eta, \eta\rangle + O(\varepsilon^2).$$

Here we used the fact that ζ is a broken Jacobi field and hence

$$\delta^2 S(\gamma)(\zeta, \eta) = -\sum_\tau \langle L_{vv}\Delta\dot\zeta(\tau), \eta(\tau)\rangle,$$

where the sum is taken over all discontinuity points of $\dot\zeta$. For small $\varepsilon > 0$ the positive definiteness condition yields a contradiction. □

The end of the proof of Proposition 4.2 is similar to the proof of Lemma 3.2. By Lemma 4.4, there exist a tubular neighborhood U of V in $M \times \mathbb{T}$ and smooth functions $(x, t) \in U \to y_\pm(x, t) \in T^*_x M$ such that near Γ the invariant manifolds $W^{s,u} \subset P^*$ are given by the equations

$$W^{s,u} \supset \{(x, p, t) \in P^* : (x, t) \in U,\ p = y_\pm(x, t)\}.$$

We have $y_+(x, t) = y_-(x, t)$ if and only if $(x, t) \in V$. The differential forms $\alpha^\pm = \langle y_\pm(x, t), dx\rangle - H(x, y_\pm(x, t), t)\,dt$ on U are closed. By Lemma 4.3, the 1-form $\alpha^+|_V = \alpha^-|_V$ is exact. Hence, the forms $\alpha^\pm|_U$ are exact, i.e., $\alpha^\pm = dS^\pm$, where $S^\pm$ are smooth functions on U. Extend them to smooth functions on $M \times \mathbb{T}$.

In U we have $y_\pm(x, t) = S^\pm_x(x, t)$ and the Hamilton–Jacobi equations are satisfied:

$$S^\pm_t(x, t) + H(x, S^\pm_x(x, t), t) = 0. \tag{4.10}$$

Let $W = (S^+ + S^-)/2$. Perform a canonical transformation $(x, p, t) \to (x, \hat{p}, t)$, $\hat{p} = p - W_x(x, t)$. Then the new Hamiltonian is

$$\widehat{H}(x, \hat{p}, t) = H(x, \hat{p} + W_x(x, t), t) + W_t(x, t). \tag{4.11}$$

Using the convexity of H in p and (4.10), for $(x, t) \in U$, we get

$$\begin{aligned} \widehat{H}(x, 0, t) &= H(x, (S_x^+ + S_x^-)/2, t) + (S_t^+ + S_t^-)/2 \\ &\leqslant (S_t^-(x, t) + H(x, S_x^-(x, t), t))/2 \\ &\quad + (S_t^+(x, t) + H(x, S_x^+(x, t), t))/2 = 0. \end{aligned} \tag{4.12}$$

Inequality (4.12) is an equality only if $S_x^+ = S_x^-$, i.e., if $(x, t) \in V$. Since V is a nondegenerate critical manifold for $S^+ - S^-$, it is a nondegenerate critical manifold for the function $u = \widehat{H}(x, 0, t)$. In the new variables, the torus $\Gamma \subset P^*$ is given by the formula $\Gamma = \{(x, 0, t) : (x, t) \in V\}$.

Let $\widehat{L}$ be the Lagrangian corresponding to the Hamiltonian $\widehat{H}$. Then

$$\widehat{L}(x, v, t) = \max_p(\langle p, v\rangle - H(x, p + W_x, t) - W_t) = L(x, v, t) - \dot{W}.$$

Since $\min_v \widehat{L}(x, v, t) = -u$, inequality (4.12) implies that $\widehat{L}$ is equal to zero on Γ and strictly positive on $\pi^{-1}(U) \setminus \Gamma$. Since V is a nondegenerate critical manifold for u, Γ is a nondegenerate critical manifold for $\widehat{L}$. Proposition 4.2 is proved.

By Proposition 4.2, in the proof of Theorem 4.2 we can assume that $L|_\Gamma \equiv 0$ and Γ is a nondegenerate critical manifold for L with zero index.

COROLLARY 4.2. *There exists a function S on a neighborhood U of the torus V in $M \times \mathbb{T}$ such that S is zero on V, positive on $U \setminus V$, V is a nondegenerate critical manifold for S, and*

$$\pm S_t(x, t) + H(x, \pm S_x(x, t), t) = 0. \tag{4.13}$$

This follows from the proof of Proposition 4.2. Let $S = (S^+ - S^-)/2$. Then $S^\pm = W \pm S$ and (4.13) is a consequence of (4.10) and (4.11). □

§5. Semiasymptotic orbits to minimal invariant sets

The goal of this section is to prove Proposition 4.3. Let $\Gamma = \Lambda_c$, $c \in H^1(M, \mathbb{R})$, be an arbitrary c-minimal invariant set of the Lagrangian system. We can assume that $c = 0$ and $\min_{\mathfrak{M}} A = 0$.

Denote by $G \subset H_1(M, \mathbb{R})$ the image of the fundamental group $\pi_1(M)$ under the Hurewicz homomorphism $\pi_1(M) \to H_1(M, \mathbb{R})$ [42], and let $p\colon \widehat{M} \to M$ be the covering of M with the group of covering transformations G. This covering satisfies the condition that the image of the fundamental group $\pi_1(\widehat{M})$ in $\pi_1(M)$ under the map p_* is the kernel of the Hurewicz homomorphism.

DEFINITION [32]. An orbit $\gamma\colon [\tau, \infty) \to M$ is called *minimizing*, if any covering orbit $\widehat{\gamma}\colon [\tau, \infty) \to \widehat{M}$, $p(\widehat{\gamma}) = \gamma$ is minimizing with respect to the Hamiltonian action $S = \int L\,dt$ between any two points.

This is a generalization of the "class A" geodesics of Morse. Every orbit contained in a minimal invariant set is minimizing [32].

PROPOSITION 5.1. *For each* $(y, \tau) \in M \times \mathbb{T}$ *there exists a minimizing trajectory* $\gamma\colon [\tau, \infty) \to M$, $\gamma(\tau) = y$, *which is semiasymptotic to* Γ *as* $t \to \infty$.

In particular, $\liminf_{t\to\infty} d(\gamma'(t), \Gamma) = 0$. Here we denote by $\gamma'\colon [\tau, \infty) \to P$, $\gamma'(t) = (\gamma(t), \dot{\gamma}(t), t)$ is the corresponding orbit in the phase space P, and d is a metric in P. The same result holds for orbits that are semiasymptotic to Γ as $t \to -\infty$.

PROOF. Proposition 5.1 is a straightforward consequence of the results of [32], but we shall include all the details, since some parts of the proof are used later.

LEMMA 5.1. *For each* $y \in M$, $\tau \in \mathbb{R}$, *there exists a minimizing trajectory* $\gamma\colon [\tau, \infty) \to M$, $\gamma(\tau) = y$, *such that*

$$\lim_{T\to\infty} S(\gamma|_{[\tau, \tau+T]})/T = 0. \tag{5.1}$$

PROOF OF LEMMA 5.1. The minimal set Γ is the union of supports of minimizing invariant measures $\mu \in \mathfrak{M}$. Taking an ergodic component of one of them, we obtain an ergodic measure $\mu \in \mathfrak{M}$ such that $A(\mu) = 0$ and $\operatorname{supp}(\mu) \subset \Gamma$. By the Birkhoff ergodic theorem, for almost all trajectories $x\colon \mathbb{R} \to M$ such that the corresponding orbit $x'\colon \mathbb{R} \to P$ is contained in $\operatorname{supp}(\mu)$, the following limit exists

$$\lim_{T\to\infty} S(x|_{[\tau, \tau+T]})/T = A(\mu) = 0. \tag{5.2}$$

For simplicity, we shall use the same notation for some covering trajectory $x\colon \mathbb{R} \to \widehat{M}$. Let $y \in \widehat{M}$ be a point in the preimage of the given point in $M \times \mathbb{T}$ such that its distance from $x(\tau + 1)$ is not greater than $\operatorname{diam} M$ (the Riemannian metric on M is lifted to an invariant metric on $\widehat{M}$). For any $T > 1$ denote by $x_T\colon [\tau, \tau + T] \to \widehat{M}$ the minimizing curve with the end points $x_T(\tau) = y$, $x_T(\tau + T) = x(\tau + T)$. We claim that there exist constants $C, D > 0$, depending only on the Lagrangian, such that

$$-C \leqslant S(x_T) \leqslant S(x|_{[\tau, \tau+T]}) + D. \tag{5.3}$$

To prove (5.3), we take a minimizing curve $\alpha\colon [\tau, \tau+1] \to \widehat{M}$ such that $\alpha(\tau) = y$ and $\alpha(\tau + 1) = x(\tau + 1)$. Let $\beta = \alpha|_{[\tau, \tau+1]} \cdot x|_{[\tau+1, \tau+T]}$, i.e., $\beta(t) = \alpha(t)$ for $\tau \leqslant t \leqslant \tau + 1$, and $\beta(t) = x(t)$ for $\tau + 1 \leqslant t \leqslant \tau + T$. Since x_T is minimizing,

$$S(x_T) \leqslant S(\beta) = S(x|_{[\tau, \tau+T]}) + S(\alpha) - S(x|_{[\tau, \tau+1]}).$$

Now (5.3) follows from

LEMMA 5.2. *There exists a* $C > 0$ *such that for any curve* $\alpha: [a, b] \to M$, *we have* $S(\alpha) \geqslant -C$.

PROOF. Let $k = [b - a] + 1$ and $\beta: [b, a + k] \to M$ be a minimizing trajectory with endpoints $\beta(b) = \alpha(b)$ and $\beta(a + k) = \alpha(a)$. Since M is compact and the length of the interval $[b, a + k]$ is between 1 and 2, we have $S(\beta) \leqslant C$, where the constant C does not depend on α and β. Let $\gamma: [a, a + k] \to M$ be the closed curve $\gamma = \alpha|_{[a, b]} \cdot \beta|_{[b, a+k]}$. Let σ be a closed trajectory minimizing the action S in the class of curves homotopic to γ, and let $\mu \in \mathfrak{M}$ be the invariant probability measure on P evenly distributed along σ. Then

$$0 \leqslant kA(\mu) = S(\sigma) \leqslant S(\gamma) = S(\alpha) + S(\beta),$$

where $S(\beta) \leqslant C$. □

Let $\lambda > 0$. By (5.2) and (5.3), $S(x_T) \leqslant \lambda T$ for all sufficiently large T. Now we shall use the following result [32].

LEMMA 5.3. *For arbitrary* $A, B > 0$ *there exists a* $K > 0$ *such that for any minimizing trajectory* $\gamma: [a, b] \to \widehat{M}$ *satisfying the inequality* $S(\gamma) \leqslant A(b - a) + B$, *we have*

$$\|\dot\gamma(t)\| \leqslant K, \qquad t \in [a, b]. \tag{5.4}$$

Hence, $\|\dot x_T(t)\| \leqslant K$ for all $\tau \leqslant t \leqslant \tau + T$, where the constant K does not depend on T. Thus there exists a sequence $T_k \to \infty$ such that the sequence of vectors $\dot x_{T_k}(\tau) \in T_y\widehat{M}$ has a limit $v \in T_y\widehat{M}$, $\|v\| \leqslant K$. Let $\gamma: [\tau, \infty) \to \widehat{M}$ be the solution of the Lagrangian equations with the initial conditions $\gamma(\tau) = y$, $\dot\gamma(\tau) = v$. Then for any $T > 0$, the sequence $x_{T_k}|_{[\tau, \tau+T]}$ is uniformly convergent to $\gamma|_{[\tau, \tau+T]}$ in the C^1 metric. Hence, γ is minimizing. By (5.3), for all $T > 0$

$$S(\gamma|_{[\tau, \tau+T]}) \leqslant S(x|_{[\tau, \tau+T]}) + D. \tag{5.5}$$

Now (5.1) follows from (5.2) and (5.5). □

LEMMA 5.4. *Let* $\gamma: [\tau, \infty) \to M$ *be a trajectory satisfying* (5.1) *and* (5.4) *for all* $t \geqslant \tau$. *Then* γ *is semiasymptotic to* Γ *as* $t \to \infty$.

The minimality of γ is not needed here. We shall use this in the proof of Theorem 4.3. Proposition 5.1 is a corollary of Lemmas 5.1 and 5.4.

PROOF OF LEMMA 5.4. Suppose that (5.1) and (5.4) are satisfied. Then the closure of the orbit $\gamma': [\tau, \infty) \to P$ is compact. Thus in order to prove that γ is semiasymptotic to Γ as $t \to +\infty$, it is sufficient to prove that every minimal (in Birkhoff's sense) compact invariant subset N of the ω-limit set Ω_+ of the orbit γ' is contained in Γ.

The set $N \subset P$ is compact and invariant and there exists a sequence $t_k \to \infty$ such that $d(\gamma'(t_k), N) \to 0$ as $k \to \infty$. We claim that there exists a sequence $T_k \to \infty$ such that

$$\max\{d(\gamma'(t), N) : t_k \leqslant t \leqslant t_k + T_k\} \to 0 \quad \text{as } k \to \infty. \tag{5.6}$$

Suppose this is false. Then there exists ε, $T > 0$ and a subsequence t_j of the sequence t_k satisfying the following condition: for every j and some $s_j \in [t_j, t_j+T]$, we have $d(\gamma'(s_j), N) \geqslant \varepsilon$. Since the closure of the orbit $\gamma': [\tau, \infty) \to P$ is compact, for some subsequence t_i of the sequence t_j the sequences $\gamma'(t_i)$ and $\gamma'(s_i)$ are convergent to some points $x \in N$ and $y \in P$, where $d(y, N) \geqslant \varepsilon$. Since solutions of Lagrangian differential equations continuously depend on initial conditions, we obtain $g^t x = y$, where $t = \lim(s_i) - \lim(t_i) \leqslant T$. This contradicts the invariance of the set N. The claim is proved.

Let μ_k be the probability measure on P evenly distributed along the trajectory $\gamma'|_{[t_k, t_k+T_k]}$. Then

$$A(\mu_k) = S(\gamma|_{[t_k, t_k+T_k]})/T_k. \tag{5.7}$$

By (5.6), for any $\varepsilon > 0$ and sufficiently large k, we have

$$\operatorname{supp}(\mu_k) \subset U_\varepsilon(\Omega_+) = \{x \in P : d(x, \Omega_+) < \varepsilon\}. \tag{5.8}$$

Lemma 5.2 implies that for $T > T_k$

$$\begin{aligned} -C \leqslant S(\gamma|_{[t_k, t_k+T_k]}) &= S(\gamma|_{[\tau, T]}) - S(\gamma|_{[\tau, t_k]}) - S(\gamma|_{[t_k+T_k, T]}) \\ &\leqslant S(\gamma|_{[\tau, T]}) + 2C. \end{aligned}$$

By (5.1) and (5.7), we have $\lim_{k\to\infty} A(\mu_k) = 0$. The set of probability measures is weakly compact. Hence, the sequence μ_k contains a subsequence that is weakly convergent to some measure μ. Since the measure μ_k is evenly distributed along the trajectory $\gamma'|_{[t_k, t_k+T_k]}$ and $T_k \to \infty$, the Krylov–Bogolyubov theorem implies that the limit measure μ is invariant and thus $\mu \in \mathfrak{M}$. Since the action functional is semicontinuous [32], we have $A(\mu) = \lim_{k\to\infty} A(\mu_k) = 0$. Hence, μ is a minimizing measure. The definition of a minimal set implies that $\operatorname{supp}(\mu) \subset \Gamma$. By (5.8), we have $\operatorname{supp}(\mu) \subset \overline{U}_\varepsilon(N)$ for any $\varepsilon > 0$. Thus $\operatorname{supp}(\mu) \subset N$. The set N is minimal in Birkhoff's sense and hence it contains no nontrivial closed invariant subsets. Thus $\operatorname{supp}(\mu) = N$ and $N \subset \Gamma$. □

COROLLARY 5.1. *Any minimal invariant set Γ such that $\pi(\Gamma) \neq M \times \mathbb{T}$ is unstable in Lyapunov's sense.* □

REMARK. Let μ be an ergodic measure with $\Lambda = \operatorname{supp}(\mu) \subset \Gamma$. If $\pi(\Lambda) \neq M \times \mathbb{T}$, then the set Λ is unstable in Lyapunov's sense, though it is possible that there exist no semiasymptotic orbits to Λ. The proof follows the method used in the proof of Theorem 2 in [15] (see also the proof of Lemma 8.4).

Let $\Lambda^+ \subset P$ be the set consisting of the trajectories $t \to (\gamma(t), \dot\gamma(t), t) \in P$, $\tau \leqslant t < \infty$, constructed in Proposition 5.1. The set Λ^- is defined in a similar way. Then $\Lambda^\pm$ are invariant under the semigroups $g^{\pm t}$, $t \geqslant 0$. If $W^\pm$ are the unions of orbits that are semiasymptotic to Γ as $t \to \pm\infty$, then $\Lambda^\pm \subset W^\pm$.

COROLLARY 5.2. *Let $\pi: P \to M \times \mathbb{T}$ be the projection. Then $\pi(\Lambda^\pm) = M \times \mathbb{T}$.* □

This proves Proposition 4.3. Proposition 4.4 implies

COROLLARY 5.3 [10]. *Suppose that the system is time-reversible and γ is a minimum point for the Hamiltonian action I on $C^\infty([0, 1/2], M)$. Then the corresponding periodic orbit is unstable.* □

Corollaries 5.1 and 5.3 are generalizations of the Hagedorn theorem on the inversion of Routh's and Lagrange's stability theorems (see [15]). Results of this section generalize the results of [10–12, 29], where the case of periodic orbits ($n = 0$) was studied. See also [38, 39] for additional information on asymptotic orbits in this case.

§6. Variational problem

The aim of this section is to prove by global variational methods an existence result (Proposition 6.1) which provides the main step in the proof of Theorem 4.2. Suppose that assumptions of Theorem 4.2 are satisfied. By Proposition 4.2, without loss of generality we can assume that $\min_{\mathfrak{M}} A = 0$, $L|_\Gamma = 0$, and Γ is a nondegenerate critical manifold for L with zero index.

PROPOSITION 6.1. *There exists $c > 0$ such that for each $h > 0$ there exist $\tau \in \mathbb{R}$, $T > 0$, and a trajectory $\gamma: [\tau, \tau + T] \to M$ with the following properties.*

(1) *The point $\gamma(\tau)$ belongs to the torus $V_\tau = f_\tau(\mathbb{T}^n)$ and the point $\gamma(\tau + T)$ to $V_{\tau+T} = f_{\tau+T}(\mathbb{T}^n)$: $\gamma(\tau) = f_\tau(x)$ and $\gamma(\tau + T) = f_{\tau+T}(y)$, $x, y \in \mathbb{T}^n$.*

(2) *Let $p(t) \in T^*_{\gamma(t)}M$ be the momentum of the trajectory γ, and $H(t) = \langle p(t), \dot\gamma(t)\rangle - L$ be the energy. Then*

$$p(\tau)\perp T_{\gamma(t)}V_\tau, \qquad p(\tau + T)\perp T_{\gamma(\tau+T)}V_{\tau+T}, \tag{6.1}$$

$$H(\tau) - \langle p(\tau), D_\tau f_\tau(x)\rangle = H(\tau + T) - \langle p(\tau + T), D_\tau f_{\tau+T}(y)\rangle = h. \tag{6.2}$$

(3) *Let $S(\gamma)$ be the action of the trajectory γ. Then*

$$S(\gamma) + hT \leqslant c. \tag{6.3}$$

Here $p(\tau)\perp T_{\gamma(\tau)}V_\tau$ means that $\langle p(\tau), v\rangle = 0$ for all $v \in T_{\gamma(t)}V_\tau$. For simplicity, we write D_τ instead of $\partial/\partial\tau$.

REMARK. Let γ': $[\tau, \tau + T] \to P$ be the corresponding orbit. For any $\varepsilon > 0$ and sufficiently small $h > 0$, (6.1)–(6.2) imply that $d(\gamma'(t), \Gamma) < \varepsilon$ for $t = \tau$ and $t = \tau + T$. Thus, γ' is almost a homoclinic orbit. However, the limit as $h \to 0$ may not exist.

The proof of Proposition 6.1 is a modification of the proof of an analogous result for the case of periodic orbits [11]. It is convenient to assume that the Lagrangian $L(x, v, t)$ is quadratic in velocity for large $\|v\|$. We can achieve this by modifying L for $\|v\| \geqslant K$. We shall prove later that for sufficiently large K the trajectory γ constructed in Proposition 6.1 is contained in the region $\|v\| \leqslant K$. Hence, it is a solution of the initial Lagrangian equations.

Now we shall state the variational problem. Let W be the Hilbert manifold of absolutely continuous curves $w: [0, 1] \to M$ such that

$$\int_0^1 \|\dot w(t)\|^2\, dt < \infty$$

(see [34]). Define the space Ω as the following subset in $W \times \mathbb{T}^{2n+1} \times \mathbb{R}^+$:

$$\Omega = \{(w, x, y, \tau, T) : w \in W,\ x, y \in \mathbb{T}^n,\ \tau \in \mathbb{T},\ T > 0,\ w(0) = f_\tau(x),\ w(1) = f_{\tau+T}(y)\}.$$

Then Ω is a Hilbert submanifold in $W \times \mathbb{T}^{2n+1} \times \mathbb{R}^+$. To each point in Ω there corresponds a curve $\gamma: [\tau, \tau + T] \to M$, $\gamma(t) = w((t-\tau)/T)$, where $\tau \in \mathbb{R}$ is some moment of time corresponding to $\tau \in \mathbb{T} = \mathbb{R}/\mathbb{Z}$. The curve γ is defined up a time shift $\gamma(t) \to \gamma(t+k)$, $k \in \mathbb{Z}$. Since $V_t = f_t(\mathbb{T}^n)$, we have $\gamma(\tau) \in V_\tau$ and $\gamma(\tau+T) \in V_{\tau+T}$. Up to a time shift, we can identify a point $(w, x, y, \tau, T) \in \Omega$ with a curve $\gamma: [\tau, \tau+T] \to M$. We shall simply write $\gamma = (w, x, y, \tau, T)$, hoping there will be no misunderstanding. For each $h > 0$, let S_h be the modified action functional on Ω:

$$S_h(\gamma) = S_h(w, x, y, \tau, T) = \int_\tau^{\tau+T} L(\gamma(t), \dot\gamma(t), t)\, dt + hT = \int_0^1 L\left(w(s), \frac{w'(s)}{T}, \tau + Ts\right) ds + hT. \tag{6.4}$$

LEMMA 6.1. *The functional S_h is C^1 on Ω and its differential satisfies the local Lipschitz condition. The curve $\gamma: [\tau, \tau+T] \to M$ from Ω is a critical point of S_h if it satisfies statements* (1)–(2) *of Proposition* 6.1.

PROOF. The proof of the first statement follows the usual pattern (see, for example, [1, 34]). Note that it is essential here that L is quadratic in velocity for large velocity. If γ is a solution of the Lagrangian equations, then by the first variation formula [2],

$$dS_h(\gamma) = \langle p(\tau+T), df_{\tau+T}(y)\rangle - \langle p(\tau), df_\tau(x)\rangle - H(\tau+T)\, d(\tau+T) + H(\tau)\, d\tau + h\, dT.$$

If $dS_h(\gamma) = 0$, then we obtain equations (6.1) and (6.2). □

It follows that to prove Proposition 6.1, we need to find a critical point of the functional S_h. The first step is to study the topology of Ω. Denote $B = \mathbb{T}^{2n+1} \times \mathbb{R}^+$ and define the map $p: \Omega \to B$ by the formula $p(\gamma) = (x, y, \tau, T) \in B$ for $\gamma = (w, x, y, \tau, T) \in \Omega$.

LEMMA 6.2. *The map p is a Serre fibration* [42].

This means that any homotopy in the base B can be lifted to a homotopy in Ω. We omit the proof. □

For any $b > 0$, let $\Lambda = \Lambda_b \subset \Omega$ be the set of points $(w, x, y, \tau, T) \in \Omega$ such that

$$w(s) = f_{\tau+sT}(x + \omega s T), \quad 0 \leqslant s \leqslant 1, \qquad y = x + \omega T, \quad 0 < T \leqslant b. \tag{6.5}$$

Then Λ is a smooth submanifold in Ω with boundary $\partial\Lambda = \{T = b\}$. Let $\gamma: [\tau, \tau+T] \to M$ be the curve corresponding to a point (w, x, y, τ, T) in Λ.

Condition (6.5) means that $\gamma(t) = f_t(x + \omega(t - \tau))$. Hence, γ is a trajectory of the Lagrangian system contained in the invariant torus Γ. Let $p\colon \Omega \to B$ be the projection. Then the image $A = p(\Lambda) \subset B$ is

$$A = \{(x, y, \tau, T) \in B : y = x + \omega T, \ 0 < T \leqslant b\}, \tag{6.6}$$

and p is a diffeomorphism from Λ onto A. The manifold A is diffeomorphic to $\mathbb{T}^{n+1} \times (0, b]$.

LEMMA 6.3. *Suppose that the inclusion $\Lambda \subset \Omega$ is a weak homotopy equivalence. Then $\pi_i M = 0$ for $i \geqslant 2$.*

PROOF. The fiber F of the Serre fibration $p\colon \Omega \to B$ is the loop space of M. Hence $\pi_i F = \pi_{i+1} M$ [42]. Consider the exact homotopy sequence of the fibration p:

$$\begin{array}{ccccccccc} \cdots & \longrightarrow & \pi_{i+1}B & \longrightarrow & \pi_i F & \longrightarrow & \pi_i \Omega & \longrightarrow & \pi_i B \longrightarrow \cdots \\ & & & & \downarrow & & \downarrow & & \\ & & & & \pi_i\Lambda & \longrightarrow & \pi_i A & & \end{array} \tag{6.7}$$

Suppose that the homomorphism $\pi_i\Lambda \to \pi_i\Omega$, which is induced by the inclusion, is an isomorphism for all i. Since $p_*\colon \pi_i\Lambda \to \pi_i A$ is an isomorphism and $\pi_i A = \pi_i(\mathbb{T}^{n+1}) = 0$ for $i \geqslant 2$, we have $\pi_i F = 0$ for $i \geqslant 2$, and, consequently, $\pi_{i+1} M = 0$. This leaves only the case $i = 1$. Let j be the inclusion of A into B given by (6.6). Since the pair (B, A) is homotopically equivalent to $(\mathbb{T}^n \times \mathbb{T}^n \times \mathbb{T}, \Delta(\mathbb{T}^n) \times \mathbb{T})$, where $\Delta(\mathbb{T}^n) = \{(x, x) : x \in \mathbb{T}^n\}$ is the diagonal, the map $j_*\colon \pi_1 A \to \pi_1 B$ is monomorphic. Therefore the image of the group $\pi_1 A = \mathbb{Z}^n$ in $\pi_1 B = \mathbb{Z}^n \times \mathbb{Z}^n \times \mathbb{Z}$ is $\Delta \times \mathbb{Z}$, where Δ is the diagonal subgroup $\Delta = \{(k, k) : k \in \mathbb{Z}^n\} \subset \mathbb{Z}^n \times \mathbb{Z}^n$. From the commutative diagram (6.7), we see that the map $p_*\colon \pi_1\Omega \to \pi_1 B$ is a monomorphism. Since the sequence is exact, $\pi_1 F = 0$. □

LEMMA 6.4. *If the inclusion of Λ into Ω is a weak homotopy equivalence, then M is weakly homotopically equivalent to $\mathbb{T}^n$.*

PROOF. By Lemma 6.3, we have $\pi_2 M = \pi_1 F = 0$. If the inclusion of Λ into Ω is a weak homotopy equivalence, then $\pi_0\Omega = 0$. For $i = 0$, (6.7) yields an exact sequence $0 \to \pi_1\Omega \to \pi_1 B \to \pi_0 F \to 0$. Thus $\pi_0 F = \pi_1 B / p_*(\pi_1\Omega)$. The image of the group $\pi_1\Omega = \pi_1\Lambda = \pi_1 A = \mathbb{Z}^n \times \mathbb{Z}$ in $\pi_1 B = \mathbb{Z}^n \times \mathbb{Z}^n \times \mathbb{Z}$ is equal to $\Delta \times \mathbb{Z}$. Consequently, $\pi_1 M = \pi_0 F = \mathbb{Z}^n$. Hence, $M = K(\mathbb{Z}^n, 1)$ and thus, M is weakly equivalent to $\mathbb{T}^n$ [42]. □

COROLLARY 6.1. *If M is compact and $n < m$, then the inclusion of Λ into Ω is not a weak homotopy equivalence.* □

Since Λ is a retract of Ω, we obtain

COROLLARY 6.2. *There exists a continuous map*

$$g\colon (S^i, x_0) \to (\Omega, \gamma_0), \qquad \gamma_0 = g(x_0) \in \Lambda, \ x_0 \in S^i,$$

that is not homotopic to any map with image in Λ in the class of maps form (S^i, x_0) to (Ω, γ_0). □

COROLLARY 6.3. *For any* $a \in (0, b]$, *we can assume that* $\gamma_0 \in \Lambda_a = \Lambda \cap \{T \leqslant a\}$. *There exists* $c > 0$, *not depending on* a *and* h, *such that*

$$\sup\{S_h(\gamma) : \gamma \in g(S^i)\} \leqslant c. \tag{6.8}$$

This follows from $S|_\Lambda \equiv 0$ and $T|_\Lambda \leqslant b$. □

REMARK. We can assume that the image $g(S^i) \subset \Omega \cap C^1$ and g is continuous in the C^1 topology.

For the proof of Lemma 3.5, we shall need an estimate of the constant in (6.8).

ADDENDUM. *Suppose that the assumptions of Lemma* 3.5 *are satisfied. Then there exists a* $C > 0$ *such that for* $0 < h \leqslant \varepsilon^{1/2}$ *the constant in* (6.8) *is not greater than* $C\varepsilon^{1/2}$.

PROOF. Let $p: M \to \widetilde{M} = M/\mathbb{T}^n$ be the projection. There exists a noncontractible C^1 map $\varphi: (S^{i+1}, x_0) \to (\widetilde{M}, y_0)$, $y_0 = p(\Gamma_0)$. Hence, there exists a noncontractible map $\psi: (S^i, x_0) \to (\Omega(\widetilde{M}), \gamma_0)$. Here $\Omega(\widetilde{M})$ is the loop space [42] and $\gamma_0(t) \equiv y_0$.

We can assume that for each $x \in S^i$ the endpoints of the curve $\gamma = \psi(x) \in \psi(S^i)$ coincide with y_0. Moreover, γ is smooth and parametrized in such a way that $\gamma: [0, T] \to \widetilde{M}$, $T = \varepsilon^{-1/2}$, and $\|\dot\gamma(t)\| = l(\gamma)\varepsilon^{1/2}$. The length is measured in the metric on $\widetilde{M}$ that is the projection of the invariant metric $\|\cdot\|$ on M. We have $\max_\gamma l(\gamma) \leqslant A < \infty$.

Let us lift each curve $\gamma = \psi(x) \in \psi(S^i)$ to a horizontal curve $\beta: [0, T] \to M$, $p\beta(t) = \gamma(t)$, with endpoints in Γ_0. (A curve is horizontal if its velocity is everywhere orthogonal to a fiber of the bundle $p: M \to \widetilde{M}$). Then $\|\dot\beta\| \equiv \|\dot\gamma\|$.

Denote $\alpha(t) = g_{\omega t}\beta(t)$. We obtain a map $g: (S^i, x_0) \to (\Omega, \Lambda)$, $g(x) = \alpha$. To estimate the constant in (6.8), let us calculate $S(\alpha)$. By (3.10), we have

$$\begin{aligned} S(\alpha) &= \int_0^T (\|\dot\alpha - v(\alpha)\|^2/2 - \varepsilon V(\alpha) + O(\varepsilon^2) \\ &\qquad + \varepsilon O(\|\dot\alpha - v(\alpha)\|) + O(\|\dot\alpha - v(\dot\alpha)\|^3))\, dt \\ &= \int_0^T (\|\dot\beta\|^2/2 - \varepsilon V(\beta) + O(\varepsilon^2) + \varepsilon O(\|\dot\beta\|) + O(\|\dot\beta\|^3))\, dt \\ &\leqslant l(\gamma)^2\varepsilon^{1/2}/2 + \max(-V)\varepsilon^{1/2} + O(\varepsilon). \end{aligned}$$

Here we used the $\mathbb{T}^n$-invariance of $\|\cdot\|$ and V. Since $hT \leqslant \varepsilon^{1/2}$, for small ε the constant c in (6.8) is not greater than $C\varepsilon^{1/2}$, where $C > A^2/2 + \max(-V) + 1$. □

Define a Riemannian metric on the Hilbert manifold Ω by the formula

$$\langle\gamma', \gamma'\rangle = \|\gamma'\|^2 = \int_0^1 \|w'(t)\|^2\, dt + \|x'\|^2 + \|y'\|^2 + \tau'^2 + T'^2. \tag{6.9}$$

Here the tangent vector $\gamma' \in T_\gamma\Omega$, $\gamma = (w, x, y, \tau, T)$, is identified with the quintet $\gamma' = (w', x', y', \tau', T')$, where $w' \in T_w W$ is a vector field along the curve w; $x', y' \in \mathbb{R}^n$, and $\tau', T' \in \mathbb{R}$. Unfortunately, the Riemannian metric (6.9) is not complete. This leads to some technical difficulties.

LEMMA 6.5. *For all c, $a > 0$ the set*

$$\Omega^{c,a} = \{\gamma \in \Omega : S_h(\gamma) \leqslant c \text{ and } T(\gamma) \geqslant a\} \subset \Omega$$

is complete in the metric (6.9).

Lemma 6.5 is a corollary of the next two lemmas.

LEMMA 6.6. *For all $0 < a < b$ the set $\{\gamma \in \Omega : a \leqslant T(\gamma) \leqslant b\}$ is complete.*

SKETCH OF THE PROOF. Let γ_k be a fundamental sequence. Then the sequences $T(\gamma_k)$, $\tau(\gamma_k) \in \mathbb{T}$, $x(\gamma_k)$, $y(\gamma_k) \in \mathbb{T}^n$, are obviously convergent. The proof that the sequence $w(\gamma_k) \in W$ is convergent is close to the proof of the completeness of W [34]. We omit the details. □

LEMMA 6.7. *Let c, $h > 0$. There exists $b > 0$ such that for any curve $\gamma \in \Omega$ satisfying the inequality $S_h(\gamma) \leqslant c$, we have $T(\gamma) \leqslant b$.*

PROOF. Since $\min_{\mathfrak{M}} A = 0$, by Lemma 5.2, we have $S \geqslant -C$, $C > 0$. Hence, $S_h(\gamma) \geqslant -C + hT(\gamma)$ and $T(\gamma) \leqslant (c + C)/h$ (this is the only place in the proof of Proposition 6.1 where the minimality of Γ is used). □

Fix $h > 0$ and denote $J = S_h$. Define the vector field $\xi = -\operatorname{grad} J$ on the Hilbert manifold Ω by the formula

$$\langle \xi(\gamma), \eta \rangle = -dJ(\gamma)(\eta) \quad \text{for all } \eta \in T_\gamma \Omega. \tag{6.10}$$

The Riemannian metric is given by (6.9). If L is not quadratic in velocity, then the vector field ξ is not smooth, but by Lemma 6.1 it is locally Lipschitz. This is sufficient for the local existence of integral curves for ξ.

COROLLARY 6.4. *Let $a > 0$. Each integral curve γ_t of the vector field ξ exists for all $t \geqslant 0$ while $T(\gamma_t) \geqslant a$.*

PROOF. It is standard. Suppose that $T(\gamma_t) \geqslant a$ for all $t \geqslant 0$ such that γ_t exists. By (6.10),

$$J(\gamma_0) - J(\gamma_t) = -\int_0^t dJ(\gamma_t)(\xi(\gamma_t))\, dt = \int_0^t \|\xi(\gamma_t)\|^2\, dt .$$

Since J is nonnegative,

$$\int_0^t \|\xi(\gamma_s)\|^2\, ds \leqslant C \quad \text{for all } t. \tag{6.11}$$

By Lemma 6.5, the orbit γ_t is contained in the complete set $\Omega^{c,a}$. Suppose that γ_t is defined for $0 \leqslant t < t_0 < \infty$. Denote by d the distance function in Ω corresponding to the Riemannian metric (6.9). For $0 < t_1 < t_2 < t_0$, we obtain from the Schwartz inequality and (6.11):

$$d(\gamma_{t_1}, \gamma_{t_2})^2 \leqslant \left(\int_{t_1}^{t_2} \|\gamma_t'\|\, dt \right)^2 \leqslant (t_2 - t_1) \int_{t_1}^{t_2} \|\gamma_t'\|^2\, dt \leqslant (t_2 - t_1)\, C .$$

Since the metric on $\Omega^{c,a}$ is complete, $\lim_{t \to t_0} \gamma_t$ exists. The local existence theorem implies that γ_t is defined for small $t - t_0 > 0$. This is a contradiction. □

Fix $c > 0$ satisfying inequality (6.8) and denote $\Pi^{c,a} = \{T \leqslant a\} \cap \Omega^c$. Since the vector field ξ is not complete in this region, we shall construct another vector field on $\Pi^{c,a}$.

LEMMA 6.8. *Suppose that* $\delta > 0$ *and let* U *be the neighborhood of the torus* $V = \pi(\Gamma)$ *in* $M \times \mathbb{T}$ *given by the inequality*

$$U = \{(q, t) \in M \times \mathbb{T} : d(q, V_t) < \delta\}.$$

If $a > 0$ *is sufficiently small, then for any curve* $\gamma \in \Pi^{c,a}$, *we have* $(\gamma(t), t) \in U$ *for all* t.

PROOF. For any t let $U_t = \{q \in M : (q, t) \in U\}$. If $a > 0$ is small enough, then for all $\tau \in \mathbb{T}$ we have

$$d(V_\tau, \partial U_t) \geqslant \delta/2 \quad \text{for } \tau \leqslant t \leqslant \tau + a. \tag{6.12}$$

Suppose that the statement of Lemma 6.8 is false. Then for any $a > 0$ there exists a curve $\gamma \in \Pi^{c,a}$, $\gamma : [\tau, \tau+T] \to M$, $0 < T \leqslant a$, such that for some $t \in [\tau, \tau+T]$, we have $\gamma(t) \in \partial U_t$. Since the Lagrangian is superlinear in velocity, for arbitrary $A > 0$ and large enough $B > 0$, the inequality $L(x, v, t) \geqslant A\|v\| - B$ is satisfied for all $(x, v, t) \in P$. Hence

$$d(V_\tau, \partial U_t) \leqslant \int_\tau^{\tau+T} \|\dot\gamma(s)\|\, ds \leqslant \frac{S(\gamma) + BT}{A} \leqslant \frac{J(\gamma) + BT}{A} \leqslant \frac{c + Ba}{A}.$$

If $A > 0$ is large and $a > 0$ sufficiently small, then this contradicts (6.12). □

LEMMA 6.9. *If* $a > 0$ *is sufficiently small, then there exists a smooth vector field* η *on* $\Pi = \Pi^{c,a}$ *with the following properties. The field* η *is nowhere zero on* $\Pi \setminus \Lambda$, $\eta|_{\Lambda\cap\Pi} = 0$, *and* $\eta \cdot T = 0$, $\eta \cdot \tau = 0$. *There exists a* $K > 0$ *such that everywhere in* Π *we have*

$$\eta \cdot J = \eta \cdot S \geqslant (K/T)\|\eta\|^2. \tag{6.13}$$

PROOF. For any $x \in \mathbb{T}^n$, $t \in \mathbb{T}$, denote by $N_{x,t} \subset T_{f(x,t)}M$ the normal space to $T_{f(x,t)}V_t$ defined by the given Riemannian metric on M. We obtain a vector bundle $N \to \mathbb{T}^{n+1}$ with fibers $N_{x,t}$. If $\delta > 0$ is small enough, then, using the exponent map, we can identify the neighborhood U of the torus V in $M \times \mathbb{T}$ with the set

$$U = \{(x, t, u) : (x, t) \in \mathbb{T}^{n+1},\ u \in N_{x,t},\ \|u\| < \delta\}.$$

Lemma 6.8 implies that for small $a > 0$ any curve $\gamma \in \Pi$ is contained in U. The curve γ is defined by $x, y \in \mathbb{T}^n$, $\tau \in \mathbb{T}$, $T \in (0, a]$ and a curve $w : [0, 1] \to M$. We have $w(s) = (v(s), u(s)) \in V_{\tau+sT}$, $0 \leqslant s \leqslant 1$. Here $v(s) \in \mathbb{T}^n$ and $u(s) \in N_{v(s),\tau+sT}$, $\|u(s)\| \leqslant \delta$, where $v(0) = x$, $v(1) = y$, and $u(0) = u(1) = 0$. We shall use the notation

$$\gamma = (x, y, \tau, T, v(s), u(s)). \tag{6.14}$$

Choose some flat Riemannian connection ∇ for the vector bundle $N \to \mathbb{T}^{n+1}$ (it exists for any vector bundle on a torus [27]). For any section $u(t) \in N$ let

$\nabla u(t) \in N$ be the covariant derivative. Suppose that $\gamma_t \in \Pi$, $\gamma_0 = \gamma$ is a smooth family of curves (a path in Π):

$$\gamma_t = (x_t\,,\, y_t\,,\, \tau_t\,,\, T_t\,,\, v_t(s)\,,\, u_t(s))\,, \qquad -\varepsilon < t < \varepsilon\,.$$

We shall identify the tangent vector $\eta = D_t|_{t=0}\gamma_t \in T_\gamma\Omega$ with the sextet

$$\eta = (x'\,,\, y'\,,\, \tau'\,,\, T'\,,\, v'(s)\,,\, u'(s))\,, \tag{6.15}$$

where $x' = D_t x_t$, $y' = D_t y_t \in \mathbb{R}^n$, $\tau' = D_t\tau_t$, $T' = D_t T_t \in \mathbb{R}$, $v'(s) = D_t v_t(s) \in \mathbb{R}^n$, and $u'(s) = \nabla_t u_t(s) \in N_{v(s)\,,\,\tau+sT}$. All derivatives are taken at $t = 0$. Define a vector field $\eta = \eta(\gamma)$ on Π by formula (6.15), where

$$\begin{gathered} x' = 0\,, \quad y' = \int_0^1 \dot v(r)\,dr - \omega T\,, \quad \tau' = 0\,, \quad T' = 0\,, \\ v'(s) = \int_0^s \dot v(r)\,dr - \omega T s\,, \quad u'(s) = u(s)\,. \end{gathered} \tag{6.16}$$

We shall prove that the vector field η satisfies the conditions of Lemma 6.9. The smoothness of η is an immediate consequence of the definition. By (6.16), we have $\eta(\gamma) = 0$ if and only if $\dot v(s) \equiv \omega T$ and $u(s) \equiv 0$. Let us prove inequality (6.13).

In the subsequent calculation, we can assume that the vector bundle $N \to \mathbb{T}^{n+1}$ and the connection ∇ are trivial. Suppose this is not so. For any closed curve α in $\mathbb{T}^{n+1}$ let R_α be the corresponding holonomy operator [27] of the flat connection ∇. Then R_α is an involution of the fiber $N_{x\,,\,t}$. Let $p\colon \mathbb{T}^{n+1} \to \mathbb{T}^{n+1}$, $p(\theta) = 2\theta$, be the two-sheet covering. Then for any closed curve α in $\mathbb{T}^{n+1}$ and the corresponding holonomy operator $\widetilde R_\alpha$ of the connection $p^*\nabla$ on the bundle p^*N, we have $\widetilde R_\alpha = R^2_{p(\alpha)} = \mathrm{id}$. Hence the connection $p^*\nabla$ is trivial. Since the calculation is local, our assumption is justified.

Now we can assume that $U \subset \mathbb{T}^{n+1} \times \mathbb{R}^{m-n}$ and in (6.14) and (6.15), we have $u(s) \in \mathbb{R}^{m-n}$ and $u'(s) \in \mathbb{R}^{m-n}$. In the neighborhood U, the Lagrangian L is given by

$$L = L(v\,,\, u\,,\, \dot v\,,\, \dot u\,,\, t)\,, \qquad v \in \mathbb{T}^n\,,\ u \in \mathbb{R}^{m-n}\,,\ t \in \mathbb{T}.$$

Let us calculate the derivative $\eta(\gamma)\cdot S$, where $\gamma \in \Pi$ is given by (6.14) and $\eta(\gamma)$ by (6.15) and (6.16). By (6.4),

$$\begin{aligned} \eta(\gamma)\cdot S = \frac{\partial}{\partial\varepsilon}\Big|_{\varepsilon=0} \int_0^1 L\Bigg(& v(s) + \varepsilon\left(\int_0^s \dot v(r)\,dr - \omega T s\right),\ u(s) + \varepsilon u(s)\,, \\ & \frac{\dot v(s)}{T} - \omega\,,\ \frac{\dot u(s) + \varepsilon \dot u(s)}{T}\,,\ \tau + Ts\Bigg)\,ds\,. \end{aligned}$$

It is convenient to reparametrize the functions v and u by introducing the new parameter $t = sT$, $0 \leqslant t \leqslant T$. For simplicity, we denote the obtained functions by $v(t)$, $u(t)$. Then

$$\begin{aligned} \eta(\gamma)\cdot S = \int_0^T \big(&\langle L_v\,,\, v(t) - v(0) - \omega t\rangle + \langle L_u\,,\, u(t)\rangle \\ &+ \langle L_{\dot v}\,,\, \dot v(t) - \omega\rangle + \langle L_{\dot u}\,,\, \dot u(t)\rangle\big)\,dt\,, \end{aligned} \tag{6.17}$$

where $L_v = L_v(v(t)\,,\, u(t)\,,\, \dot v(t)\,,\, \dot u(t)\,,\, \tau + t)$ and all other derivatives of L are calculated at the same point. Let

$$R(v\,,\, u\,,\, \dot v\,,\, \dot u\,,\, t) = \langle L_u\,,\, u\rangle + \langle L_{\dot v}\,,\, \dot v - \omega\rangle + \langle L_{\dot u}\,,\, \dot u\rangle\,. \tag{6.18}$$

LEMMA 6.10. *Suppose that $B > 0$ and $\delta > 0$ are small enough. If $|u| < \delta$, then*

$$R(v, u, \dot v, \dot u, t) \geqslant B(|u|^2 + |\dot v - \omega|^2 + |\dot u|^2). \tag{6.19}$$

PROOF. By the assumption at the beginning of this section, for fixed v and t the function $L(v, u, \dot v, \dot u, t)$ has a local nondegenerate minimum 0 when $u = \dot u = 0$ and $\dot v = \omega$. By (6.18), the same is true for the function R. Thus, inequality (6.19) holds for sufficiently small $|u|$, $|\dot v - \omega|$, and $|\dot u|$. Therefore, it is sufficient to prove (6.19) for the case $|\dot v - \omega|^2 + |\dot u|^2 \geqslant \varepsilon > 0$, where ε is small but fixed, and $\delta < \sqrt{\varepsilon}$ arbitrarily small. For small $|u|$ and some $A > 0$ we have:

$$\begin{aligned} R(v, u, \dot v, \dot u, t) \geqslant A(|\dot v - \omega|^2 + |\dot u|^2) &+ \langle R_{\dot v}(v, u, \omega, 0, t), \dot v - \omega\rangle \\ &+ \langle R_{\dot u}(v, u, \omega, 0, t), \dot u\rangle + R(v, u, \omega, 0, t). \end{aligned}$$

Indeed, L is uniformly convex in $\dot u$ and $\dot v$ and quadratic in velocity for large velocities. By (6.18), for small u the function R is uniformly convex on any ray of type $\{(\lambda \dot u, \omega + \lambda(\dot v - \omega)) : \lambda \geqslant 0\}$ in the velocity space $\{\dot v, \dot u\}$. The desired inequality follows immediately.

Since

$$R_{\dot v}(v, 0, \omega, 0, t) = R_{\dot u}(v, 0, \omega, 0, t) = 0$$

and $u = 0$ is a nondegenerate minimum for $R(v, u, \omega, 0, t)$, there exist $C, D > 0$ such that

$$R(v, u, \dot v, \dot u, t) \geqslant A(|\dot v - \omega|^2 + |\dot u|^2) - C(|u| \cdot |\dot v - \omega| + |u| \cdot |\dot u|) + D|u|^2.$$

The inequality $2|ab| \leqslant k^2a^2 + b^2/k^2$ yields

$$R(v, u, \dot v, \dot u, t) \geqslant (A - Ck^2/2)(|\dot v - \omega|^2 + |\dot u|^2) - (C/k^2 - D)|u|^2.$$

Let k be so small that $3B = A - Ck^2/2 > 0$. Since $|\dot v - \omega|^2 + |\dot u|^2 \geqslant \varepsilon$, for small $\delta > 0$ and $|u| < \delta$, we have

$$|u|^2 \leqslant |\dot v - \omega|^2 + |\dot u|^2, \qquad (Ck^{-2} - D)|u|^2 \leqslant B(|\dot v - \omega|^2 + |\dot u|^2).$$

It follows that

$$R(v, u, \dot v, \dot u, t) \geqslant 2B(|\dot v - \omega|^2 + |\dot u|^2) \geqslant B(|\dot v - \omega|^2 + |\dot u|^2 + |u|^2).$$

Inequality (6.19) is proved. □

Using (6.17)–(6.19), we obtain

$$\eta \cdot S \geqslant \int_0^T (\langle L_v, v(t) - v(0) - \omega t\rangle + B(|\dot v - \omega^2| + |\dot u|^2 + |u|^2))\, dt. \tag{6.20}$$

Recall that we assumed that, for large velocity, L is a positive definite quadratic form in velocity. Since $L_v = 0$ for $u = \dot u = \dot v - \omega = 0$ and the same is true for the derivatives of L_v, there exists a constant $A > 0$ such that

$$|L_v| \leqslant A(|\dot v - \omega|^2 + |\dot u|^2 + |u|^2). \tag{6.21}$$

Since the function L is superlinear in velocity, for any $C > 0$ and some $D > 0$, we have $L \geqslant C\,|\dot{v} - \omega| - D$. Hence, for $0 \leqslant t \leqslant T$ and $0 < T \leqslant a$

$$|v(t)-v(0)-\omega t| = \left|\int_0^t (\dot{v}(s)-\omega)\,ds\right| \leqslant \int_0^T |\dot{v}(s)-\omega|\,ds \leqslant \frac{S(\gamma)+DT}{C} \leqslant \frac{c+Da}{C}.$$

If $C > 0$ is large enough and $a > 0$ is small, then $|v(t) - v(0) - \omega t| \leqslant B/(2A)$ for $0 \leqslant t \leqslant T$. Now (6.20) and (6.21) imply

$$\eta \cdot S \geqslant \frac{B}{2}\int_0^T (|\dot{v} - \omega|^2 + |\dot{u}|^2 + |u|^2)\,dt \geqslant \frac{B}{2}\int_0^T (|\dot{v}(t) - \omega|^2 + |\dot{u}(t)|^2)\,dt\,.$$

Returning to the initial parametrization $s = t/T$, we obtain

$$\eta \cdot S \geqslant \frac{B}{2T}\int_0^1 (|\dot{v}(s) - \omega T|^2 + |\dot{u}(s)|^2)\,ds\,.$$

The definitions (6.9) of the Riemannian metric and (6.16) of the vector field η imply

$$\|\eta(\gamma)\|^2 = \int_0^1 (|\dot{v}(s) - \omega T|^2 + |\dot{u}(s)|^2)\,ds + \left|\int_0^1 (\dot{v}(s) - \omega T)\,ds\right|^2.$$

Inequality (6.13) follows immediately. Lemma 6.9 is proved.

COROLLARY 6.5. *For each curve $\gamma = (x\,,\, y\,,\, \tau\,,\, T\,,\, v(s)\,,\, u(s)) \in \Pi$ define the curve $r(\gamma) \in \Pi \cap \Lambda$ by the formula*

$$r(\gamma) = (x\,,\, x + \omega T\,,\, \tau\,,\, T\,,\, x + \omega T s\,,\, 0) \in \Lambda\,.$$

Then for some $C > 0$, we have $\|\eta(\gamma)\| \geqslant Cd\,(\gamma\,,\, r(\gamma))$ for all $\gamma \in \Pi$.

This follows from the last formula in the proof of Lemma 6.9. □

The map r is a retraction of $\Pi^{c\,,a}$ onto $\Pi^{c\,,a} \cap \Lambda$. The vector field $-\eta$ on Π defines a semigroup of transformations $\psi_t\colon \Pi \to \Pi$, $t \geqslant 0$. Explicit formulas follow from (6.16): if $\gamma \in \Pi$ is given by $\gamma = (x\,,\, y\,,\, \tau\,,\, T\,,\, v(s)\,,\, u(s))$, then $\psi_t(\gamma) = (x\,,\, y_t\,,\, \tau\,,\, T\,,\, v_t(s)\,,\, u_t(s))$, where

$$\begin{aligned}
y_t &= e^{-t}\left(\int_0^1 \dot{v}(s)\,ds - \omega T\right) + x + \omega T\,,\\
v_t(s) &= e^{-t}\left(\int_0^s \dot{v}(r)\,dr - \omega T s\right) + x + \omega T s\,,\\
u_t(s) &= e^{-t} u(s)\,.
\end{aligned}$$

In the latter formula we used the trivialization of the bundle $N \to \mathbb{T}^{n+1}$ which was introduced in the proof of Lemma 6.9. The flow ψ_t does not depend on the choice of the trivialization and is identical on $\Pi \cap \Lambda$. We obtain

COROLLARY 6.6. *The vector field* $-\eta$ *on* $\Pi^{c,a}$ *defines a semigroup of transformations* $\psi_t\colon \Pi^{c,a} \to \Pi^{c,a}$, $t \geqslant 0$, *retracting* $\Pi^{c,a}$ *onto* $\Lambda \cap \Pi^{c,a}$. *For any* $\gamma \in \Pi^{c,a}$, *we have* $\lim_{t\to\infty} \psi_t(\gamma) = r(\gamma)$. □

We shall combine the vector fields ξ on $\Omega^{c,a}$ and $-\eta$ on $\Pi = \Pi^{c,a}$ to obtain a vector field generating a semigroup of transformations retracting Ω^c onto $\Lambda \cap \Omega^c$. Choose a constant $B > 0$ such that the statement of Lemma 6.9 is satisfied on $\Pi^{c,B}$ and fix $0 < a < b < B$. Denote $Y = \{g \in \Omega : 0 < T(\gamma) < b\}$ and $X = \{\gamma \in \Omega : T(\gamma) > a\}$. Then $X \cup Y = \Omega$. Let f be a smooth function on $(0, \infty)$ such that $f(T) = 0$ if $0 < T \leqslant a$, $f(T) = 1$ if $b \leqslant T < \infty$ and $f'(T) > 0$ if $a < T < b$. Define smooth functions F and G on Ω by the formulas $F(\gamma) = f(T(\gamma))$ and $G = 1 - F$. Then $F + G \equiv 1$, $F \equiv 0$ on Y and $G \equiv 0$ on X, $F \equiv 1$ on $X \setminus Y$ and $G \equiv 1$ on $Y \setminus X$.

The vector field ξ is defined on Ω and the vector field η on $\Pi^{c,b} = Y \cap \Omega^c$. Let

$$\zeta = F\xi - G\eta\,. \tag{6.22}$$

Then ζ is a Lipschitz vector field on Ω^c and $\zeta = \xi$ on $X \setminus Y$, $\zeta = -\eta$ on $Y \setminus X$.

LEMMA 6.11. *There exists a* $C > 0$ *such that everywhere in* $\Pi^{c,B}$

$$\zeta \cdot J \leqslant -C\,\|\zeta\|^2. \tag{6.23}$$

PROOF. By (6.10) and (6.22),

$$\zeta \cdot J = \langle \operatorname{grad} J\,,\, \zeta\rangle = \langle \operatorname{grad} J\,,\, -F \operatorname{grad} J - G\eta\rangle = -F\,\|\operatorname{grad} J\|^2 - G\eta \cdot J\,.$$

Inequality (6.13) implies

$$\zeta \cdot J \leqslant -(F\|\xi\|^2 + GK\|\eta\|^2/T) \leqslant -(F\|\xi\|^2 + GK\|\eta\|^2/B)\,, \tag{6.24}$$

since $T < B$ on $\Pi^{c,B}$. Further

$$\begin{aligned}\|\zeta\|^2 &= F^2\|\xi\|^2 + G^2\|\eta\|^2 - 2FG\langle\xi\,,\,\eta\rangle\\ &\leqslant F^2\|\xi\|^2 + G^2\|\eta\|^2 + 2FG\|\xi\|\cdot\|\eta\| \leqslant 2(F^2\|\xi\|^2 + G^2\|\eta\|^2)\,.\end{aligned}$$

Since $0 \leqslant F\,,\, G \leqslant 1$, for $0 < 2C < \min(1\,,\, K/B)$, we get (6.23). □

LEMMA 6.12. *All trajectories of the vector field* ζ *on* Ω^c *are defined for all* $t \geqslant 0$. *Hence,* ζ *generates a semigroup of transformations* $\varphi_t\colon \Omega^c \to \Omega^c$, $t \geqslant 0$.

PROOF. Since the vector field ζ satisfies the local Lipschitz condition, we can use the local existence theorem. Let $\gamma \in \Omega^c$. On $Y \setminus X$, we have $\zeta = -\eta$. Thus if $\gamma \in Y \setminus X$, then the trajectory $\gamma_t = \varphi_t\gamma = \psi_t\gamma$ is defined for all $t \geqslant 0$. If $\gamma \in X \setminus Y$, then Corollary 6.4 implies that the trajectory γ_t can be continued until the exit from $\Pi^{c,A}$, $b < A < B$.

Thus, it is sufficient to prove that the trajectory γ_t, $0 \leqslant t < t_0$, such that $\gamma_t \in \{a \leqslant T \leqslant B\}$ for $0 \leqslant t < t_0$, has a limit as $t \to t_0$. This can be deduced from the completeness of the metric $\|\cdot\|$ on $\{a \leqslant T \leqslant B\}$ and the inequality (6.23), as in the proof of Corollary 6.6. □

The next lemma shows that the functional J satisfies the Palais–Smale condition on $\Omega^{c,a}$.

LEMMA 6.13. *Let $\gamma_k \in \Omega^{c,a}$ be a sequence such that $\|\operatorname{grad} J(\gamma_k)\| \to 0$ as $k \to \infty$. Then it contains a subsequence converging to a critical point of J.* □

SKETCH OF THE PROOF. Lemma 6.7 implies that for $\gamma \in \Omega^{c,a}$ the length of the interval on which the curve γ is defined is bounded: $0 < a \leqslant T(\gamma) \leqslant B < \infty$. Thus, the sequence $T(\gamma_k)$ contains a converging subsequence. The same is true for the sequence $\tau(\gamma_k)$. Now the proof of Lemma 6.13 closely follows the proof of the Palais–Smale condition for the standard variational functional [34]. We omit the details. □

Let $g\colon (S^i, x_0) \to (\Omega^c, \gamma_0)$, where $\gamma_0 \in \Lambda \cap \Pi^{c,a}$, be a continuous map.

PROPOSITION 6.2. *Suppose that the functional J has no critical points in $\Omega^{c,a}$. Then the map g is homotopic, in the class of maps $(S^i, x_0) \to (\Omega^c, \gamma_0)$, to a map with image in Λ.*

Proposition 6.1 follows from Proposition 6.2. Corollary 6.3 implies that the functional J has a critical point in $\Omega^{c,a}$. Lemma 6.1 shows that this critical point satisfies the statement of Proposition 6.1.

PROOF OF PROPOSITION 6.2. Suppose that there are no critical points of the functional J in $\Omega^{c,a}$. Lemma 6.12 implies that there exists $\delta > 0$ such that $\|\operatorname{grad} J(\gamma)\| \geqslant \delta$ for all $\gamma \in \Omega^{c,a}$. Let ψ_t, $t \geqslant 0$, be the semigroup corresponding to the vector field $-\eta$. Then $\zeta = -\eta$ and $\varphi_t = \psi_t$ on $\Pi = \Pi^{c,a}$. Hence, the set Π is φ_t-invariant and the semigroup φ_t is retracting Π on Λ_a. Thus, the set $\Omega^{c,a} = \overline{\Omega^c \setminus \Pi^{c,a}}$ is also φ_t-invariant for $t \geqslant 0$.

LEMMA 6.14. *Let $A \subset \Omega^{c,a}$ be a compact set. For any $\alpha > 0$ there exists $R > 0$ such that for $t \geqslant R$ and all $\gamma \in \varphi_t(A)$, we have $T(\gamma) \leqslant a + \alpha$. In short, $\varphi_t(A) \subset \Pi^{c,a+\alpha}$.*

PROOF. Let $\gamma \in A$. The trajectory $\gamma_t = \varphi_t\gamma$ is defined for all $t \geqslant 0$ and contained in $\Omega^{c,a}$. We claim that $F(\gamma_t) \to 0$ as $t \to \infty$ uniformly in $\gamma \in A$.

Lemma 6.11 implies that $J(\gamma_t)$ is not increasing. Since J is bounded from below, we obtain

$$\int_0^t \zeta \cdot J(\gamma_t)\,dt = \int_0^t \frac{dJ(\gamma_t)}{dt}\,dt = J(\gamma_t) - J(\gamma) \geqslant -D, \tag{6.25}$$

where the constant $D > 0$ does not depend on $\gamma \in \Omega^{c,a}$. Since $\|\xi(\gamma_t)\| \geqslant \delta$, inequalities (6.24) and (6.25) imply that

$$\int_0^\infty F(\gamma_t)\delta^2\,dt \leqslant D. \tag{6.26}$$

From (6.23) and (6.25) it follows that

$$\int_0^\infty \|\zeta(\gamma_t)\|^2\,dt \leqslant \frac{D}{C}. \tag{6.27}$$

Using (6.26) and (6.27), we obtain that for any $\varepsilon > 0$ there exists an $N > 0$ such that for all $k > N$

$$\int_k^{k+1} F(\gamma_t)\,dt < \varepsilon, \qquad \int_k^{k+1} \|\zeta(\gamma_t)\|^2\,dt < \varepsilon. \tag{6.28}$$

Since A is compact, N can be taken independently of $\gamma \in A$. By (6.28), there exists an $s \in [k, k+1]$ such that $F(\gamma_s) < \varepsilon$. Since $F(\gamma) = f(T(\gamma))$, for any $t \in [k, k+1]$ we get

$$|F(\gamma_t) - F(\gamma_s)| = \left| \int_s^t D_u F(\gamma_u)\, du \right| = \left| \int_s^t f'(T(\gamma_u)) D_u T(\gamma_u)\, du \right|$$
$$\leqslant \max(f') \left| \int_s^t |\langle \zeta, \operatorname{grad} T(\gamma_u)\rangle|\, du \right|.$$

By the definition of the metric (6.9), we have $\|\operatorname{grad} T\| \equiv 1$. Using the Schwartz inequality and (6.28), we obtain

$$|F(\gamma_t) - F(\gamma_s)| \leqslant \max(f') \int_k^{k+1} \|\zeta(\gamma_u)\|\, du$$
$$\leqslant \max(f') \left(\int_k^{k+1} \|\zeta(\gamma_u)\|^2\, du \right)^{1/2} \leqslant \max(f') \varepsilon^{1/2}.$$

Hence, $0 \leqslant F(\gamma_t) \leqslant \varepsilon + \max(f')\varepsilon^{1/2}$ for $k \leqslant t \leqslant k+1$. This proves the claim:

$$\max\{F(\gamma) : \gamma \in \varphi_t(A)\} \to 0$$

as $t \to \infty$. Since $F(\gamma) = f(T(\gamma))$ and for any $\alpha > 0$ the function f is bounded away from zero on $[a + \alpha, \infty)$, Lemma 6.14 is proved.

Now we can complete the proof of Proposition 6.2. Let a, $\alpha > 0$ be sufficiently small. Then we can assume that Corollary 6.6 holds if we replace a with $a' = a+\alpha$. The semigroup ψ_t retracts $\Pi^{c, a'}$ onto $\Lambda_{a'}$. Denote $A = g(S^i)$. Lemma 6.14 implies that $\varphi_t(A \cap \Omega^{c, a}) \subset \Pi^{c, a+\alpha}$ for $t \geqslant R$. Since $\varphi_t(\Pi^{c, a}) \subset \Pi^{c, a}$, we have $\varphi_t(A) \subset \Pi^{c, a+\alpha}$ for $t \geqslant R$. The semigroups φ_t and ψ_t are trivial on $\Lambda_{a'}$. Hence, the map $g\colon (S^i, x_0) \to (\Omega^c, \gamma_0)$ is homotopic to the map $r \circ \varphi_R \circ g$ into $\Lambda_{a'}$. The homotopy g_s is given by the formula

$$g_s = \begin{cases} \varphi_{2sR} \circ g, & 0 \leqslant s \leqslant 1/2, \\ \psi_{-\ln(2-2s)} \circ \varphi_R \circ g, & 1/2 \leqslant s \leqslant 1. \end{cases} \quad \square$$

If the assumptions of Proposition 4.1 are satisfied (only this case is needed to prove Theorems 4.1 and 1.1–1.3), then Theorem 4.2 is an easy consequence of Proposition 6.1. In the general case some additional work is needed.

§7. Proof of Theorem 4.2

In this section we shall complete the proof of Theorem 4.2. For given $h > 0$, let $\gamma\colon [\tau, \tau + T] \to M$ be the trajectory constructed in Proposition 6.1.

PROPOSITION 7.1. *There exists a neighborhood U of the torus V in $M \times \mathbb{T}$ such that*

(1) *for all sufficiently small $h > 0$ there exists a $t \in (\tau, \tau+T)$ such that $(\gamma(t), t) \in (M \times \mathbb{T}) \setminus U$;*

(2) *let* $t_- = \inf\{t \in (\tau, \tau+T) : (\gamma(t), t) \in \partial U\}$, $t_+ = \sup\{t \in (\tau, \tau+T) : (\gamma(t), t) \in \partial U\}$. *Then* $t_- - \tau \to \infty$ *and* $\tau + T - t_+ \to \infty$ *as* $h \to 0$;

(3) $\lim_{h\to 0} d(\gamma'(t_-), W^u(\Gamma)) = 0$ *and* $\lim_{h\to 0} d(\gamma'(t_+), W^s(\Gamma)) = 0$.

The proof is based on simple estimates for the behavior of the solution $\gamma(t)$ of the Hamiltonian equations in special canonical coordinates in a neighborhood of the invariant torus Γ. As in the proof of Lemma 6.9, while performing local calculations in a neighborhood of Γ, we can assume that the normal bundle of the torus $V \subset M \times \mathbb{T}$ is trivial. Thus we can introduce local coordinates $x \in \mathbb{T}^n$, $t \in \mathbb{T}$, $u \in \mathbb{R}^{m-n}$ in a neighborhood of $V \subset M \times \mathbb{T}$ such that $V = \{u = 0\}$. We take $U = \{(x, u, t) : |u| \leqslant \delta\}$, where $\delta > 0$ is sufficiently small. The variables $x \in \mathbb{T}^n$ and $u \in \mathbb{R}^{m-n}$ are local Lagrangian coordinates. The torus $\Gamma \subset P$ is given by the equations

$$\Gamma = \{(x, u, \dot{x}, \dot{u}, t) : \dot{x} = \omega,\ u = \dot{u} = 0\}.$$

Denote by $y \in \mathbb{R}^n$ and $v \in \mathbb{R}^{m-n}$ the conjugate momenta. Since the invariant torus Γ consists of critical points of the Lagrangian, in the symplectic coordinates x, u, y, v, we have $\Gamma = \{(x, u, y, v, t) : y = 0,\ u = v = 0\}$, and the Hamiltonian is

$$H(x, u, y, v, t) = \langle \omega, y\rangle + O(|y|^2 + |u|^2 + |v|^2). \tag{7.1}$$

LEMMA 7.1. *It is possible to choose the coordinates* $x \in \mathbb{T}^n$ *and* $u \in \mathbb{R}^{m-n}$ *in such a way that in the corresponding symplectic variables the Hamiltonian takes the form*

$$\begin{aligned} H = {} & \langle \omega, y\rangle + \langle Ay, y\rangle/2 + \langle (C + O(|y| + |u| + |v|))(v+u), v-u\rangle/2 \\ & + \langle O(|u| + |v|), y\rangle + O(|y|^3). \end{aligned} \tag{7.2}$$

Here $A = A(x, t)$ *and* $C = C(x, t)$. *The matrices* A *and* $B = (C + C^*)/2$ *are symmetric and positive definite.*

PROOF. Let $S(x, u, t)$, $|u| \leqslant \delta$, be the function defined in Corollary 4.2. By (4.13), S satisfies the Hamilton–Jacobi equations

$$\pm S_t + H(x, u, \pm S_x, \pm S_u, t) = 0, \qquad S(x, 0, t) \equiv 0, \tag{7.3}$$

and $u = 0$ is a nondegenerate minimum for S. By Morse's lemma with parameters, there exists a change of variables $F(x, u, t) = (x, f(x, u, t), t)$ such that $\widehat{S} = S \circ F = |u|^2/2$. Perform the corresponding time-dependant symplectic transformation preserving the Poincaré–Cartan 1-form $\alpha = \langle y, dx\rangle + \langle v, du\rangle - H\,dt$. Then, in the new variables, the new Hamiltonian will be $\widehat{H} = (H - \langle v, f_t\rangle)\circ(F^*)^{-1}$. The Hamilton–Jacobi equations are still satisfied. Thus, (7.3) becomes

$$\widehat{H}(x, u, 0, \pm u, t) \equiv 0.$$

Taking into account (7.1), we see that $\widehat{H}$ is of type (7.2). Since the Lagrangian L is positive definite in velocity, the matrices A and B are positive definite. $\square$

Let us introduce the symplectic variables x, y, z_+, z_-, where $z_\pm = (v \pm u)/\sqrt{2}$. Then by (7.2),

(7.4)

$$H = \langle \omega, y\rangle + \langle Ay, y\rangle/2 + \langle (C + O(|y| + |z|))z_+, z_-\rangle + \langle O(|z|), y\rangle + O(|y|^3).$$

The hyperbolic torus and its stable and unstable manifolds are given by the equations

$$\begin{aligned} \Gamma &= \{(x, y, z_+, z_-) : y = 0,\ z = 0\}, \\ W^{s,u} &= \{(x, y, z_+, z_-) : y = 0,\ z_\pm = 0\}. \end{aligned} \tag{7.5}$$

REMARK. In a neighborhood of an ergodic hyperbolic invariant torus, there exists a Birkhoff's normalizing symplectic transformation which takes the Hamiltonian to a simpler form than (7.6) (see, for example, [23, 43]). However, these transformations destroy the Lagrangian structure of the phase space and thus they are forbidden. Admissible symplectic transformations are generated by transformations of the Lagrange variables and by adding full derivatives to the Lagrangian.

PROOF OF PROPOSITION 7.1. We shall prove only the part of Proposition 7.1 related to the left boundary point $t = \tau$ of the trajectory $\gamma\colon [\tau, \tau + T] \to M$ which was constructed in Proposition 6.1. For any $h > 0$ denote by $x(t) \in \mathbb{T}^n$, $u(t) \in \mathbb{R}^{m-n}$, $y(t) \in \mathbb{R}^n$, $v(t) \in \mathbb{R}^{m-n}$, the solution of the Hamiltonian equations corresponding to the trajectory $\gamma(t)$. Of course, these functions are defined only when $|u(t)|$ is sufficiently small, or else our coordinates do not make sense. From (6.1) it follows that $u(\tau) = 0$ and $y(\tau) = 0$. By (6.2),

$$H(x(\tau), 0, 0, v(\tau), \tau) = h. \tag{7.6}$$

Substituting (7.4) into (7.6), we get

$$\langle B(x(\tau), \tau)v(\tau), v(\tau)\rangle + O(|v(\tau)|^3) = 2h.$$

Since the matrix B is positive definite:

$$2a\,|v|^2 \leqslant \langle Bv, v\rangle \leqslant b\,|v|^2/2 \quad \text{for all } v \text{ with } a, b > 0, \tag{7.7}$$

for small $h > 0$, we have

$$2h/b \leqslant |v(\tau)|^2 \leqslant 2h/a. \tag{7.8}$$

Let

$$\begin{cases} \dot{x} = H_y = \omega + Ay + O(|z|), \\ \dot{y} = -H_x = O(|y|^2) + O(|z_+|)O(|z_-|) + O(|y|)O(|z|), \\ \dot{z}_+ = H_{z_-} = (C + O(|y| + |z|))z_+ + O(|y|), \\ \dot{z}_- = -H_{z_+} = -(C^* + O(|y| + |z|))z_- + O(|y|), \end{cases} \tag{7.9}$$

be the Hamiltonian system with the Hamiltonian (7.4). Denote $r = |y|$ and $r_\pm = |z_\pm|$. Equations (7.9) imply that

$$\begin{aligned} \dot{r} &= \langle e, \dot{y}\rangle = O(r^2) + O(r_+r_-) + O(r)\,O(r_+ + r_-), \\ \dot{r}_\pm &= \langle e_\pm, \dot{z}_\pm\rangle = \pm\langle (B + O(r + r_+ + r_-))e_\pm, e_\pm\rangle + O(r), \end{aligned} \tag{7.10}$$

where $e = y/|y|$ and $e_\pm = z_\pm/|z_\pm|$. By (7.5), we have $z_+(\tau) = z_-(\tau) = v(\tau)/\sqrt{2}$. Thus, (7.8) yields

$$r(\tau) = 0, \qquad r_\pm(\tau) = \varepsilon, \qquad h/b \leqslant \varepsilon^2 \leqslant h/a. \tag{7.11}$$

For given $\delta, \varepsilon, \lambda > 0$, consider the compact subset $D = \{r_- \leqslant \varepsilon,\ r \leqslant \lambda r_+,\ |u| \leqslant \delta\}$ of the phase space P.

LEMMA 7.2. *Let $\delta > 0$ be sufficiently small and $\Lambda > 0$ sufficiently large. Then for any sufficiently small $\varepsilon > 0$ and $\lambda = \Lambda\varepsilon$, every trajectory of the Hamiltonian system* (7.9), *starting in D at $t = \tau$, for $t > \tau$ cannot leave D through the part $D \cap \{r_- = \varepsilon\} \cup \{r = \lambda r_+\}$ of the boundary ∂D.*

PROOF. Let δ and ε be small. Then λ and $r_+ = |u\sqrt{2} + z_-| \leqslant \delta\sqrt{2} + \varepsilon$ are also small. By (7.7) and (7.10), everywhere in D any solution of the system (7.9) satisfies the inequalities

$$
\begin{cases} \dot r_+ \geqslant (3a/2) r_+ - O(\lambda r_+) \geqslant a r_+ \,, \\ \dot r_- \leqslant -a r_- + O(\lambda r_+) \leqslant -a r_- + c\lambda r_+ \,, \\ \dot r \leqslant O(\lambda^2 r_+) + O(\varepsilon r_+) + O(\lambda r_+(\varepsilon + r_+)) \,, \end{cases} \tag{7.12}
$$

where c is a positive constant. If $\lambda = \Lambda\varepsilon$ and $\varepsilon \to 0$, then the last inequality (7.12) implies

$$
\dot r \leqslant C\varepsilon r_+ \,, \qquad C > 0 \,.
$$

The constants c and C do not depend on δ, ε, and Λ. Using (7.12), we see that everywhere in D

$$
\dot r - \lambda \dot r_+ \leqslant (C - \Lambda a)\varepsilon < 0
$$

for $\Lambda > C/a$. If $r = r_-$, then

$$
\dot r_- \leqslant -a\varepsilon + C\Lambda\varepsilon\,(\delta\sqrt{2} + \varepsilon) < 0
$$

for $\delta\sqrt{2} + \varepsilon < a/(\Lambda C)$. Lemma 7.2 is proved.

The first inequality (7.12) and (7.10) imply that while the trajectory $\gamma(t)$ stays in D, we have

$$
r_+(t) \geqslant \varepsilon e^{a(t-\tau)} \,, \qquad t \geqslant \tau \,.
$$

Thus, for some $t > \tau$, the trajectory $\gamma(t)$ exits D. By Lemma 7.2, $|u(t)| = \delta$ and hence $(\gamma(t)\,,\, t) \in \partial U = \{|u| = \delta\}$. The exit time is less than $\tau + T$, since $r_+(\tau + T) = \varepsilon$. The definition of the set D implies that $|z_-| \leqslant \varepsilon$ and $|y| \leqslant \Lambda\varepsilon\,(\delta\sqrt{2} + \varepsilon)$. By (7.5) and (7.11), we get that $\gamma'(t) \to W^u$ as $h \to 0$. The last statement of Proposition 7.1 is evident from (7.10) and (7.11). □

PROOF OF THEOREM 4.2. By Proposition 6.1, there exists a $c > 0$, not depending on h, such that $S(\gamma) \leqslant c$. Lemma 5.4 implies that there exists a constant $C > c$, not depending on h, such that

$$
S(\gamma|_{[a\,,\,b]}) \leqslant C \quad \text{for all } [a\,,\,b] \subset [\tau\,,\,\tau + T]. \tag{7.13}
$$

Choose a sequence $h_k \to 0$, $h_k > 0$, and denote by $\gamma_k \colon [\tau_k\,,\,\tau_k + T] \to M$ the corresponding sequence of trajectories constructed in Proposition 6.1. Let $t_k = t_-$, where t_- is defined in (2) of Proposition 7.1. Since the Lagrangian is 1-periodic in time, we can assume that $t_k \in [0\,,\,1]$ and thus some subsequence of t_k has a limit $t_0 \in [0\,,\,1]$. Then (2) of Proposition 7.1 implies that $\tau_k \to -\infty$ and $\tau_k + T_k \to \infty$ as $k \to \infty$. We claim that the sequence γ_k contains a subsequence that is convergent on any finite time interval to an orbit which is semihomoclinic to the invariant torus Γ.

The superlinearity of L and (7.13) imply that the sequence γ_k is uniformly bounded: $\|\dot\gamma_k(t)\| \leqslant K$, where the constant K does not depend on k (see Lemma 7.3 below). Since M is compact, there exists a subsequence γ_j such that the sequence $\gamma'(t_j) = (\gamma_j(t_j), \dot\gamma_j(t_j), t_j)$ is convergent to some point $(x_0, v_0, t_0) \in P$. We have $(x_0, t_0) \in \partial U$. Let $x: \mathbb{R} \to M$ be the trajectory with initial condition (x_0, v_0, t_0). Then for any time interval $[a, b]$, the sequence $\gamma_j|_{[a,b]}$ is uniformly convergent to $x|_{[a,b]}$ in the C^1-metric. Hence, (7.13) implies that $S(x|_{[a,b]}) \leqslant C$ and x is stable in Lagrange's sense. By Lemma 5.4, x is semiasymptotic to Γ as $t \to \pm\infty$. Since $(x(t_0), t_0) \in \partial U$, the orbit x' is not contained in Γ and thus it is a nontrivial semihomoclinic orbit. The claim is proved.

In fact, we have obtained that x is asymptotic to Γ as $t \to -\infty$. This follows from (3) of Proposition 7.1: $x'(t_0) \in W^u(\Gamma)$. Thus, $W^u \cap W^+ \setminus \Gamma \neq \varnothing$. Replacing t_- with t_+, we get $W^- \cap W^s \setminus \Gamma \neq \varnothing$. This proves part (1) of Theorem 4.2.

To complete the proof, it remains to get rid of the assumption (in the proof of Proposition 6.1) that the function $L(x, v, t)$ is quadratic in velocity for sufficiently large $\|v\|$. Let us take a sufficiently large $K > 0$ and modify L for $\|v\| \geqslant K$ in such a way that the new Lagrangian L' is quadratic in velocity for large $\|v\|$, satisfies the Mather conditions, and the inequality

$$L'(x, v, t) \geqslant A\|v\| - B, \tag{7.14}$$

where the positive constants A and B do not depend on K. Then Proposition 6.1 is valid. We claim that if K is large enough, then the constant $c > 0$ in Proposition 6.1 can be taken not depending on K and on the function L'.

The constant c is introduced in the inequality (6.8). By the remark after Corollary 6.3, c is the maximum of the functional S_h on some C^1-bounded set of piecewise smooth curves in Ω. If $K > 0$ is so large that $\|\dot\gamma(t)\| \leqslant K$ for all curves γ in this set, then c does not depend on the choice of L'. The claim is proved.

The constant C in (7.13) also does not depend on L'. Analyzing the proof of Lemma 5.2, we see that $C = c + 2D$, where D is such a constant that for all $a < b$, $1 \leqslant b - a \leqslant 2$, any two points of M can be connected by a curve $\alpha: [a, b] \to M$ satisfying the inequality $S(\alpha) \leqslant D$. If, for instance, α is a shortest geodesic, then $\|\dot\alpha(t)\| \leqslant \operatorname{diam}(M)/(b-a)$. Hence, if $K > \operatorname{diam}(M)$, then $D = \sup S(\alpha)$ does not depend on L'.

Now we shall use the following obvious consequence of the completeness condition [32].

LEMMA 7.3. *Let $N > 0$ and suppose that $K > 0$ is large enough. If $x: [a, a+1] \to M$ is a solution of the Lagrangian system with the Lagrangian L satisfying the inequality $\|\dot x(a)\| \leqslant N$ for some $t \in [a, a+1]$, then $\|\dot x(t)\| \leqslant K$ for all $t \in [a, a+1]$.*

Let $N > (B + C)/A$, where A and B are defined in (7.14) and C in (7.13). For a given N, take K as in Lemma 7.3. Let $\gamma: [\tau, \tau + T] \to M$ be the trajectory of the Lagrangian system with the Lagrangian L' constructed in Proposition 6.1. Then by Proposition 6.1, $\|\dot\gamma(\tau)\|$ is uniformly bounded by a constant that depends only on the original Lagrangian and is less than K if K is large enough. Suppose that $\|\dot\gamma(t)\| \geqslant K$ for some $t \in (\tau, \tau + T)$. Then there exists an $a \in [\tau, \tau + T]$ such that

$\|\dot\gamma(a+1)\| = K$ and $\|\dot\gamma(t)\| \leqslant K$ for $t \in [a, a+1]$. Then $\gamma|_{[a,a+1]}$ is a solution of the initial Lagrangian system with Lagrangian L. By Lemma 7.3, we see that $\|\dot\gamma(t)\| \geqslant N$ for all $t \in [a, a+1]$. Then by (7.14), we get $S(\gamma|_{[a,a+1]}) > C$, which contradicts (7.13). Hence γ is a solution of the original Lagrangian system.

This completes the proof of statement (1) of Theorem 4.2. Now suppose that assumptions of Proposition 4.1 are satisfied. We can assume that $L|_\Gamma = 0$ and $L|_{P\setminus\Gamma} > 0$. In this case, Proposition 6.1 implies that the constructed semihomoclinic orbit $x\colon \mathbb{R} \to M$ satisfies the inequality

$$\int_{-\infty}^{+\infty} L(x(t), \dot x(t), t)\, dt \leqslant c\,. \tag{7.15}$$

Hence, x is homoclinic to Γ. We omit the simple proof based on the completeness condition. For $n = 0$, see, for example, [15, 37]. Theorem 4.2 is proved.

PROOF OF LEMMA 3.5. Suppose that the assumptions of Lemma 3.5 are satisfied. By the Addendum to Corollary 6.3, the constant c in (7.15) is equal to $C\sqrt{\varepsilon}$. Now (1) of Lemma 3.5 follows from statement (2) of Theorem 4.2. The rest of Lemma 3.5 follows in a similar way from Proposition 4.3 and Theorem 4.3. □

Now to complete the proofs of Theorems 1.1–1.3, it remains to prove Theorem 4.3.

§8. Semihomoclinic orbits to minimal invariant sets

In this section we shall prove Theorem 4.3.

PROOF OF PART (1) OF THEOREM 4.3. The proof is similar to the proof of Theorem 4.2 in §6, and in a certain respect even simpler, since the constructed semihomoclinics are minimizing. The difficulty is that now it can happen that the minimal invariant set Γ is topologically complicated. For example, in the one-dimensional case Γ is usually a Cantor set. Without loss of generality, we can assume that for any invariant measure $\mu \in \mathfrak{M}$ with support in Γ, we have $A(\mu) = \min_{\mathfrak{M}} A = 0$.

By the assumption of Theorem 4.3, the Čech homology group $G = H_1(M \times \mathbb{T}, V, \mathbb{R})$ is not zero. Let $i\colon V \to M \times \mathbb{T}$ be the inclusion. Consider the following exact sequence [42]:

$$i_*\colon H_1(V, \mathbb{R}) \to H_1(M \times \mathbb{T}, \mathbb{R}) \to H_1(M \times \mathbb{T}, V, \mathbb{R}) \to \widetilde{H}_0(V, \mathbb{R})\,.$$

We shall assume that Γ is connected, i.e., $\widetilde{H}_0(V, \mathbb{R}) = 0$. If this is not so, then the proof is simpler: no covering spaces are needed. The exact sequence implies

$$G = H_1(M \times \mathbb{T}, V, \mathbb{R}) = H_1(M \times \mathbb{T}, \mathbb{R})/i_*(H_0(V, \mathbb{R}))\,.$$

By the assumption of Theorem 4.3, we have $G \neq 0$. For small $\delta > 0$, let

$$U = U_\delta = \{x \in M \times \mathbb{T} : d(x, V) < \delta\}\,.$$

Then U is connected and by the definition of the Čech homology group,

$$G = \lim_{\delta\to 0} H_1(M \times \mathbb{T}, \mathbb{R})/i_*(H_0(U, \mathbb{R}))\,.$$

Thus, for small $\delta > 0$ the map $i_*\colon H_1(U, \mathbb{R}) \to H_1(M \times \mathbb{T}, \mathbb{R})$ is not epimorphic. It is sufficient to prove

PROPOSITION 8.1. *If U is connected and the homomorphism i_* is not epimorphic, then there exists a semihomoclinic orbit to Γ.*

PROOF. Let $p\colon \widehat{M} \to M$ be the covering with the transformation group

$$\widehat{H}_1(M, \mathbb{Z}) = \operatorname{Im}(\pi_1(M) \to H_1(M, \mathbb{R})),$$

which was already considered in §5. Denote by $\varphi\colon \widehat{M} \times \mathbb{R} \to M \times \mathbb{T}$ the corresponding covering with transformation group

$$H = \widehat{H}_1(M \times \mathbb{T}, \mathbb{Z}) = \widehat{H}_1(M, \mathbb{Z}) \times \mathbb{Z}.$$

Let $K = i_*(\widehat{H}_1(U, \mathbb{Z})) \subset H$. By the assumption, $K \neq H$.

LEMMA 8.1. *The set $\varphi^{-1}(U) \subset \widehat{M} \times \mathbb{R}$ is connected if and only if $K = H$.*

This is an easy consequence of the theory of coverings (see, for example, [42]). □

By the assumption of Proposition 8.1, the set $\varphi^{-1}(U)$ is not connected. Let $\widehat{U}$ be one of its connected components. The group $K \subset H$ acts on $\widehat{M} \times \mathbb{R}$ and $\widehat{U}$ is K-invariant: for all $g \in K$, we have $g\widehat{U} = \widehat{U}$. If $g \in H \setminus K$, then $g\widehat{U} \cap \widehat{U} = \varnothing$. The connected components of $\varphi^{-1}(U)$ are in one-to-one correspondence with the elements of the group H/K:

$$\varphi^{-1}U = \bigcup_{g \in H/K} g(\widehat{U}).$$

The next lemma is also a standard fact of the theory of coverings.

LEMMA 8.2. *The map $\varphi\colon \widehat{U} \to U$ is a covering with the transformation group K. Thus $U = \widehat{U}/K$.* □

Denote $\widehat{V} = \varphi^{-1}(V) \cap \widehat{U}$. By Lemma 8.2, the quotient $\widehat{V}/K = V$ is compact. Let $\pi\colon \widehat{P} = T\widehat{M} \times \mathbb{R} \to \widehat{M} \times \mathbb{R}$ be the projection and $\widehat{\Gamma} \subset \widehat{P}$ the corresponding invariant set: $\pi(\widehat{\Gamma}) = \widehat{V}$. Then $\pi^{-1}\colon \widehat{V} \to \widehat{\Gamma}$ is a Lipschitz map. For fixed $g \in H \backslash K$, let $g\widehat{U}$, $g\widehat{V}$, and $g\widehat{\Gamma}$, be the translations of $\widehat{U}$, $\widehat{V}$, and $\widehat{\Gamma}$.

Now let us formulate the variational problem. If Γ is isotropic (see the definition in §4) and $i^*\colon H^1(M \times \mathbb{T}, \mathbb{R}) \to H^1(U, \mathbb{R})$ is surjective, then after subtracting from the Lagrangian the local full derivative $\alpha/dt = \langle y(x, t), \dot{x}\rangle - H(x, y(x, t), t)$ corresponding to the closed 1-form (4.8), we can assume that $L|_\Gamma = 0$. In this case, the variational problem is similar to the variational problem in §6. This simplifies the proof. Of course, the new Lagrangian contains a term which is linear in velocity and only Lipschitz in x, but this is no difficulty, since the trajectories we are looking for are minimizers. However, since in general it is not proved that any minimal invariant set is isotropic, we cannot use this in the proof. Thus, we do not assume that $L|_\Gamma = 0$. Therefore, the definition of a suitable function space is a bit intricate. Since we are looking for minimizers, the space of curves under consideration can be narrowed.

Let $T > 0$. For each point $x \in \widehat{V}$ denote by $\gamma_x\colon [\tau, \tau + T] \to \widehat{M}$ a minimizing trajectory with the boundary conditions

$$(8.1) \qquad (\gamma_x(\tau), \tau) = x, \qquad (\gamma_x(\tau + T), \tau + T) = g\psi^T x.$$

Here $\psi^t = \pi \circ g^t \circ \pi^{-1} \colon \widehat{V} \to \widehat{V}$, where $g^t \colon \widehat{P} \to \widehat{P}$ is the phase flow. As in the proof of Proposition 4.2, by using Lemma 5.2 it can be shown that there exists a constant $C > 0$ such that for all $T \geqslant 1$ and any $x \in \widehat{V}$, we have

$$\|\dot{\gamma}_x(t)\| \leqslant C \quad \text{for all } t \in [\tau, \tau + T].$$

The group H acts on the space of curves $\gamma \colon [a, b] \to \widehat{M}$: the curve γ with the graph $G_\gamma = \{(\gamma(t), t) \in \widehat{M} \times \mathbb{R} : a \leqslant t \leqslant b\}$ is mapped by a translation $f \in H$, $f \colon \widehat{M} \times \mathbb{R} \to \widehat{M} \times \mathbb{R}$, to the curve $f(\gamma)$ with the graph $f(G_\gamma)$. The functional S is H-invariant. We can assume that $\gamma_{f(x)} = f(\gamma_x)$ for all $f \in K$. Then the function $x \in \widehat{V} \to S(\gamma_x) \in \mathbb{R}$ is invariant under the action of the group K on $\widehat{V}$. Hence it defines a function on V. Standard variational methods imply that $S(\gamma_x)$ is continuous in $T > 0$ and $x \in V$. Let μ be an arbitrary minimizing measure with $\operatorname{supp}(\mu) \subset \Gamma$. Denote

$$F(T) = \int_V S(\gamma_x)\, d\mu(x). \tag{8.2}$$

Here the integral is defined by the measure on V which is the projection of the invariant measure μ on Γ (for simplicity we write $d\mu(x)$ instead of $d\mu(\pi^{-1}(x))$). The continuous function $y \in \Gamma \to S(\gamma_{\pi(y)})$ is μ-integrable and therefore, the integral is well defined.

LEMMA 8.3. *The function F on $\mathbb{R}^+$ is continuous and $F(T) \to +\infty$ as $T \to 0$. For arbitrary $h > 0$, we have $F_h(T) = F(T) + hT \to +\infty$ as $T \to \infty$.*

PROOF. The first statement is evident. Let us lift the Riemannian metric on M to an H-invariant Riemannian metric on $\widehat{M} \times \mathbb{R}$. Since the quotient $\widehat{V}/K$ is compact and $\widehat{V} \cap g\widehat{V} = \varnothing$, we get

$$d(\widehat{V}, g\widehat{V}) > c > 0. \tag{8.3}$$

By the superlinear property of the Lagrangian and the boundary conditions (8.1), we see that $S(\gamma_x) \to \infty$ as $T \to 0$ uniformly in $x \in V$ (the proof is similar to the proof of Lemma 6.8). Hence $F(T) \to \infty$.

Let $T \to +\infty$. Then $\liminf_{T\to\infty} S(\gamma_x)/T \geqslant 0$. The proof is based on the assumption that Γ is a minimal invariant set and $\min A = 0$. Since the proof is close to that of Lemma 6.7, we omit the details. By (8.2) and the definition of F_h, we obtain $\lim_{T\to\infty} F_h(T) = \infty$. □

COROLLARY 8.1. *For all $h > 0$, the function $F_h \colon \mathbb{R}^+ \to \mathbb{R}$ assumes its minimum.* □

For given $h > 0$, let $T > 0$ be a point of minimum for F_h and for each $x \in \widehat{V}$ let $\gamma_x \colon [\tau, \tau + T] \to \widehat{M}$ be the corresponding curve.

LEMMA 8.4. *There exists a $B > 0$ such that for all sufficiently small $h > 0$, we have*

$$\int_V \|\dot{\gamma}_x(\tau) - u(x)\|^2\, d\mu(x) \leqslant Bh,$$

$$\int_V \|\dot{\gamma}_x(\tau + T) - dg \cdot u(\psi^T(x))\|^2\, d\mu(x) \leqslant Bh.$$

Here for $x = (y, t) \in \widehat{V}$ we denote by $u(x) \in T_y\widehat{M}$ the velocity vector of the trajectory in $\widehat{V}$ passing through x (the velocity component of $\pi^{-1}(x) \in \widehat{\Gamma}$).

PROOF. Let us take a small variation $T \to T + \varepsilon$. For each $x \in \widehat{V}$ let $\gamma_{x,\varepsilon}$ be a smooth family of curves satisfying the boundary conditions (8.1), where T is replaced by $T + \varepsilon$. Again, we can assume that for any $f \in K$ the curve $\gamma_{f(x),\varepsilon}$ is the translation $f(\gamma_{x,\varepsilon})$ of the curve $\gamma_{x,\varepsilon}$. By (8.1), (8.2), and the definition of T, we have

$$\int_V S(\gamma_{x,\varepsilon})\, d\mu(x) + h(T+\varepsilon) \geqslant \int_V S(\gamma_x)\, d\mu(x) + hT.$$

Differentiating with respect to ε at $\varepsilon = 0$ and using the free-boundary first variation formula [2, 4], we obtain

$$\text{(8.4)} \quad \int_V (\langle p, D_\varepsilon|_{\varepsilon=0}\gamma_{x,\varepsilon}(\tau+T+\varepsilon)\rangle - H(\gamma_x(\tau+T), p, \tau+T))\, d\mu(x) + h = 0.$$

Here $p = L_v(\gamma_x(\tau+T), \dot\gamma_x(\tau+T), \tau+T)$ is the momentum of the trajectory γ_x and H the Hamiltonian. By (8.1), we have

$$D_\varepsilon|_{\varepsilon=0}\gamma_{x,\varepsilon}(\tau+T+\varepsilon) = dg\cdot u(\psi^T(x)) = u(g\psi^T(x)).$$

Since the measure μ is minimizing and invariant,

$$0 = \int_\Gamma L\, d\mu = \int_V L(\pi^{-1}(x))\, d\mu(x) = \int_V L(\pi^{-1}(g\psi^T(x)))\, d\mu(x).$$

Thus, (8.4) yields

$$\int_V (\langle p, u(g\psi^T(x))\rangle - H(\gamma_x(\tau+T), p, \tau+T) - L(\pi^{-1}g\psi^T(gx)))\, d\mu(x) + h = 0.$$

Let $u = u(g\psi^T(x))$ and $y = L_v(\gamma_x(\tau+T), u, \tau+T)$. Consider the integrand as a function $f(p)$. Then

$$f(p) - \langle p, u\rangle - H(p) - L(u).$$

There exists $A > 0$, independent of h and x, such that $\|p\| \leqslant A$ and $\|y\| \leqslant A$. By the positive definiteness assumption, there exists $a > 0$, such that for any $\|p\| \leqslant A$ we have $\langle H_{pp}\xi, \xi\rangle \geqslant a\|\xi\|^2$ for all ξ. The definition of the Legendre transform implies

$$f(p) \geqslant -\frac{a}{2}\|p-y\|^2 + \min_p(\langle p, u\rangle - H(p)) - L(u) = -\frac{a}{2}\|p-y\|^2.$$

Thus

$$\int_V \|p-y\|^2\, d\mu(x) \leqslant \frac{2h}{a}.$$

Rewriting this inequality in terms of velocities, we obtain the second inequality in the statement of Lemma 8.4.

To prove the first inequality, we carry out the transformation of variables $x = g^{-1}\psi^{-\varepsilon}g(y)$ in the integral on the left-hand side of (8.4). Since the measure μ is invariant, we obtain

$$\int_V S(\gamma_{g^{-1}\psi^{-\varepsilon}g(y),\varepsilon})\,d\mu(y) + h(T+\varepsilon) \geqslant \int_V S(\gamma_x)\,d\mu(x) + hT.$$

When ε varies, the right-boundary point $g\psi^T y \in g(\widehat{V})$ of the curve

$$\gamma_{g^{-1}\psi^{-\varepsilon}g(y)} : [\tau - \varepsilon\,,\, \tau + T] \to \widehat{M}$$

is fixed, and the left-boundary point varies with ε. We obtain an inequality like (8.4), but the initial and final point are interchanged. Now after differentiating with respect to ε, the proof of the second inequality follows the same lines as the proof of the first. $\square$

REMARK. Since the measure μ is invariant and $A(\mu) = 0$, in the case $g = 0$ the corresponding function (8.2) is identically zero. Thus for any $g \in H \setminus K$ there exists $D > 0$ such that for sufficiently small $h > 0$ we have

$$\int_V S(\gamma_x)\,d\mu(x) \leqslant D\,, \qquad S(\gamma_x) \geqslant -C \quad \text{for all } x.$$

It is easy to see that for any $R > 0$ the constant D can be estimated as

$$D \leqslant \max_{x\in V} S(\alpha_x).$$

Here for any $x \in V$, $\alpha_x : [\tau\,,\, \tau + R] \to \widehat{M}$ is any curve such that $(\alpha_x(\tau)\,,\, \tau) = \widehat{x}$ and $(\alpha(\tau + R)\,,\, \tau + R) = g\psi^R\widehat{x}$, where $\widehat{x}$ is a point in $\widehat{V}$ projected into x. We shall use this estimate in §9.

For simplicity, denote $a = \tau$ and $b = \tau + T$. The remark and Lemma 8.4 imply that for any $h > 0$ there exists an $x \in \operatorname{supp}(\mu)$ such that the corresponding minimizing trajectory $x(t) = \gamma_x(t)$, $a \leqslant t \leqslant b$, satisfies the following conditions: $(x(a)\,,\, a) \in \widehat{V}$, $(x(b)\,,\, b) \in g(\widehat{V})$,

$$\text{(8.5)} \qquad \|\dot{x}(a) - u(x(a)\,,\, a)\|^2 \leqslant 3Bh\,, \qquad \|\dot{x}(b) - u(x(b)\,,\, b)\|^2 \leqslant 3Bh\,,$$

and $S(x) \leqslant 3D + 2C$. The proof is obvious: the μ-measure of the sets where any of these inequalities is not satisfied is less than $1/3$.

Lemma 5.2 implies

$$\text{(8.6)} \qquad S(x|_{[c\,,\,d]}) \leqslant 3D + 4C \quad \text{for all } [c\,,\, d] \subset [a\,,\, b].$$

Inequalities (8.5) imply that $d\,(x'(a)\,,\, \widehat{\Gamma}) \leqslant Bh$ and $d\,(x'(b)\,,\, g\widehat{\Gamma}) \leqslant Bh$. The distance in $\widehat{P}$ is measured by the H-invariant Riemannian metric on $\widehat{M}$. Since the sets $\widehat{\Gamma}$ and $g\widehat{\Gamma}$ are g^t-invariant, we obtain

COROLLARY 8.2. *For any $T > 0$ and sufficiently small $h > 0$, we have $b - a \geqslant 2T$, $(x(t), t) \in \widehat{U}$ for $a \leqslant t \leqslant a + T$, and $(x(t), t) \in g(\widehat{U})$ for $b - T \leqslant t \leqslant b$.* □

Now we can complete the proof of Proposition 8.1. Since the points $(x(a), a)$ and $(x(b), b)$ belong to different components of $\varphi^{-1}(U)$, for any $h > 0$ there exists a $c \in [a, b]$ such that $(x(c), c) \in \partial\widehat{U}$. Let $h_i \to 0$. Replacing the curve x_i by an appropriate translation $g_i(x_i)$, $g_i \in K$, we can assume that the sequence $(x_i(c_i), c_i)$ is contained in a compact subset of $\widehat{M} \times \mathbb{R}$ and the sequence $x'(c_i) = (x(c_i), \dot{x}(c_i), c_i)$ is convergent in $\widehat{P}$ to a point (x_0, v_0, t_0). By Corollary 8.2, $c_i - a_i \to +\infty$ and $b_i - c_i \to +\infty$ as $t \to \infty$. Let $x\colon \mathbb{R} \to \widehat{M}$ be the trajectory with the initial condition (x_0, v_0, t_0). Corollary 8.2 implies that for all $a < b$ the restriction $x|_{[a,b]}$ is minimizing and satisfies (8.6). Hence

$$\lim_{b-a\to\infty} S(x|_{[a,b]})/(b-a) = 0.$$

By Lemma 5.4, x is semiasymptotic to Γ as $t \to \pm\infty$. Since $(x_0, t_0) \in \partial\widehat{U}$, the orbit x is not contained in Γ. Proposition 8.1 and, consequently, part (1) of Theorem 4.3, are proved.

PROOF OF PART (2) OF THEOREM 4.3. As in the proof of part (1), for simplicity, suppose that Γ is connected. The disconnected case requires only small modifications. Replacing L by $L - \dot{S} - \min A$, we can assume that

$$L|_\Gamma = 0, \qquad L|_{P\setminus\Gamma} > 0. \tag{8.7}$$

We shall continue to use the notations of the proof of Proposition 8.1. It is sufficient to prove

PROPOSITION 8.2. *Suppose that the neighborhood $U = U_\delta$ is small enough. Then there exist at least $2\dim H/K$ homoclinic orbits to Γ.*

PROOF. We follow [14], where the similar case of brake orbits (librations) was considered (see also [29, 37, 39] for the case of homoclinics to an equilibrium and [10] for the case of homoclinics to periodic orbits of reversible systems).

Let $g \in K$ and let $\Omega = \Omega_g$ be the set of piecewise smooth curves $x\colon [a, b] \to \widehat{M}$ such that $a < b$ are arbitrary and $(x(a), a) \in \widehat{V}$, $(x(b), b) \in g(\widehat{V})$. Denote

$$R(g) = \inf_{\gamma\in\Omega_g} S(\gamma). \tag{8.8}$$

LEMMA 8.5. *The function $R\colon H \to \mathbb{R}$ is zero on $K \subset H$, constant on every conjugate class $g + K$, and*

$$\inf_{H\setminus K} R > \varepsilon > 0. \tag{8.9}$$

The set $\{g \in H/K : R(g) \leqslant C\}$ is finite for all $C > 0$.

PROOF. The first two statements are evident. Let us prove (8.9). There exists $\lambda > 0$ such that $L|_{\pi^{-1}(\varphi^{-1}(U))} > \lambda$. Hence $S(\gamma) \geqslant T\lambda$, where $T > 0$ is the time that the curve $\gamma \in \Omega_g$ spends outside of $\varphi^{-1}(U)$. The superlinearity of L and (8.3) imply that if the T is small, then the action $S(\gamma)$ is large. Thus, $S(\gamma)$ is bounded from below by a universal positive constant. The proof of the last statement is similar. □

We define a set of independent generators $g_1, \dots, g_k$ of the semigroup $G = H/K$ by induction. By Lemma 8.5, the function R assumes the minimum on $G \setminus \{0\}$ at some point $g_1 \in G$. Suppose that the elements $g_1, \dots, g_{i-1} \in G$ are already defined. Denote by $[g_1, \dots, g_{i-1}] \subset G$ the generated semigroup, i.e., the set

$$\left\{\sum_{j=1}^{i-1} n_j g_j : n_j \in \mathbb{Z},\ n_j \geqslant 0\right\}.$$

If it is not equal to G, we define an element $g_i \in G$ by the formula

$$R(g_i) = \min_{G \setminus [g_1, \dots, g_{i-1}]} R. \tag{8.10}$$

By Lemma 8.5, the minimum exists. It is easy to see that $k = 2 \operatorname{rank} G$.

LEMMA 8.6. *Let $g = g_i$. If $g = f + f'$ for some $f, f' \in H$, then $R(f) \geqslant R(g)$ or $R(f') \geqslant R(g)$.*

This is an immediate corollary of (8.10): we have either $f \notin [g_1, \dots, g_{i-1}]$ or $f' \notin [g_1, \dots, g_{i-1}]$. □

Let $\gamma: \mathbb{R} \to M$ be a homoclinic trajectory to Γ. One of the covering curves $\widehat{\gamma}: \mathbb{R} \to \widehat{M}$ is asymptotic to $\widehat{\Gamma}$ as $t \to -\infty$ and to $g(\widehat{\Gamma})$, $g \in H$, as $t \to +\infty$. The element g is determined up to adding an element from K, and thus defines an element $[\gamma] \in G = \widehat{H}_1(M \times \mathbb{T}, U, \mathbb{Z})$. The corresponding element in $H_1(M \times \mathbb{T}, V, \mathbb{R})$ is the homology class of the homoclinic orbit γ. This definition is equivalent to the definition in §4. Proposition 8.2 is a corollary of the following

LEMMA 8.7. *Suppose that $g \in G$ satisfies the condition of Lemma 8.6. Then there exists a homoclinic orbit γ such that $[\gamma] = g$.*

PROOF. For any $h > 0$ define the functional S_h on $\Omega = \Omega_g$ by the formula

$$S_h(x) = S(x) + h(b - a), \qquad x: [a, b] \to \widehat{M}.$$

Since V is compact, there exists an S_h-minimizing curve $x \in \Omega_g$. Taking a small variation of the points $(x(a), a)$ and $(x(b), b)$ along minimizing trajectories in $\widehat{V}$ and $g(\widehat{V})$, as it was done in the proof of Lemma 8.4, and using the first variation formula and (8.7), we obtain that for small $h > 0$ the statement of Corollary 8.2 holds. Now (8.6) and (8.7) imply that $S(x) \leqslant C$, where the constant C does not depend on h. Let

$$c = \sup\{t \in [a, b] : x([a, t]) \subset \widehat{U}\}, \qquad d = \inf\{t \in [a, b] : x([t, c]) \subset g(\widehat{U})\}.$$

By Corollary 8.2, $c - a \to \infty$ and $b - d \to \infty$ as $h \to 0$.

LEMMA 8.8. *Suppose that the neighborhood* $W \subset U$ *of the set* $V \subset M \times \mathbb{T}$ *is small enough. Then for each sufficiently small* $h > 0$, *we have* $(x(t), t) \notin \varphi^{-1}(W)$ *for* $c \leqslant t \leqslant d$.

PROOF. Let

$$A = \min_{g \in H \setminus \{0\}} R(g). \tag{8.11}$$

Denote $\widehat{W} = \varphi^{-1} W \cap \widehat{U}$.

LEMMA 8.9. *There exist* $B > 0$ *and* $\varepsilon < \min(A/2, B)$ *such that*

(1) *if* $\gamma: [u, v] \to \widehat{M}$ *is a curve connecting* $\partial \widehat{U}$ *and* $\partial \widehat{W}$, *i.e.*, $(\gamma(u), u) \in \partial \widehat{U}$ *and* $(\gamma(v), v) \in \partial \widehat{W}$ *or vice versa, then* $S(\gamma) \geqslant B$;

(2) *for each point* $(x_0, t_0) \in \partial \widehat{W}$ *there exist two curves* $\alpha: [u, t_0] \to \widehat{M}$ *and* $\beta: [t_0, v] \to \widehat{M}$ *such that* $\alpha(t_0) = \beta(t_0) = x_0$, $(\alpha(u), u) \in \widehat{V}$, $(\beta(v), v) \in \widehat{V}$, *and* $S(\alpha) < \varepsilon$, $S(\beta) < \varepsilon$.

We omit the proof. □

Suppose that Lemma 8.8 is false. Then there exists a $t_0 \in [c, d]$ such that the point $(x(t_0), t_0)$ is in $\partial \varphi^{-1} W$. Hence, $(x(t_0), t_0) \in \partial f(\widehat{W})$ for some $f \in H$. By the definitions (8.2) and (8.8), for each $\lambda > 0$ and sufficiently small $h > 0$, we have

$$S(x) \leqslant R(g) + \lambda. \tag{8.12}$$

Let α and β be the curves defined in (2) of Lemma 8.9. Since $x|_{[a, t_0]} \cdot \beta \in \Omega_f$ and $\alpha \cdot x|_{[t_0, b]} \in \Omega_{g-f}$, the definition of the function R implies

$$S(x) = S(x|_{[a, t_0]} \cdot \beta) - S(\beta) + S(\alpha \cdot x|_{[t_0, b]}) - S(\alpha) \geqslant R(f) + R(g - f) - 2\varepsilon.$$

Lemma 8.6 yields $R(f) \geqslant R(g)$ or $R(g - f) \geqslant R(g)$. If $f \neq 0$ and $f \neq g$, then (8.11) implies $S(x) > R(g) + A - 2\varepsilon$. If $\lambda > 0$ is sufficiently small, then this contradicts (8.12).

Now suppose, for example, that $f = 0$. Then by (1) and (2) of Lemma 8.9,

$$S(x) \geqslant S(x|_{[a, t_0]}) + S(\alpha \cdot x|_{[t_0, b]}) - \varepsilon \geqslant B + R(g) - \varepsilon.$$

This contradicts (8.12). □

COROLLARY 8.3. *There exists* $T > 0$ *such that* $d - c \leqslant T$ *for all* $h > 0$.

Now we can complete the proof of Lemma 8.7. Let $h_i \to 0$ and let $x_i: [c_i, d_i] \to \widehat{M}$ be the corresponding minimizer. Performing a suitable sequence of translations of $\widehat{M} \times \mathbb{R}$, we can assume that the sequence x_i contains a converging subsequence $x_j: [c_j, d_j] \to \widehat{M}$. Denote by $\gamma: \mathbb{R} \to \widehat{M}$ the maximal solution of the Lagrangian system corresponding to the limiting curve. Then $S(\gamma) < \infty$ and $(\gamma(t), t) \in \widehat{U}$ for $t < \lim c_j$, $(\gamma(t), t) \in g(\widehat{U})$ for $t > \lim d_j$. Thus, the corresponding orbit $\gamma': \mathbb{R} \to P$ is homoclinic to Γ and $[\gamma] = g$. Lemma 8.7 and Proposition 8.2 are proved.

This completes the proof of Theorem 4.3.

§9. Minimal invariant sets of perturbed Hamiltonian systems

In this section we shall prove Theorems 1.1′–1.3′. In §§1–2 it was shown that it is sufficient to consider only the case of autonomous Hamiltonian systems (Theorem 1.2′).

PROOF OF THEOREM 1.2′. According to §2, the problem is reduced to the case of a positive definite autonomous Lagrangian system with the Lagrangian function (3.10′). Since the assumptions of Mather are satisfied, to each cohomology class $c \in H^1(M, \mathbb{R})$ there corresponds a minimal invariant set $\Gamma = \Lambda_c$. In the present autonomous case, this set is t-invariant. Thus we can assume that $\Gamma \subset TM$. Suppose that c is an $O(\varepsilon)$-small cohomology class: $\|c\| \leqslant C\varepsilon$, where $C > 0$ is fixed.

PROPOSITION 9.1. *Let $\delta > 0$. Then for all sufficiently small $\varepsilon > 0$, the minimal invariant set Γ is contained in $TU \subset TM$, where $U = U_\delta = \{x \in M : d(x, \Gamma_0) \leqslant \delta\}$.*

REMARK. This shows that as $\varepsilon \to 0$, the projection of the set $\Gamma \subset TM$ to M tends to Γ_0. Making the calculations more carefully, it is possible to estimate the size δ of the neighborhood U by a function of ε. From our proof it follows only that $\delta \leqslant C\alpha^{1/2}$, where $\alpha > 0$ was introduced in Lemma 2.3′. The estimate depends on the Diophantine properties of the frequency vector ω. If the assumptions of Theorem 1.2 are satisfied, then the proof yields $\delta = o(\varepsilon^{1/4})$. Probably in this case $\delta = o(\varepsilon^{1/2})$.

Proposition 9.1 and definition (4.4) imply that for every invariant measure μ such that $\operatorname{supp}(\mu) \subset \Gamma$, we have $\rho(\mu) = i_* \nu$, $\nu \in H_1(U, \mathbb{R}) = H_1(\Gamma_0, \mathbb{R}) = \mathbb{R}^n$. Thus

$$A_c(\mu) = A(\mu) - \langle c, \rho(\mu)\rangle = A(\mu) - \langle I, \nu(\mu)\rangle,$$

where $I = i^* c \in E \subset \mathbb{R}^n$. Hence, for $c = O(\varepsilon)$ the set Λ_c depends only on $I \in E$.

Let $c = \varepsilon[\lambda]$, where λ, $\|\lambda\|_{C^1} \leqslant C$, is a closed 1-form on M. Averaging λ with respect to the $\mathbb{T}^n$-action, we can assume that λ is invariant. Then $\langle \lambda, v\rangle \equiv \text{const}$, since $d\langle \lambda, v\rangle = v \cdot \lambda - i_v d\lambda = 0$. Replace L by

$$\widehat{L} = L - \varepsilon\langle \lambda, \dot{x} - v\rangle. \tag{9.1}$$

The Lagrangian (9.1) is of same type (3.10′) and the corresponding Lagrangian equations are the same. Thus, from now on we can assume that $c = 0$.

The normal form (3.10′) depends on the parameter $\alpha > 0$ which was introduced in Lemma 2.3′. In what follows we shall always assume that α is sufficiently small but fixed, and $\varepsilon \to 0$. By A, B, C, and D we shall denote arbitrary positive constants independent of α and ε.

LEMMA 9.1. *If $\alpha > 0$ is sufficiently small, then for small $\varepsilon > 0$ any two points $x, y \in \Gamma_0$ can be connected by a curve $\gamma\colon [0, T] \to \Gamma_0$, $\varepsilon^{-1/2} \leqslant T \leqslant 2\varepsilon^{-1/2}$, with Hamiltonian action*

$$S(\gamma) \leqslant C\alpha\varepsilon^{1/2}. \tag{9.2}$$

PROOF. Let $g_\theta\colon M \to M$, $\theta \in \mathbb{T}^n$, be the group action on M which was introduced in §1. Let $\beta\colon [0, T] \to \Gamma_0$ be the shortest geodesic joining x and $g_{-\omega T}(y)$.

Then

$$\|\dot{\beta}(t)\| = \rho/T, \qquad \rho = d(x, y). \tag{9.3}$$

Since the frequency vector ω is nonresonant, the curve $t \to g_{\omega t}(x)$ is dense in Γ_0. Hence, for small ε it is possible to choose T in such a way that the distance ρ in (9.3) is arbitrary small.

Let $\gamma(t) = g_{\omega t}(\beta(t))$, $0 \leqslant t \leqslant T$. Then

$$\dot{\gamma}(t) = v(\gamma(t)) + dg_{\omega t}(\beta(t))\,\dot{\beta}(t).$$

The metric in M is $\mathbb{T}^n$-invariant. Hence by (9.3), $\|\dot{\gamma}(t) - v(\gamma(t))\| = \rho/T$. Since $V|_{\Gamma_0} = 0$, (3.13′) implies

$$S(\gamma) = \int_0^T L\,dt \leqslant \int_0^T \left(\frac{(\rho/T)^2}{2a} + \varepsilon\alpha + K\varepsilon^2\right) dt = \frac{\rho^2}{2aT} + \varepsilon T(\alpha + K\varepsilon).$$

For $\varepsilon^{-1/2} \leqslant T \leqslant 2\varepsilon^{-1/2}$ and $\rho \leqslant \alpha^{1/2}$, we get

$$S(\gamma) \leqslant \varepsilon^{1/2}(\alpha/(2a) + 2(1 + K(\alpha)\varepsilon)).$$

For $\varepsilon \to 0$, we obtain (9.2). □

REMARK. The estimate for ρ in (9.3) depends on the Diophantine properties of ω. If ω satisfies the Diophantine condition, then ρ can be estimated by a power of ε. Thus, inequality (9.2) can be sharpened. This shows one way how the Diophantine properties of ω enter into estimates for $\delta(\varepsilon)$. Another way is through small divisors in the proof of Lemma 2.3′. However, we shall not bother about getting the best possible estimates.

COROLLARY 9.1. *If ε is small enough, then $-C'\alpha\varepsilon \leqslant \min_{\mu\in\mathfrak{M}} A(\mu) \leqslant C\alpha\varepsilon$.*

PROOF. Since $V \leqslant 0$, the left-hand side inequality is evident from (3.13′). Apply Lemma 9.1, taking $x = y$. Then $\gamma\colon [0, T] \to M$, $T \geqslant \varepsilon^{-1/2}$, is a closed curve and by (9.2), $S(\gamma)/T \leqslant C\alpha\varepsilon$. Let $\beta\colon [0, T] \to M$ be a minimizing closed curve homotopic to γ. The orbit β is the support of an invariant measure μ. Thus by the definition, $A(\mu) \leqslant C\alpha\varepsilon$. □

Replacing L by $L - \min_{\mathfrak{M}} A$, we can assume that $\min_{\mathfrak{M}} A = 0$. By Corollary 9.1, the new function u in (3.10′) differs from the old one by less than $\varepsilon\alpha\max(C, C')$, and thus it also satisfies (3.17) (with different α, equal to $\alpha\max(C, C', 1)$). Thus, (9.2) still holds.

Let $0 < r < \delta$ and $W = U_r$.

LEMMA 9.2. *If $\alpha < \inf_{M\setminus W}(-V)/2$ and $\varepsilon > 0$ is small enough, then for any curve $\gamma\colon [0, T] \to \overline{U \setminus W}$ connecting ∂U with ∂W or vice versa, we have*

$$S(\gamma) \geqslant \varepsilon^{1/2}\rho(\partial U, \partial W),$$

where ρ *is the distance in the Jacobi metric* $\sqrt{-V(x)/b}\,\|\dot{x}\|$ *on* $M\setminus\Gamma_0$. *If* Γ_0 *is a nondegenerate critical manifold for* V, *then* $\rho(\partial U, \partial W)\geqslant C(\delta^2-r^2)$, $C>0$.

PROOF. First let us prove the inequality

$$L(x,\dot{x},\varepsilon)\geqslant\sqrt{-\varepsilon V(x)/b}\,\|\dot{x}-v(x)\|+B\alpha\varepsilon,\qquad x\in M\setminus W,\ b,B>0. \tag{9.4}$$

Rewrite (3.13′) as follows:

$$\begin{aligned}L(x,\dot{x},\varepsilon)\geqslant{}&\sqrt{-\varepsilon V(x)/b}\,\|\dot{x}-v(x)\|+\Big(\|\dot{x}-v(x)\|/\sqrt{2b}\\&-\sqrt{-\varepsilon V(x)/2}\Big)^2-\varepsilon V(x)/2-\alpha\varepsilon-K(\alpha)\varepsilon^2.\end{aligned} \tag{9.5}$$

We have

$$2\alpha<\inf_{M\setminus W}V\leqslant\max_W V\leqslant Dr^2,\qquad D>0. \tag{9.6}$$

If ε is small enough, then the sum of the last three terms in (9.5) is not less than $B\alpha\varepsilon$, $B>0$. This proves (9.4).

Take the integral of (9.4) from $t=0$ to $t=T$. Since the distance in M is invariant under the group action of the torus $\mathbb{T}^n$, the sets U and W are $\mathbb{T}^n$-invariant. Let $\beta(t)=g_{-\omega t}(\gamma(t))$, $0\leqslant t\leqslant T$. The integral of the first term on the right-hand side of (9.4) equals $\varepsilon^{1/2}l(\beta)$, where $l(\beta)$ is the length of β in the Jacobi metric $\sqrt{-V(x)/b}\,\|\dot{x}\|$. Hence $l(\beta)\geqslant\rho(\partial U,\partial W)$. □

LEMMA 9.3. *Suppose that* α *satisfies* (9.6) *and* $\varepsilon>0$ *is small enough. Then each point in* ∂W *can be connected with* Γ_0 *by a curve* γ *such that* $S(\gamma)\leqslant B\varepsilon^{1/2}r^2$ *and* $S(\gamma^{-1})\leqslant B\varepsilon^{1/2}r^2$, $B>0$.

PROOF. Let $\beta\colon[0,\varepsilon^{-1/2}]\to M$ be the shortest geodesic connecting $x\in\partial W$ with Γ_0 and let $\gamma(t)=g_{\omega t}(\beta(t))$. Since $\|\dot{\gamma}(t)-v(\gamma(t))\|=\|\dot{\beta}(t)\|\equiv r\varepsilon^{1/2}$, by (3.13′) and (9.6) we have

$$S(\gamma)\leqslant\varepsilon^{1/2}\big(r^2/(2a)+\max_W(-V)+\alpha+K\varepsilon\big)\leqslant B\varepsilon^{1/2}r^2.\quad\square$$

PROOF OF PROPOSITION 9.1. Let $\mu\in\mathfrak{M}$ be an ergodic minimal invariant measure such that $\operatorname{supp}(\mu)\subset\Gamma$. It is sufficient to show that $\operatorname{supp}(\mu)\subset TU$.

Suppose this is not so. Since $A(\mu)=0$, almost all trajectories $x\colon\mathbb{R}\to M$ such that the corresponding orbit $x'\colon\mathbb{R}\to TM$ is contained in $\operatorname{supp}(\mu)$, have the following property. There exist arbitrary large $a, b>0$ such that $S(x|_{[-a,0]})\leqslant 0$ and $S(x|_{[0,b]})\leqslant 0$.

This is a corollary of the Birkhoff ergodic theorem. We can assume that the trajectory x is not contained in U. By (9.4), if a and b are large enough, then $x([-a,0])\cap W\neq\varnothing$ and $x([0,b])\cap W\neq\varnothing$. Let

$$c=\sup\{t : x([-a,t])\subset M\setminus W\},\qquad d=\inf\{t : x([t,b])\subset M\setminus W\}.$$

Then $x(c)$, $x(d) \in \partial W$ and (9.4) implies $S(x|_{[c,d]}) \leqslant 0$. By our assumption, there exist $c \leqslant p \leqslant q \leqslant d$ such that $x(p)$, $x(q) \in \partial W$, $x([p,q]) \subset M \setminus W$, and $x(t) \in M \setminus U$ for some $t \in [p, q]$. By Lemma 9.3,

$$S(x|_{[p,q]}) \geqslant 2\varepsilon^{1/2}\rho(\partial U, \partial W). \tag{9.7}$$

By Lemmas 9.1 and 9.3, the pairs of points $x(p)$, $x(q)$, and $x(d)$, $x(c)$ can be joined by curves β and γ respectively, with the Hamiltonian actions

$$S(\beta) \leqslant \varepsilon^{1/2}(C\alpha + 2Br^2), \qquad S(\gamma) \leqslant \varepsilon^{1/2}(C\alpha + 2Br^2). \tag{9.8}$$

Consider the closed curve $\sigma = x|_{[c,p]} \cdot \beta \cdot x|_{[q,d]} \cdot \gamma$, where for each curve the independent variable is appropriately shifted. Then by (9.7) and (9.8), we have

$$\begin{aligned} S(\sigma) &= S(x|_{[c,d]}) - S(x|_{[p,q]}) + S(\beta) + S(\gamma) \\ &\leqslant 0 - 2\varepsilon^{1/2}\rho(\partial U, \partial W) + 2\varepsilon^{1/2}(C\alpha + 2Br^2). \end{aligned}$$

Since the minimum of A on $\mathfrak{M}$ is zero, the closed curve σ has a nonnegative action $S(\sigma) \geqslant 0$. Hence by (9.6),

$$\rho(\partial U, \partial W) \leqslant (CD/2 + 2B)r^2.$$

Since r is arbitrary small, the size δ of U is arbitrary small. In the nondegenerate case, $\delta = O(\alpha^{1/2})$. □

PROOF OF PART (1) OF THEOREM 1.2′. We have already showed in Proposition 9.1 that as $\varepsilon \to 0$, the projection of Γ into M is contained in an arbitrarily small neighborhood U_δ of Γ_0. To complete the proof, it is sufficient to make sure that for small $\varepsilon > 0$ and every trajectory $t \to (x(t), p(t)) \in \Gamma$, the norm $\|p(t)\|$ is $o(\delta\varepsilon^{1/2})$-small.

LEMMA 9.4. *Let* $\Delta, \Lambda > 0$. *For sufficiently small* $\alpha, \varepsilon > 0$ *and any trajectory* $t \to (x(t), p(t)) \in T^*M$ *such that* $\|p(\tau)\|^2 \leqslant \Delta\varepsilon$, *the function* $E(t) = \|p\|^2/2 + \varepsilon V$ *satisfies the inequality*

$$|E(t) - E(\tau)| \leqslant A\alpha\varepsilon \quad \text{for } |t - \tau| \leqslant \Lambda\varepsilon^{-1/2}, \qquad A > 0. \tag{9.9}$$

PROOF. As in the proof of Lemma 3.1, differentiating the function $E(t)$ along a trajectory $(x(t), p(t))$ of the Hamiltonian system (3.1′), for sufficiently small ε and $\|p(t)\|$ we obtain

$$\dot{E} = O(\varepsilon^2) + \varepsilon O(\|p\|^2) + O(\|p\|^4) + P,$$

where

$$P = \{\langle Ap, p\rangle/2 + \varepsilon V, \langle wp, p\rangle/2 + \varepsilon u\}.$$

Calculating the Poisson bracket and using (3.17), we get

$$\begin{aligned} P &= O(\|w\| + \|\nabla w\|)\,O(\|p\|^3) + \varepsilon O(\|w\|)\,O(\|p\|) + \varepsilon O(\|\nabla u\|)\,O(\|p\|) \\ &= \alpha(O(\|p\|^3) + \varepsilon O(\|p\|)). \end{aligned}$$

Suppose that

$$\|p(t)\|^2 \leqslant 2\Delta^2\varepsilon \quad \text{for } |t-\tau| \leqslant \Lambda\varepsilon^{-1/2}. \tag{9.10}$$

Then $\dot{E} = O(\varepsilon^2) + \alpha O(\varepsilon^{3/2})$. Thus, there exist constants $C, D > 0$ such that $|\dot{E}| \leqslant C\varepsilon^2 + D\alpha\varepsilon^{3/2}$. Integrating this inequality, we obtain that while (9.10) holds,

$$|E(t) - E(\tau)| \leqslant \Lambda(C\varepsilon^{3/2} + D\alpha\varepsilon), \qquad |t-\tau| \leqslant \Lambda\varepsilon^{-1/2}.$$

If $A > D$ and ε is so small that $\Lambda(C\varepsilon^{1/2} + D\alpha) < A\alpha$, then we get (9.9). This inequality implies that the assumption (9.10) is justified. □

LEMMA 9.5. *Let $c > 0$ and $v \geqslant c\delta^2$. If $\Delta = 2(\max_U(-V) + v)$ and $\alpha, \varepsilon > 0$ are sufficiently small, then for any trajectory $t \to (x(t), p(t)) \in \Gamma$, we have $\|p(t)\|^2 \leqslant \Delta\varepsilon$.*

PROOF. Suppose this is not so. It is sufficient to consider a generic trajectory in Γ. We know that $x(t) \in U$. By (3.10′),

$$L \geqslant \|p\|^2/(2a) - \varepsilon(V + \alpha) - K\varepsilon^2.$$

Since $V \leqslant 0$, for $\|p(t)\|^2 \geqslant v\varepsilon$ and sufficiently small $\alpha, \varepsilon > 0$, we have

$$L(x(t), \dot{x}(t), \varepsilon) > \lambda\varepsilon, \qquad \lambda = av/3. \tag{9.11}$$

Thus, $\|p(t)\|^2 \leqslant v\varepsilon$ for some t. Our assumption implies that there exists $\tau \in \mathbb{R}$ such that $\|p(\tau)\|^2 = \Delta\varepsilon$. Hence $E(\tau) \geqslant v\varepsilon$. If α and ε are small, then by Lemma 9.4, $E(t) \geqslant v\varepsilon/2$ for $\tau \leqslant t \leqslant \Lambda\varepsilon^{-1/2}$. Thus $\|p(t)\|^2 \geqslant v\varepsilon$. By (9.11),

$$S(x|_{[\tau,\tau+\Lambda\varepsilon^{-1/2}]}) \geqslant \Lambda\lambda\varepsilon^{1/2}. \tag{9.12}$$

As in the proof of Proposition 8.3, by the Birkhoff ergodic theorem, we can assume that there exists arbitrarily large $T > \tau$ such that $S(x|_{[\tau,\tau+T]}) \leqslant 0$. We shall take $T > \tau + \varepsilon^{-1/2}$.

By Lemma 9.3, the points $x(\tau)$ and $x(\tau + \varepsilon^{-1/2})$ can be connected by a curve β, and the points $x(T)$ and $x(\tau)$ by a curve γ, such that the actions of both curves are less than $B\delta^2\varepsilon^{1/2}$. By (9.12), the action of the closed curve

$$x_{[\tau,\tau+\varepsilon^{-1/2}]} \cdot \beta \cdot x_{[\tau+\varepsilon^{-1/2},T]} \cdot \gamma$$

is less than $\varepsilon^{1/2}(2B\delta^2 - \Lambda\lambda)$. If $\Lambda > 2B\delta^2/\lambda \geqslant 8B/(ac)$, then we get a contradiction. □

Part (1) of Theorem 1.2′ is proved.

We omit the proof of (2) of Theorem 1.2′, since it is similar to the proof of part (3).

PROOF OF PART (3) OF THEOREM 1.2′. First let us show that there exists at least one semihomoclinic orbit. By the assumption, the map $i_*: H_1(U, \mathbb{R}) \to H_1(M, \mathbb{R})$ is not epimorphic. Hence by Proposition 8.1, there exists a trajectory $t \to (x(t), p(t)) \in T^*M$ which is semihomoclinic to Γ. We must only show that

$$\|p(t)\|^2 \leqslant \Delta\varepsilon, \qquad \Delta = 2(\max(-V) + v), \quad v > 0. \tag{9.13}$$

In particular, the orbit is contained in the set where the Hamiltonian was not modified.

LEMMA 9.6. *There exists a constant $B > 0$ independent of ε, such that for all τ and $T > 0$ we have*

$$S(x|_{[\tau,\tau+T]}) \leqslant B\varepsilon^{1/2}. \tag{9.14}$$

Suppose that Lemma 9.6 is proved. Let $\Lambda > 2B/v$ and $T = \Lambda\varepsilon^{-1/2}$. Then, as in the proof of Lemma 9.5, inequality (9.14) implies that for arbitrary τ and some $t \in [\tau, \tau + T]$, we have

$$L(x(t), \dot{x}(t), \varepsilon) \leqslant \varepsilon B/\Lambda < \varepsilon v/2.$$

By (3.13′), if α and ε are small, then $\|p(t)\|^2 \leqslant \varepsilon v$ and $E(t) \leqslant \varepsilon v/2$. Lemma 9.4 yields that if $\alpha \leqslant v/(2A)$, then $E(t) \leqslant \varepsilon v$ for all $t \in [\tau, \tau + T]$. Hence $\|p(t)\|^2 \leqslant 2(\varepsilon v - \varepsilon V) = \varepsilon\Delta$, which proves (9.13).

PROOF OF LEMMA 9.6. Let us return to the proof of Proposition 8.1. In the present case, the Lagrangian system is autonomous. We claim that inequality (8.6) takes the form

$$-C'\varepsilon^{1/2} \leqslant S(x|_{[c,d]}) \leqslant (3D' + 4C')\varepsilon^{1/2}. \tag{9.15}$$

By (3.10′), any two points of M can be joined by a curve with Hamiltonian action less than $C'\varepsilon^{1/2}$. Since $\min A = 0$, as in the proof of Lemma 5.2, it follows that for any curve β in M, we have $S(\beta) \geqslant -C'\varepsilon^{1/2}$. To prove the right-hand inequality (9.15), we shall use the estimate for the constant D in (8.6) given in the remark before inequalities (8.5). Let $R = \varepsilon^{-1/2}$. Lift the $\mathbb{T}^n$-action on M to $\widehat{M}$. Then $g \circ g_{\omega t} \equiv g_{\omega t} \circ g$. For any $x \in \widehat{V}$ let $\alpha_x : [\tau, \tau + \varepsilon^{-1/2}] \to \widehat{M}$ be the curve defined by the formula $\alpha_x(t) = g_{\omega t}(\beta(t))$, where $\beta : [\tau, \tau + \varepsilon^{-1/2}] \to \widehat{M}$ is a shortest geodesic joining x and $y = g_{-\omega\varepsilon^{-1/2}}(g\psi^{\varepsilon^{-1/2}}x)$: $\beta(\tau) = x$ and $\beta(\tau + \varepsilon^{-1/2}) = y$. Then the constant D in (8.6) is bounded by $\sup_x S(\alpha_x)$. We claim that in the present case $\sup_x S(\alpha_x) \leqslant D'\varepsilon^{1/2}$, where $D' > 0$ is independent of $x \in \widehat{V}$ and ε.

Let $x(t) = \psi^t(x) \in \widehat{V}$. Since by (2) of Theorem 1.2′, the set Γ is $o(\varepsilon^{1/2})$-close to M, there exists a $K > 0$, independent of ε and α, such that $\|\dot{x}(t) - v(x(t))\| \leqslant K\varepsilon^{1/2}$. Let $z(t) = g_{-\omega t}(x(t))$. Since $g_{\omega t}$ is an isometry,

$$\|\dot{z}(t)\| = \|dg_{-\omega t}(x(t))(\dot{x}(t) - v(x(t)))\| = \|\dot{x}(t) - v(x(t))\| \leqslant K\varepsilon^{1/2}.$$

Hence

$$d(g_{\omega t}x, \psi^t x) = d(x, z(t)) \leqslant K\varepsilon^{1/2}t \quad \text{for } \tau \leqslant t \leqslant \tau + \varepsilon^{-1/2}.$$

Thus, there exists a constant $A > 0$ independent of ε and x, such that $d(x, y) \leqslant A$. Hence, $\|\dot{\beta}(t)\| \leqslant A\varepsilon^{1/2}$ for $\tau \leqslant t \leqslant \tau + \varepsilon^{-1/2}$. Therefore, $\|\dot{\alpha}_x(t) - v(\alpha_x(t))\| \leqslant A\varepsilon^{1/2}$. By (3.10′), we obtain $S(\alpha_x) \leqslant D'\varepsilon^{1/2}$. The claim is proved.

Now inequality (9.15) follows from (8.6). If $B = 3D + 4C$, then (9.15) implies (9.14). □

Now the proof of the existence of a semihomoclinic orbit in the region (9.13) is complete.

The proof of the estimate for the number of semihomoclinic orbits is a modification of the proof of Proposition 8.2. In the present case, assumption (8.7), which was essential for the proof of Proposition 8.2, is not satisfied. Let

$$L_0 = \|\dot{x} - v(x)\|^2/2 - \varepsilon V .$$

Define the function R and the generators g_i as in (8.8) and (8.9), replacing everywhere L by L_0 and Γ by Γ_0. Since $L_0|_{\Gamma_0} = 0$, the definition is correct. Thus, Lemmas 8.5–8.6 hold. Let g be an element of $H \setminus K$ satisfying the condition of Lemma 8.6. As in the proof of Proposition 8.1, associate with g a semihomoclinic orbit $x\colon \mathbb{R} \to \widehat{M}$. It is sufficient to make sure that if α and ε are small, then $x(t) \in \widehat{U}$ for large negative t and $x(t) \in g(\widehat{U})$ for large positive t. The proof is a combination of the proofs of Lemma 8.8 and Proposition 9.1. We omit the details. □

This completes the proof of Theorem 1.2′ and thus of Theorems 1.1′–1.3′.

References

1. A. Ambrosetti, *Critical points and nonlinear variational problems*, Scuola Normale Superiore, Pisa, 1991.
2. V. I. Arnold, *Mathematical methods in classical mechanics*, "Nauka", Moscow, 1974; English transl., Springer-Verlag, New York, Heidelberg, and Berlin, 1978.
3. ______, *Instability of dynamical systems with many degrees of freedom.*, Dokl. Akad. Nauk SSSR **156** (1964), 9–12; English transl. in Soviet Math. Dokl. **5** (1964).
4. V. I. Arnold, V. V. Kozlov, and A. I. Neishtadt, *Dynamical systems* 3, Sovremennye Problemy Matematiki. Fundamental′nye Napravleniya, vol. 3, VINITI, Moscow, 1985; English transl., Encyclopedia of Math. Sciences, vol. 3, Springer-Verlag, New York, Heidelberg, and Berlin, 1988.
5. S. Aubry and P. Y. Le Daeron, *The discrete Frenkel–Kontorova model and its extensions*, Phys. D **8** (1983), 381–422.
6. V. Benci and F. Giannoni, *Homoclinic orbits on compact manifolds*, J. Math. Anal. Appl. **157** (1991), 568–576.
7. D. Bernstein and A. Katok, *Birkhoff periodic orbits for small perturbations of completely integrable Hamiltonian systems with convex Hamiltonians*, Invent. Math. **88** (1987), no. 2, 225–241.
8. G. D. Birkhoff, *Dynamical systems*, Amer. Math. Soc., New York, 1927.
9. S. V. Bolotin, *Libration motions of natural dynamical systems*, Vestnik Moskov. Univ. Ser. I Mat. Mekh. **1978**, no. 6, 72–77; English transl. in Moscow Univ. Math. Bull. **32** (1978).
10. ______, *Libration motions of reversible Hamiltonian systems.*, PhD Thesis, Moscow State University, Moscow, 1981. (Russian)
11. ______, *Existence of homoclinic motions*, Vestnik Moskov. Univ. Ser. I Mat. Mekh. **1983**, no. 6, 98–103; English transl. in Moscow Univ. Math. Bull. **38** (1983).
12. ______, *Motions that are doubly asymptotic to invariant tori in the theory of the perturbations of Hamiltonian systems*, Prikl. Mat. Mekh. **54** (1992), no. 3, 497–501; English trans. in J. Appl. Math. Mech. **54** (1990).
13. ______, *Homoclinic orbits to minimal tori of Lagrangian systems*, Vestnik Moskov. Univ. Ser. I Mat. Mekh. **1992**, no. 6, 34–41; English transl. in Moscow Univ. Math. Bull. **47** (1992).
14. S. V. Bolotin and V. V. Kozlov, *Libration in systems with many degrees of freedom*, Prikl. Mat. Mekh. **42** (1978), no. 2, 245–250; English trans. in J. Appl. Math. Mech. **42** (1978).
15. ______, *On asymptotic solutions of the equations of dynamics*, Vestnik Moskov. Univ. Ser. I Mat. Mekh. **1980**, no. 4, 84–89; English transl. in Moscow Univ. Math. Bull. **34** (1980).
16. L. Chierchia and G. Galavotti, *Drift and diffusion in phase space*, CARR Reports on Mathematical Physics, Rome, Italy, 1992.
17. C. C. Conley and E. Zehnder, *The Birkhoff–Lewis fixed point theorem and a conjecture by V. I. Arnold*, Invent. Math. **73** (1983), 33–49.

18. V. Coti Zelati, I. Ekeland, and E. Sere, *A variational approach to homoclinic orbits in Hamiltonian systems*, Math. Ann. **288** (1990), 133–160.

19. V. Coti Zelati and P. H. Rabinowitz, *Homoclinic orbits for second order Hamiltonian systems possessing superquadratic potentials*, J. Amer. Math. Soc. **4** (1991), 693–727.

20. R. Douadi, *Stabilité ou instabilité des points fixes elliptiques*, Ann. Sci. École Norm. Sup. (4) **21** (1988), 1–46.

21. F. Giannoni, *On the existence of homoclinic orbits on Riemannian manifolds.*, Preprint, Centre for the Mathematical Sciences, Univ. of Wisconsin, Madison, WI, 1991.

22. F. Giannoni and P. H. Rabinowitz, *On the multiplicity of homoclinic orbits on Riemannian manifolds for a class of second order Hamiltonian systems*, Nonlinear Differential Equations and Applications **1** (1993), 1–49.

23. S. M. Graff, *On the conservation of hyperbolic tori for Hamiltonian systems*, J. Differential Equations **15** (1974), no. 1, 1–69.

24. G. Hedlund, *Geodesics on a two-dimensional Riemannian manifold with periodic coefficients*, Ann. of Math. (2) **33** (1932), 719–739.

25. M. R. Herman, *Existence et non existence de tores invariants par des diffeomorphismes symplectiques*, Seminaire sur les Equations aux Derivées Partielles, Exp. XIV, Ecole Politechnique, Palaiseau, 1988.

26. P. J. Holmes and J. E. Marsden, *Melnikov's method and Arnold diffusion for perturbations of integrable Hamiltonian systems*, J. Math. Phys. **23** (1982), 669–675.

27. S. Kobayashi and S. Nomizu, *Foundations of differential geometry,* Vols. 1 and 2, Interscience, New York, 1963–1969.

28. V. V. Kozlov, *Integrability and noninitegrability in classical mechanics*, Uspekhi. Mat. Nauk **38** (1983), no. 1, 3–67; English transl. in Russian Math. Surveys **38** (1983).

29. ———, *Calculus of variations in large and classical mechanics*, Uspekhi. Mat. Nauk **40** (1985), no. 2, 33–60; English transl. in Russian Math. Surveys **40** (1985).

30. J. N. Mather, *Existence of quasiperiodic orbits for twist homeomorphisms of the annulus*, Topology **21** (1982), 457–467.

31. ———, *Variational construction of orbits of twist diffeomorphisms*, J. Amer. Math. Soc. **4** (1991), no. 2, 207–263.

32. ———, *Action minimizing invariant measures for positive definite Lagrangian systems*, Math. Z. **227** (1991), 169–207.

33. J. Moser, *Monotone twist mappings and the calculus of variations*, Ergodic Theory Dynamical Systems **6** (1986), 401–413.

34. R. Palais, *Morse theory on Hilbert manifolds*, Topology **2** (1963), 299–340.

35. I. C. Percival, *A variational principle for invariant tori of fixed frequency*, J. Phys. A **12** (1979), no. 3, 57–60.

36. H. Poincaré, *Les méthodes nouvelles de la mécanique céléste.* II, Gauthier–Villars, Paris, 1893.

37. P. H. Rabinowitz, *Periodic and heteroclinic solutions for a periodic Hamiltonian system*, Ann. Inst. H. Poincaré Anal. Non Linéaire **6** (1989), 331–346.

38. ———, *Homoclinic and heteroclinic orbits for a class of Hamiltonian systems*, Calc. Var. Partial Differential Equations **1** (1992 3), no. 1, 1–36.

39. P. H. Rabinowitz and K. Tanaka, *Some results on connecting orbits for a class of Hamiltonian systems*, Math. Z. **206** (1991), 473–499.

40. D. Salamon, E. Zehnder, *KAM theory in configuration space*, Comment. Math. Helv. **64** (1989), 84–132.

41. E. Sere, *Existence of infinitely many homoclinics in Hamiltonian systems*, Math. Z. **209** (1992), 27–42.

42. E. Spanier, *Algebraic topology*, McGraw Hill, New York, 1966.

43. D. V. Treshchev, *The fracture mechanism of resonant tori of Hamiltonian systems*, Mat. Sb. **180** (1989), no. 10, 1325–1346; English transl. in Math. USSR-Sb. **68** (1991).

44. A. Weinstein, *Lagrangian submanifolds and Hamiltonian systems*, Ann. of Math. (2) **98** (1973), 377–410.

45. S. Wiggins, *Global bifurcations and chaos*, Springer-Verlag, New York, Heidelberg, and Berlin, 1991.
46. E. Zehnder, *Generalized implicit function theorems with applications to some small divisor problems.* II, Comm. Pure Appl. Math. **29** (1976), no. 1, 49–111.

Translated by THE AUTHOR

DEPARTMENT OF MATHEMATICS AND MECHANICS, MOSCOW STATE UNIVERSITY, MOSCOW 119899, RUSSIA

Amer. Math. Soc. Transl.
(2) Vol. **168**, 1995

An Estimate of Irremovable Nonconstant Terms in the Reducibility Problem

DMITRY V. TRESHCHEV

§1. Introduction

Let us consider the system of ordinary differential equations

$$\begin{cases} \dot{x} = \omega, \\ \dot{y} = A(x)y, \end{cases} \tag{1.1}$$

where x lies on the m-dimensional torus $\mathbb{T}^m = \mathbb{R}^m/2\pi\mathbb{Z}^m$, y belongs to the space $\mathbb{R}^n$, $\omega \in \mathbb{R}^m$ is a constant vector, A is a matrix-valued function on $\mathbb{T}^m$; dots denote derivatives with respect to the time t.

Equations (1.1) are called *reducible* if there exists a C^1-smooth change of variables

$$y = B(x)z, \qquad B \text{ is nondegenerate,} \tag{1.2}$$

which transforms the equations to the following form:

$$\begin{cases} \dot{x} = \omega, \\ \dot{z} = \widehat{A}z, \end{cases} \tag{1.3}$$

with some constant matrix $\widehat{A}$.

Note that the reducibility problem is usually formulated in a slightly different form, with the equation $\dot{y} = A(t)y$ and the change of variables $y = B(t)z$ instead of (1.1) and (1.2). Here the matrices A and B are quasiperiodic in their arguments. If in the new coordinates the system becomes $\dot{z} = \widehat{A}z$ with a constant matrix $\widehat{A}$, then the system is called reducible. Our definition of the reducibility is fairly close to the usual one.

In accordance with Floquet–Lyapunov theory [1], the reducibility problem always has a solution for $m = 1$. One can note that sometimes the function B is not 2π-periodic, but 4π-periodic in x.

In the case $n = 1$ the matrices A, B, and $\widehat{A}$ are ordinary functions and the following criterion of the reducibility of equations (1.1) exists: the equation

$$\partial f(x) = A(x) - \widehat{A}, \qquad \widehat{A} = \text{const}, \tag{1.4}$$

1991 *Mathematics Subject Classification*. Primary 34A25.

where

$$\partial = \sum_{k=1}^{m} \omega_k \frac{\partial}{\partial x_k} \tag{1.5}$$

must have a C^1-smooth solution $f\colon \mathbb{T}^m \to \mathbb{R}$. Indeed, the change of variables (1.2) transforms the second equation (1.1) to the form $\dot z = (A - \partial \ln |B|)\, z$. Thus $B = \pm \exp f$, where f is a solution of (1.4).

Explicit solutions of equation (1.4) can be obtained using Fourier expansions. Convergence of the Fourier series f depends on the arithmetic properties of the frequencies ω and on the smoothness class of the function A. There exist far-reaching generalizations of these simple arguments (see [2]).

The problem becomes much more complicated in the case $n > 1$, $m > 1$. It is known [3] that equations (1.1) are reducible if they satisfy certain spectral conditions. We note that these conditions do not appeal to any small parameter, but are quite strong and difficult to verify for concrete systems.

Another type of results concerns the case

$$A(x) = A_0 + \varepsilon A_1(x), \tag{1.6}$$

where A_0 is a constant matrix and ε is a small parameter. Such systems are studied using rapidly convergent methods of Newton type. We formulate one of the latest results of this type.

For any periodic function $f(x)$ let the constant f^0 be the average of f:

$$f^0 = (2\pi)^{-m} \int_{\mathbb{T}^m} f(x)\, dx .$$

THEOREM (A. Jorba and C. Simo, [4]). *Consider equation* (1.1), (1.6) *with* $\varepsilon \in (0, \varepsilon_0)$. *The eigenvalues* $\lambda_1, \dots, \lambda_n$ *of the constant matrix* A_0 *are assumed to be distinct. Also assume that*

1. A_1 *is analytic on the set* $\{x \in \mathbb{C}^m/2\pi\mathbb{Z}^m : |\operatorname{Im} x_j| \leqslant \rho_0,\ j = 1, \dots, m\}$, $\rho_0 > 0$.

2. *The vector* v, *where* $v^T = (\lambda_1, \dots, \lambda_n, i\omega_1, \dots, i\omega_m)$, $i = \sqrt{-1}$, *satisfies the nonresonance conditions* $|\langle k, v\rangle| \geqslant c_v |k|^{-\gamma_v}$ *for all*

$$k \in \{(k', k'') \in \mathbb{Z}^{n+m} : k' \in \mathbb{Z}^n,\ |k'| = 0 \text{ or } |k'| = 2;\ k'' \in \mathbb{Z}^m,\ |k''| \neq 0\},$$

where c_v *and* γ_v *are positive numbers and* $|k| = \sum |k_j|$.

3. *Let* $\lambda_j^0(\varepsilon)$ $(j = 1, \dots, n)$ *be the eigenvalues of* $A^0 = A_0 + \varepsilon A_1^0$ *and the following inequalities hold*:

$$\left| \frac{d}{d\varepsilon} (\lambda_l^0(\varepsilon) - \lambda_j^0(\varepsilon))\Big|_{\varepsilon=0} \right| > 2\delta > 0 \quad \text{for all } 1 \leqslant l < j \leqslant n .$$

Then there exists a Cantor set $E \subset (0, \varepsilon_0)$ *of positive Lebesgue measure such that system* (1.1) *is reducible. If* ε_0 *is small enough, the relative measure of* E *in* $(0, \varepsilon_0)$ *is close to* 1.

The proof of the theorem is based on a certain inductive procedure. The matrices A_0 and A_1 are transformed in accordance with the procedure. After the jth step they are equal to $A_0^{(j)}$ and $A_1^{(j)}$. In order to carry out the jth step of the procedure, it is necessary to satisfy certain nonresonance conditions that coincides with condition 2 of the theorem for $j = 1$. In the case $j > 1$, one must replace in this condition the numbers $\lambda_1, \dots, \lambda_n$ with the eigenvalues $\lambda_1^{(j)}, \dots, \lambda_n^{(j)}$ of the matrix $A_0^{(j-1)}$ and the constant c_ν with a new (smaller) constant c_j. (Jorba and Simo take $c_j = \text{const}/j^2$.) Each nonresonance condition does not hold for some values of ε. To overcome this difficulty, one must consider only "good" ε's, for which all nonresonance conditions hold. The set E in the theorem is the set of such values of ε.

These arguments explain why the theorem in question does not guarantee the reducibility of system (1.1), (1.6) for all small ε. Apparently, irreducible systems of the form (1.1), (1.6) actually exist for arbitrarily small ε. From the dynamical point of view, irreducibility is generated by the resonances between the external (ω) and the internal ($\operatorname{Im}\lambda^{(j)}$) frequencies. There are no such resonances if all the magnitudes $\operatorname{Im}\lambda^{(j)}$ are equal to zero. This situation takes place if the eigenvalues of the matrix A_0 are real and pairwise distinct [5]. In this case for small values of ε_0 we have $E = (0, \varepsilon_0)$.

In the general situation, the relative measure of E is close to 1 by the nondegeneracy condition 3 of the theorem, but E is a Cantor set. In particular, 0 is a limit point of the set $(0, \varepsilon_0)/E$. We believe, that the relative measure of the set E on the interval $(0, \varepsilon_0)$ can be estimated from below by the value $1 - \text{const}\cdot\varepsilon_0$, where the constant does not depend on ε_0.

This work was partially supported by the Russian Foundation of Basic Research and the International Science Foundation.

§2. Main result

In this paper we consider the case of a small matrix A:

(2.1) $$A(x) = \varepsilon a(x), \qquad \varepsilon > 0,$$

and try to weaken the rate of dependence of the matrix a on the angles x. We do not eliminate the dependence a on x but make it exponentially small with respect to the parameter ε. Moreover, we sometimes obtain an asymptotics of irremovable (in the framework of our averaging method) nonconstant terms in the matrix $a(x)$. Apparently, our results are close to the optimal ones that can be obtained for all small ε if a^0, the average of $a(x)$ is assumed not to have a real spectrum.

To formulate the results, we introduce some notation and definitions. We denote by $|f|$ the absolute value of f if f is a real number, the standard Euclidean norm of f if f is a vector, and $\max_{j,l}\{|f_{jl}|\}$ if $f = (f_{jl})$ is a matrix. We denote by $\langle\ ,\ \rangle$ the standard scalar product.

DEFINITION 1. We call a norm $\|\cdot\|$ in $\mathbb{R}^k$ *strictly convex* if for any two vectors $\mu_1, \mu_2 \in \mathbb{R}^k$ such that $\|\mu_1\| = \|\mu_2\| = 1$ the inequality

(2.2) $$\|\mu_1\| + \|\mu_2\| \geqslant \|\mu_1 + \mu_2\| + P\|\mu_1 - \mu_2\|^2$$

holds with some positive constant P depending only on the norm $\|\cdot\|$.

The concept of strictly convex norm can be formulated in a more geometrical terms. Indeed, we note that the sphere $S = \{\mu \in \mathbb{R}^k : \|\mu\| = 1\}$ bounds a convex set because of the triangle inequality.

PROPOSITION 1. *Let the sphere S be a smooth surface. Then the norm $\|\cdot\|$ is strictly convex if and only if the normal curvature of S at any point along any direction is greater then zero.*

The proof of Proposition 1 is given in §11.

Let us fix the frequency vector ω. Further we always assume that $|\omega| = 1$. The general case is reduced to this one by the change of time: $t \mapsto t/|\omega|$.

DEFINITION 2. We call a norm $\|\cdot\|$ in $\mathbb{R}^m$ *acceptable* if for any $\mu \in \mathbb{R}^m$

$$\|\mu\| = |\langle \mu, \omega\rangle|/2 + \|\mu - \omega\langle \mu, \omega\rangle\|_\pi, \tag{2.3}$$

where $\|\cdot\|_\pi$ is a strictly convex norm on the plane $\pi = \{\tau \in \mathbb{R}^m : \langle \tau, \omega\rangle = 0\}$.

REMARK. One can define an acceptable norm with the help of the relation

$$\|\mu\| = \beta|\langle \mu, \omega\rangle| + \|\mu - \omega\langle \mu, \omega\rangle\|_\pi, \qquad \beta > 0, \tag{2.4}$$

instead of (2.3). We use Definition 2 for convenience and keep in mind that (2.4) is transformed to (2.3) by the change $\|\cdot\| \to 2\beta\|\cdot\|$, $\|\cdot\|_\pi \to 2\beta\|\cdot\|_\pi$.

Acceptable norms play a key role in our constructions.

For example, in the case $m = 2$ the norm $\|\tau\| = \max\{|\tau_1\omega_1|, |\tau_2\omega_2|\}$ as acceptable.

DEFINITION 3. The vector ω is called *strictly nonresonance* if for any nonzero vector $\tau \in \mathbb{Z}^m$ we have:

$$|\langle \tau, \omega\rangle| \geqslant (\Gamma/\|\tau\|)^{m-1+\gamma}, \qquad \Gamma, \gamma > 0. \tag{2.5}$$

REMARK. It can be shown that for any $\gamma > 0$, the relative measure of vectors ω satisfying inequalities (2.5) with some Γ is complete on the unit sphere S. For $\gamma = 0$, the set of such vectors have zero measure, but it is dense in S and in $\mathbb{R}^m$ (see [6]).

Let us expand the function $a(x)$ in the Fourier series

$$a(x) = \sum_{\tau \in \mathbb{Z}^m} a^\tau e^{i\langle \tau, x\rangle}. \tag{2.6}$$

We assume the function $a(x)$ to be real analytic. More precisely, we assume that there exist positive constants c, α such that

$$|a^\tau| \leqslant c e^{-\alpha\|\tau\|}. \tag{2.7}$$

Here $\|\cdot\|$ is some acceptable norm.

The matrix a^0 is the average of $a(x)$. Let Λ be the maximal imaginary part of the eigenvalues of the matrix a^0. Since a^0 is real, we have $\Lambda \geqslant 0$.

THEOREM 1. *Let the frequency vector* ω *be strictly nonresonance and* $\Lambda > 0$. *Then for some positive constants* c^0, c', c'', c''', θ, *and small values of* ε *there exists the change of coordinates* (1.2) *that transforms equations* (1.1), (2.1) *to the form*

$$\begin{cases} \dot{x} = \omega\,, \\ \dot{z} = \varepsilon(\hat{a} - \tilde{a}(x))z\,, \qquad \hat{a} = \text{const}\,, \end{cases} \tag{2.8}$$

where $|\hat{a} - a^0| \leqslant c'\varepsilon$ *and*

$$|\tilde{a}(x)| \leqslant c^0 \exp(-\alpha\Gamma(2\Lambda\varepsilon)^{-\frac{1}{m-1+\gamma}})\,. \tag{2.9}$$

Moreover, let us put

$$G = \{\tau \in \mathbb{Z}^m : \|\tau\| < 2\Gamma(2\Lambda\varepsilon + c'''\varepsilon^2)^{-\frac{1}{m-1+\gamma}}\,,\ |\langle\tau\,,\,\omega\rangle| \leqslant 2\Lambda\varepsilon + c'''\varepsilon^2\}\,.$$

Then $\tilde{a}(x) = \varphi(x) + \Delta_1(x) + \Delta_2(x)$, *where*

$$\varphi(x) = \sum_{\tau\in G} a^\tau e^{i\langle\tau\,,\,x\rangle}\,, \qquad |\Delta_1(x)| \leqslant c''\varepsilon^\theta \exp(-\alpha\Gamma(2\Lambda\varepsilon)^{-\frac{1}{m-1+\gamma}})\,,$$

$$|\Delta_2(x)| \leqslant c_\Delta \varepsilon^{-m-\frac{1}{m-1+\gamma}} \exp(-2\alpha\Gamma(2\Lambda\varepsilon)^{-\frac{1}{m-1+\gamma}})\,.$$

The function $\tilde{a}(x)$ *is real analytic and for any* $\alpha' < \alpha$ *there exists a constant* $\tilde{c} = \tilde{c}(\alpha')$ *such that the Fourier coefficients* $\tilde{a}^\tau$ *satisfy the inequalities* $|\tilde{a}^\tau| \leqslant \tilde{c}\, e^{-\alpha'\|\tau\|}$.

REMARKS. 1. Theorem 1 shows that the dependence of a on x can be made exponentially small. The corresponding exponentially small term $\tilde{a}$ can be estimated by (2.9). The function $\varphi(x)$ can be regarded as the asymptotics of $\tilde{a}(x)$, since, in general, the norm $|\cdot|$ of φ is much greater than the norms of Δ_1 and Δ_2.

2. It can be shown that the change $y \to z$ in Theorem 1 is close to the identity: $B(x) = I + O(\varepsilon)$, where I is the unit matrix of order n.

3. There is some arbitrariness in the choice of the norm $\|\cdot\|$. This arbitrariness can be used to increase the constants α and Γ and thus to improve estimate (2.9).

4. Let us consider the set $\mathcal{E} \subset (0\,,\,\varepsilon_0)$ of the small parameter values corresponding to reducible systems (1.1), (2.1). Apparently, the relative measure of $\mathcal{E}$ on the interval $(0\,,\,\varepsilon_0)$ is estimated from below by

$$1 - \text{const}\cdot\varepsilon \exp(-\alpha\Gamma(2\Lambda\varepsilon)^{-\frac{1}{m-1+\gamma}})\,,$$

where the constant does not depend on ε.

§3. Reducibility of invariant tori in Hamiltonian systems

The reducibility problem containing a small matrix $A(x)$ (see (2.1)) arises naturally in some problems of Hamiltonian mechanics. Indeed, consider a Hamiltonian system

$$\dot{x} = \partial H/\partial y\,, \qquad \dot{y} = -\partial H/\partial x$$

with s degrees of freedom in canonical coordinates

$$(y, x) = (y_1, \ldots, y_s, x_1 \bmod 2\pi, \ldots, x_s \bmod 2\pi).$$

The Hamiltonian function

$$H(x, y, \varepsilon) = H_0(y) + \varepsilon H_1(x, y, \varepsilon) \tag{3.1}$$

is assumed to be real-analytic. For $\varepsilon = 0$ this system is integrable and the variables y, x are the action-angle variables. Solutions of the unperturbed ($\varepsilon = 0$) system are located on the invariant tori $\mathbb{T}^s_y = \{(x, y) : y = \text{const}\}$. The angle variables x vary along the tori $\mathbb{T}^s_y$ in accordance with the equations $\dot{x} = \omega(y)$, where $\omega(y) = \partial H_0/\partial y$.

For each frequency vector $\omega = \omega(y)$ there exists a subgroup $g = g(y)$ of the group $(\mathbb{Z}^s, +)$ such that $g = \{\tau \in \mathbb{Z}^s : \langle \omega, \tau \rangle = 0\}$. Let $\operatorname{rank} g = n$, $n + m = s$. The invariant torus $\mathbb{T}^s_y$ is fibered by m-dimensional subtori $\mathbb{T}^m_{y,x}$ that are the closures of the unperturbed system trajectories passing through the points y, x.

It is well known that resonance tori $\mathbb{T}^s_y$ are destroyed by perturbation. On the other hand, a theorem due to Poincaré claims that if a resonance torus is fibered by periodic solutions (invariant one-dimensional tori $\mathbb{T}^1_{y,x}$), then as a rule a finite number of such solutions survive becomes slightly deformed by perturbation. This result can be generalized: if a resonance torus is fibered by invariant nonresonance tori of any dimension $1 \leqslant m < s$, then as a rule a finite collection of these subtori is not destroyed by a small perturbation. To formulate the corresponding theorem, we introduce some notation.

Let us fix a point y^0, a vector $\omega^0 = \omega(y^0)$, and the corresponding group $g = g(\omega^0)$. For any continuous function $f : \mathbb{T}^s \to \mathbb{R}$ we define the average $\langle f \rangle_g$ by setting

$$\langle f \rangle_g(x) = \lim_{T \to \infty} \frac{1}{T} \int_0^T f(x + \omega^0 t)\, dt.$$

Let the Fourier expansion of the function f be $f(x) = \sum_{\tau \in \mathbb{Z}^s} f_\tau e^{i\langle \tau, x \rangle}$. It is easy to show that $\langle f \rangle_g(x) = \sum_{\tau \in g} f_\tau e^{i\langle \tau, x \rangle}$.

DEFINITION 4. We call the vector ω^0 *strictly m-nonresonance* if $\operatorname{rank} g = n = s - m$ and there exist positive constants c, N such that for any vector $\tau \in \mathbb{Z}^m \setminus g$ the inequality $|\langle \omega^0, \tau \rangle| \geqslant c|\tau|^{-N}$ holds.

One can note that any strictly s-nonresonance vector is strictly nonresonance in the sense of Definition 3.

THEOREM [7]. *Let the following conditions hold*:

1) *the unperturbed system is nondegenerate, i.e.,* $\det \Pi \neq 0$, *where* Π *is the matrix* $\partial^2 H_0/(\partial y^2)(y^0)$;

2) *the frequency vector* ω^0 *is strictly m-nonresonance*;

3) *for some critical point* x^0 *of the function* $h(x) = \langle H_1(y^0, x, 0) \rangle_g$ *the Hesse matrix* $W = \partial^2 h/(\partial x^2)(x^0)$ *is such that the matrix* $W\Pi$ *has* n *eigenvalues* $\lambda_j \in \mathbb{C} \setminus \mathbb{R}^+$, $j = 1, \ldots, n$, *where* $\mathbb{R}^+ = \{\lambda \geqslant 0\}$.

Then for any small $\varepsilon \geqslant 0$ there exists an analytic in ε, $\varepsilon > 0$ and smooth in $\sqrt{\varepsilon}$ at the point $\varepsilon = 0$ family of m-dimensional nonresonance invariant tori $\mathbb{T}^m_{y^0}(\varepsilon)$ of the original Hamiltonian system. Moreover, $\mathbb{T}^m_{y^0}(0) = \mathbb{T}^m_{y^0, x^0}$.

REMARKS. 1. Since the rank of the matrix W never exceeds n, we also have rank $W\Pi \leqslant n$.

2. The tori $\mathbb{T}^m_{y^0}(\varepsilon)$, $\varepsilon > 0$, are exponentially unstable due to the hyperbolicity condition 3) of the theorem. Every torus $\mathbb{T}^m_{y^0}(\varepsilon)$, $\varepsilon > 0$, has a pair of invariant asymptotic Lagrangian surfaces Λ_+, Λ_- such that any solution located on the surface Λ_+ (Λ_-) exponential approaches the torus $\mathbb{T}^m_{y^0}(\varepsilon)$ as $t \to -\infty$ $(t \to +\infty)$. Apparently, these tori generate the Arnold diffusion in the perturbed system.

3. It is shown in paper [7] that in a vicinity of $\mathbb{T}^m_{y^0}(\varepsilon)$ the Hamiltonian (3.1) can be represented in the form

$$\begin{aligned}(3.2)\qquad &\langle \omega^*, Y\rangle + \sqrt{\varepsilon}\langle Y, \Gamma(\sqrt{\varepsilon})Y\rangle/2 + \sqrt{\varepsilon}\langle Z_-, \Omega(X, \sqrt{\varepsilon})Z_+\rangle \\ &+ \sqrt{\varepsilon}G(X, Y, Z, \sqrt{\varepsilon}) + \varepsilon G_1(X, Y, Z, \sqrt{\varepsilon}),\end{aligned}$$

where X, Y, Z are new coordinates such that $dy \wedge dx = dY \wedge dX + dZ_- \wedge dZ_+$, $Y = (Y_1, \dots, Y_m)$, $X = (X_1, \dots, X_m) \mod 2\pi$, $Z_\pm = (Z_1^\pm, \dots, Z_m^\pm)$ and the tori $\mathbb{T}^m_{y^0}(\varepsilon)$ have the form $\{(X, Y, Z) : Y = 0, Z = 0\}$. Moreover, the vector $\omega^* \in \mathbb{R}^m$ is strictly nonresonance and for small ε the matrix $\Gamma(\sqrt{\varepsilon})$ is nondegenerate; the matrix Ω is analytic in X and $\Omega(X, \sqrt{\varepsilon}) = \Omega_0 + O(\sqrt{\varepsilon})$, $\Omega_0 = \text{const}$. The function G (G_1) is of third order in Z (in Y and Z).

DEFINITION 5. An invariant torus $\mathbb{T}^m_{y^0}(\varepsilon)$ is said to be *reducible* if there exists a canonical change $X, Y, Z \to \widehat{X}, \widehat{Y}, \widehat{Z}$ with generating function $S = X\widehat{Y} + \langle Z_+, B(X)\widehat{Z}_-\rangle$ (the matrix B is nondegenerate) that transforms the Hamiltonian (3.2) to the form

$$\begin{aligned}(3.3)\qquad \widehat{H} = &\langle \omega^*, \widehat{Y}\rangle + \sqrt{\varepsilon}\langle \widehat{Y}, \widehat{\Gamma}(\sqrt{\varepsilon})\widehat{Y}\rangle/2 + \sqrt{\varepsilon}\langle \widehat{Z}_-, \widehat{\Omega}(\sqrt{\varepsilon})\widehat{Z}_+\rangle \\ &+ \sqrt{\varepsilon}\widehat{G}(\widehat{X}, \widehat{Y}, \widehat{Z}, \sqrt{\varepsilon}) + \varepsilon \widehat{G}_1(\widehat{X}, \widehat{Y}, \widehat{Z}, \sqrt{\varepsilon}),\end{aligned}$$

where the matrix $\widehat{\Omega}$ does not depend on $\widehat{X}$.

Using the relations

$$\widehat{X} = X, \qquad Y = \widehat{Y} + \frac{\partial}{\partial x}\langle Z_+, B\widehat{Z}_-\rangle, \qquad Z_- = B\widehat{Z}_-, \qquad \widehat{Z}_+ = B^T Z_+,$$

it is easy to show that the function $\widehat{G}$ $(\widehat{G}_1)$ is of third order in $\widehat{Z}$ (in $\widehat{Y}$ and $\widehat{Z}$). We obtain the equalities

$$\widehat{\Gamma} = \Gamma, \qquad \partial B = \sqrt{\varepsilon}(B\widehat{\Omega} - \Omega B).$$

Thus the reducibility of the torus $\mathbb{T}^m_{y^0}(\varepsilon)$ is equivalent to the reducibility of system (1.1) with $A(x) = -\sqrt{\varepsilon}\,\Omega(x, \sqrt{\varepsilon})$. Theorem 1 shows in particular that for small positive ε_0 the relative measure of irreducible tori $\mathbb{T}^m_{y^0}(\varepsilon)$, $\varepsilon \in (0, \varepsilon_0)$, is exponentially small in ε_0.

§4. Averaging method

We assume that the matrix B generating the change (1.2) is a solution of the differential equation

$$B'_\delta = B_\delta b_\delta\,, \qquad B_0 = I\,.$$

Here the prime denotes the derivative $\partial/\partial\delta$, δ is a real parameter, $b_\delta = b_\delta(x)$ is some square matrix. In accordance with the theory of linear differential equations, the matrix B_δ is nondegenerate for all δ.

Let the change $y = B_\delta y_\delta$ transform system (1.1) to the system

$$\begin{cases} \dot{x} = \omega\,, \\ \dot{y}_\delta = A_\delta(x)\, y_\delta\,. \end{cases}$$

Then the following relation holds:

$$\partial B_\delta = AB_\delta - B_\delta A_\delta\,, \tag{4.1}$$

where the differential operator ∂ is defined by formula (1.5).

Let us differentiate (4.1) with respect to δ. After easy calculations, we obtain the equation

$$A'_\delta = -\partial b_\delta - [b_\delta\,;\, A_\delta]\,. \tag{4.2}$$

Here $[\cdot\,,\cdot]$ is the matrix commutator: $[A_1\,,\,A_2] = A_1A_2 - A_2A_1$. Note that equation (4.2) does not contain the matrix B_δ.

We consider the case $A_\delta = \varepsilon a_\delta$ and $b_\delta = \varepsilon\xi_D a_\delta$, where the linear operator ξ_D has the following form:

$$\xi_D a_\delta = -\sum_{\tau\in D} i\,\mathrm{sgn}\,\langle\omega\,,\,\tau\rangle\, a_\delta^\tau e^{i\langle\tau\,,\,x\rangle}.$$

Here D is some subset of $\mathbb{Z}^m$ and $a_\delta = \sum_{\tau\in\mathbb{Z}^m} a_\delta^\tau e^{i\langle\tau\,,\,x\rangle}$.

Equation (4.2) takes the following form:

$$a' = -\partial\xi_D a - \varepsilon\,[\xi_D a\,,\,a]\,. \tag{4.3}$$

Here and further we usually omit the index δ. Let the set D contain the point $0\in\mathbb{Z}^m$. Putting

$$\begin{gathered} D^\pm = D\cap\{\tau\in\mathbb{Z}^m : \pm\langle\tau\,,\,\omega\rangle > 0\}\,, \\ a^* = \sum_{\tau\in\mathbb{Z}^m\setminus D} a^\tau e^{i\langle\tau\,,\,x\rangle}\,, \quad a^+ = i\xi_{D^+}a\,, \quad a^- = -i\xi_{D^-}a\,, \end{gathered} \tag{4.4}$$

we obtain the following useful relations:

$$\begin{gathered} \xi_D a = \xi_{D^+}a + \xi_{D^-}a = -ia^+ + ia^-\,, \qquad a = a^+ + a^- + a^*\,, \\ [\xi_D a\,,\,a] = 2i\,[a^-\,,\,a^+] + [\xi_D a\,,\,a^*]\,. \end{gathered} \tag{4.5}$$

Let us consider equation (4.3) without the small term $\varepsilon\,[\xi_D a\,,\,a]$:

$$a' = -\partial\xi_D a\,. \tag{4.7}$$

We have the infinite system

$$\begin{cases} (a^\tau)' = -|\langle\omega\,,\,\tau\rangle|\,a^\tau & \text{if } \tau \in D\,,\\ (a^\tau)' = 0 & \text{if } \tau \in \mathbb{Z}^m \setminus D\,. \end{cases}$$

The solution of this system has the following form:

$$\begin{cases} a_\delta^\tau = e^{-|\langle\omega\,,\,\tau\rangle|\delta} a_0^\tau & \text{if } \tau \in D\,,\\ a_\delta^\tau = a_0^\tau & \text{if } \tau \in \mathbb{Z}^m \setminus D\,. \end{cases}$$

Since the numbers $|\langle\omega\,,\,\tau\rangle|$ are positive for any $\tau \neq 0$, we see that the solution a_δ tends to the function

$$a_0^0 + \sum_{\tau\in\mathbb{Z}^m\setminus(D\cup\{0\})} a_\delta^\tau e^{i\langle\tau\,,\,x\rangle - |\langle\omega\,,\,\tau\rangle|\delta}\,.$$

The second summand in this expression is small if the set $\mathbb{Z}^m \setminus (D\cup\{0\})$ is situated outside some circle of big radius centered at the origin. In particular, if $D = \mathbb{Z}^m\setminus\{0\}$, then a_δ tends to a_0^0.

Let us complexify slightly the system (4.7) and take into account the term $\varepsilon\,[\xi_D a\,,\,a^0]$. We have the relation

$$a' = -\xi_D a - \varepsilon\,[\xi_D a\,,\,a^0]\,. \tag{4.8}$$

The equations for the Fourier coefficients a^τ remain linear and are separated again:

$$\begin{cases} (a^\tau)' = -|\langle\omega\,,\,\tau\rangle|\,a^\tau + \varepsilon i\,\operatorname{sgn}\langle\omega\,,\,\tau\rangle\,[a^\tau\,,\,a^0] & \text{if } \tau \in D\,,\\ (a^\tau)' = 0 & \text{if } \tau \in \mathbb{Z}^m \setminus D\,. \end{cases}$$

The eigenvalues of the linear operator $[\,\cdot\,,\,a^0]\colon L(n\,,\,\mathbb{C}) \to L(n\,,\,\mathbb{C})$ are $\lambda_{lj} = \lambda_l - \lambda_j$, where λ_s, $s = 1\,,\,\dots\,,\,n$, are the eigenvalues of the matrix a^0. If we assume all λ_s to be different then for the eigenvectors a_{lj}^τ, $\tau \in D$ of the operator $[\,\cdot\,,\,a^0]$, we have the equations

$$\begin{cases} (a_{lj}^\tau)' = -k_{lj}^\tau a_{lj}^\tau & \text{if } \tau \in D\,,\\ (a_{lj}^\tau)' = 0 & \text{if } \tau \in \mathbb{Z}^m \setminus D\,, \end{cases}$$

where $k_{lj}^\tau = |\langle\omega\,,\,\tau\rangle| - \varepsilon i\,\operatorname{sgn}\langle\omega\,,\,\tau\rangle\,(\lambda_l - \lambda_j)$. The real parts of the coefficients k_{lj}^τ can be nonpositive. The corresponding coefficients a_{lj}^τ cannot be eliminated. Since for some λ_l and $\lambda_j = \overline{\lambda}_l$ we have $k_{lj}^\tau = |\langle\omega\,,\,\tau\rangle| - 2\varepsilon\Lambda$, there is no hope that our averaging method removes the coefficients a^τ with τ satisfying the inequality $|\langle\omega\,,\,\tau\rangle| \leqslant 2\varepsilon\Lambda$.

It is necessary to note that these conclusions based on the investigation of equations (4.7) and (4.8) are heuristic and not rigorous. We have presented them here due to their clarity and simplicity.

The problem becomes considerably more complicated if we take into account all the terms in equation (4.3). To study this equation further we have to deal with cumbersome estimates.

We believe that various versions of our averaging method will be useful in a number of problems of perturbation theory. Some generalizations of the method are contained in §10.

§5. Main theorem

We put

$$Q = 2\Gamma(2\Lambda + c'''\varepsilon)^{-\frac{1}{m-1+\gamma}}, \qquad c''' = \text{const} > 0, \tag{5.1}$$

$$\varkappa < (3 + (m-1+\gamma)^{-1})/4 + \beta(m-2+\gamma) \leqslant 1 \tag{5.2}$$

with

$$0 < \beta < \varkappa/(m-1+\gamma) \tag{5.3}$$

and consider the following sets:

$$D_\nu = \{\mu \in \mathbb{Z}^m : \|\mu\| \leqslant Q\varepsilon^{-\frac{1}{m-1+\gamma}}, \ |\langle\tau, \mu\rangle| \geqslant \nu\}, \tag{5.4}$$

$$B_j = \{\mu \in \mathbb{Z}^m : \|\mu\| \leqslant Q\varepsilon^{-\frac{1}{m-1+\gamma}} + \tfrac{1}{2}\Gamma(j+1)\varepsilon^{-\frac{\varkappa}{m-1+\gamma}}\}, \qquad j = 0, 1, \dots,$$

$$D_* = D_{2\Lambda\varepsilon + c'''\varepsilon^2}.$$

The averaging method described in §4 allows us to prove the following (main) theorem.

THEOREM. *Let the conditions of Theorem 1 hold. Then for small values of ε there exists a change of coordinates* (1.2) *that transforms equations* (1.1), (2.1) *to the form* (2.8) *and the following estimates hold*:

1. $|\hat{a} - a^0| \leqslant c'\varepsilon$.
2. $\tilde{a}(x) = \sum_{\tau \in \mathbb{Z}^m \setminus D_*} \tilde{a}^\tau e^{i\langle\tau, x\rangle}$.
3. $|\tilde{a}^\tau - a^\tau| \leqslant c_0''\varepsilon^{\frac{\gamma(1-\varkappa)}{m-1+\gamma}} e^{-\alpha\|\tau\|}$ if $\tau \in B_0 \setminus D_*$.
4. $|a^\tau| \leqslant \varepsilon^{j(\varkappa-2)} e^{-\alpha\|\tau\|}$ if $\tau \in B_j$, $j \in \mathbb{N}$.

Here c', c_0'' are some positive constants.

Theorem 1 follows from the main theorem. Indeed, the function $\tilde{a}(x)$ satisfies the following estimate:

$$\tilde{a}(x) = \Big(\sum_{\tau \in D_0 \setminus D_*} + \sum_{\tau \in B_0 \setminus D_0} + \sum_{j \in \mathbb{N}} \sum_{\tau \in B_j \setminus B_{j-1}}\Big) \tilde{a}^\tau e^{i\langle\tau, x\rangle}.$$

Let us denote the three summands in the right-hand side of this relation by Σ_1, Σ_2, and Σ_3.

The sum Σ_1 is estimated with the help of Proposition 8 (see §12) and the nonresonance conditions (2.5): the number of terms in Σ_1 does not exceed $2K(2\Gamma)^{m-1}$ and the absolute value of each term does not exceed

$$\begin{aligned} 2c\, e^{-\alpha\|\tau\|} &\leqslant 2c \exp(-\alpha\Gamma(2\Lambda\varepsilon + c'''\varepsilon^2)^{-\frac{1}{m-1+\gamma}}) \\ &\leqslant 2c\,(1 + O(\varepsilon)) \exp(-\alpha\Gamma(2\Lambda\varepsilon)^{-\frac{1}{m-1+\gamma}}). \end{aligned}$$

Thus

$$|\Sigma_1| \leqslant 5K(2\Gamma)^{m-1} c \exp(-\alpha\Gamma(2\Lambda\varepsilon)^{-\frac{1}{m-1+\gamma}}).$$

The sums Σ_2 and Σ_3 are estimated in the following way:

$$|\Sigma_2| \leqslant 2c \sum_{\|\tau\| \geqslant Q\varepsilon^{-\frac{1}{m-1+\gamma}}} e^{-\alpha\|\tau\|} \leqslant \mathrm{const}\, \varepsilon^{-\frac{m-1}{m-1+\gamma}} \exp(-2\alpha\Gamma(2\Lambda\varepsilon)^{-\frac{1}{m-1+\gamma}}),$$

$$|\Sigma_3| \leqslant \sum_{j\in\mathbb{N}} \sum_{\tau \in B_j \setminus B_{j-1}} \varepsilon^{j(\varkappa-2)} e^{-\alpha\|\tau\|} \leqslant \mathrm{const}\, \varepsilon^{-\frac{m-1}{m-1+\gamma}} \exp(-2\alpha\Gamma(2\Lambda\varepsilon)^{-\frac{1}{m-1+\gamma}}).$$

Inequality (2.9) follows from these three estimates.

To prove the remaining part of Theorem 1, we note that $G = D_0 \setminus D_*$ and consider the sum Σ_1 in more detail:

$$\Sigma_1 = \sum_{\tau \in D_0 \setminus D_*} \tilde{a}^\tau e^{i\langle \tau, x\rangle} = \sum_{\tau \in D_0 \setminus D_*} (a^\tau + \Delta^\tau) e^{i\langle \tau, x\rangle},$$

where

$$|\Delta^\tau| = |\tilde{a}^\tau - a^\tau| \leqslant c_0'' \varepsilon^{\frac{\gamma(1-\varkappa)}{m-1+\gamma}} e^{-\alpha\|\tau\|}.$$

The sum $\Sigma' = \sum_{\tau \in D_0 \setminus D_*} \Delta^\tau e^{i\langle \tau, x\rangle}$ is estimated in the same way as Σ_1 above:

$$|\Sigma'| \leqslant 3c_0'' K(2\Gamma)^{m-1} \varepsilon^{\frac{\gamma(1-\varkappa)}{m-1+\gamma}} \exp(-\alpha\Gamma(2\Lambda\varepsilon)^{-\frac{1}{m-1+\gamma}}).$$

Thus we obtain the estimates of the functions $\Delta_1 = \Sigma'$ and $\Delta_2 = \Sigma_2 + \Sigma_3$. The analyticity of $\tilde{a}(x)$ and the estimates of $\tilde{a}^\tau$ follow from statements 1–4 of the main theorem. Theorem 1 is proved.

The proof of the main theorem is based on Lemmas 1, 2, and 3. Each lemma corresponds to some stage of averaging of system (1.1): Lemma 1 corresponds to the preliminary stage, Lemma 2 corresponds to the inductive stage, and Lemma 3 corresponds to the final stage. We carry out the averaging using the procedure that was described in §4. The sets D in the operator ξ_D are different on different stages.

In the preliminary stage $D = D_1$ (see (5.4)), Lemma 1 shows us that the averaging procedure with $D = D_1$ annihilates the coefficients a_δ^τ for $\tau \in D_1$ in the limit $\delta \to \infty$ and only slightly changes the other coefficients.

In the inductive stage, we use a finite sequence of sets D_{2-j}. After that we annihilate the coefficients a^τ for $\tau \in D_{c^\varkappa}$ and change slightly the remaining coefficients.

In the final stage we use $\widehat{D}$ which contains the set D_* for sufficiently large c'''. We annihilate all the coefficients a^τ, $\tau \in \widehat{D}$, for $\delta = \infty$. At this stage a part of the remaining coefficients may change considerably. But we succeed in obtaining a satisfactory estimates for these changes.

Thus the main theorem is a direct consequence of these three lemmas. We formulate Lemmas 1, 2, and 3 in this section and prove them in the next four sections.

LEMMA 1. *The solution a_δ of equation* (4.3) *with $D = D_1$ and $a_0^\tau \leqslant ce^{-\alpha\|\tau\|}$ satisfies the inequalities*

$$|a_\delta^\tau| \leqslant c\,|\langle\tau\,,\,\omega\rangle|^{-1}e^{-\delta/2-\alpha\|\tau\|} + c\,e^{-|\langle\tau\,,\,\omega\rangle|\delta-\alpha\|\tau\|} \qquad \text{if } \tau \in D_1\,,$$

$$|a_\delta^\tau - a_0^\tau| \leqslant w\varepsilon^{\frac{\gamma}{m-1+\gamma}}\ln\varepsilon^{-1}e^{-\alpha\|\tau\|} \qquad \text{if } \tau \in \mathbb{Z}^m \setminus D_1\,,$$

$$|a_\delta^0 - a_0^0| \leqslant w_0\varepsilon$$

with some positive constants w, w_0.

LEMMA 2. *Assume that the following statements hold*:

1) $\varepsilon < \lambda < \nu \leqslant 1$;
2) $a_0^\tau = 0 \quad$ *if* $\tau \in D_\nu$,
$|a_0^\tau| \leqslant c_1 e^{-\alpha\|\tau\|} \quad$ *if* $\tau \in \mathbb{Z}^m \setminus D_\nu$;
3) $\nu^{m-1}\varepsilon^\gamma\lambda^{-m+1-\gamma} \leqslant 2^{m-1}\varepsilon^{\vartheta(m-1+\gamma)}, \qquad \vartheta = \text{const} > 0$;
4) $p = L\varepsilon^{\frac{\gamma}{m-1+\gamma}}\lambda^{\frac{m-1}{m-1+\gamma}} \leqslant 2\nu$.

Then the solution a_δ of equation (4.3) *with $D = D_\lambda$ satisfies the inequalities*

$$|a_\delta^\tau| \leqslant c_1 p\,|\langle\tau\,,\,\omega\rangle|^{-1}e^{-\lambda\delta/2-\alpha\|\tau\|} \qquad \text{if } \tau \in D_\nu\,,$$

$$|a_\delta^\tau| \leqslant 2c_1 e^{-\lambda\delta/2-\alpha\|\tau\|} \qquad \text{if } \tau \in D_\lambda \setminus D_\nu\,,$$

$$|a_\delta^\tau - a_0^\tau| \leqslant w'\varepsilon^\vartheta e^{-\alpha\|\tau\|} \qquad \text{if } \tau \in \mathbb{Z}^m \setminus D_\lambda\,,$$

$$|a_\delta^0 - a_0^0| \leqslant w_0'\varepsilon\lambda^{-1}\big(\exp(-\alpha\Gamma\nu^{-\frac{1}{m-1+\gamma}}) + p^2\big)\,.$$

We use Lemma 2 as an inductive one. At jth step ($j = 0, 1, \dots, J$) we put $\nu = 2^{-j}$, $\lambda = 2^{-j-1}$. The constant J satisfies the relation $2^{-J-1} = \varepsilon^\varkappa$ with some $\varkappa$ defined by (5.2), (5.3). Conditions 1), 2), 4) are evidently satisfied at every step. Let us verify condition 3):

$$\begin{aligned}\nu^{m-1}\varepsilon^\gamma\lambda^{-(m-1+\gamma)} &= \varepsilon^\gamma 2^{-j(m-1)+(j+1)(m-1+\gamma)}\\ &= \varepsilon^\gamma 2^{(j+1)\gamma+m-1} \leqslant 2^{m-1}\varepsilon^{\gamma(1-\varkappa)}.\end{aligned}$$

Thus after J steps we annihilate the coefficients a^τ for $\tau \in D_{\varepsilon^\varkappa}$, and the new values a_{new}^τ of the coefficients a^τ for $\tau \in \mathbb{Z}^m\setminus D_{\varepsilon^\varkappa}$ differ only slightly from the corresponding original values a_{orig}^τ:

$$|a_{\text{new}}^\tau - a_{\text{orig}}^\tau| \leqslant \overline{w}\,\varepsilon^{\frac{\gamma(1-\varkappa)}{m-1+\gamma}}e^{-\alpha\|\tau\|}\,, \qquad \overline{w} = \text{const} > 0\,.$$

For $\tau = 0$ the estimate is better:

$$\begin{aligned}|a_{\text{new}}^0 - a_{\text{orig}}^0| \leqslant \varepsilon w_0 + \varepsilon w_0' \sum_{j=0}^{J} 2^{j+1}\Big(&\exp(-\alpha\Gamma 2^{\frac{1}{m-1+\gamma}})\\ &+ L^2\varepsilon^{\frac{2\gamma}{m-1+\gamma}}2^{-(j+1)\frac{m-1}{m-1+\gamma}}\Big) \leqslant \text{const}\,\varepsilon\,.\end{aligned}$$

LEMMA 3. *Assume that the following statements hold*:

$$a_0^\tau = 0 \qquad \text{if } \tau \in D_{\varepsilon^\varkappa}\,,$$
$$|a_0^\tau| \leqslant c_2 e^{-\alpha\|\tau\|} \qquad \text{if } \tau \in \mathbb{Z}^m \setminus D_{\varepsilon^\varkappa}\,,$$

the maximal imaginary part of eigenvalues of the matrix a_0^0 *is equal to* $\Lambda' > 0$. *We put* $\widehat{D} = D_{2\Lambda'\varepsilon+2\varepsilon^2}$, $p' = \exp(-l\varepsilon^{-q})$. *Then the solution* a_δ *of equation* (4.3) *with* $D = \widehat{D}$ *satisfies the inequalities*

$$|a_\delta^\tau| \leqslant c_2 p' |\langle\tau\,,\,\omega\rangle|^{-1} e^{-\varepsilon^2\delta-\alpha\|\tau\|} \qquad \text{if } \tau \in D_{\varepsilon^\varkappa}\,,$$
$$|a_\delta^\tau| \leqslant 2c_2 e^{-\varepsilon^2\delta-\alpha\|\tau\|} \qquad \text{if } \tau \in \widehat{D} \setminus D_{\varepsilon^\varkappa}\,,$$
$$|a_\delta^\tau - a_0^\tau| \leqslant \varepsilon e^{-\alpha\|\tau\|} \qquad \text{if } \tau \in B_0 \setminus \widehat{D}\,,$$
$$|a_\delta^\tau| \leqslant \varepsilon^{j(\varkappa-2)} e^{-\alpha\|\tau\|} \qquad \text{if } \tau \in B_j\,,\ j \in \mathbb{N}\,,$$
$$|a_\delta^0 - a_0^0| \leqslant \varepsilon^3.$$

Here c_2, l, q *are some positive constants.*

§6. Continuous induction. Existence lemmas

In order to prove Lemmas 1, 2, and 3, we study the infinite system of ordinary differential equations:

$$(6.1)\qquad \begin{cases} (a^\tau)' = -|\langle\omega\,,\,\tau\rangle| a^\tau - \varepsilon[\xi_D a\,,\,a]_\tau & \text{if } \tau \in D\,, \\ (a^\tau)' = -\varepsilon[\xi_D a\,,\,a]_\tau & \text{if } \tau \in \mathbb{Z}^m \setminus D\,, \\ a^\tau|_{\delta=0} = a_0^\tau\,, \end{cases}$$

which is the "coordinate form" of equation (4.3). Here $[\cdot\,,\,\cdot]_\tau$ denotes the Fourier coefficient of $[\cdot\,,\,\cdot]$ corresponding to the number τ. We must prove Lemmas 4 and 5 (existence lemmas), which state the existence of solutions of such systems.

We define the solution of system (6.1) as the limit of the solutions of its finite-dimensional versions:

$$(6.2)\qquad \begin{cases} (a^\tau)' = -|\langle\omega\,,\,\tau\rangle| a^\tau - \varepsilon[\xi_D a\,,\,a]_\tau & \text{if } \tau \in D \text{ and } \|\tau\| \leqslant N\,, \\ (a^\tau)' = -\varepsilon[\xi_D a\,,\,a]_\tau & \text{if } \tau \in \mathbb{Z}^m \setminus D \text{ and } \|\tau\| \leqslant N\,, \\ (a^\tau)' = 0 & \text{if } \tau > N\,, \\ a^\tau|_{\delta=0} = a_0^\tau\,. \end{cases}$$

The limit is considered as $N \to \infty$. We always use the finite sets D; thus in (6.2) we can ignore the coefficients a^τ with sufficiently large values of $\|\tau\|$.

To formulate the existence lemmas, we introduce some notation. For periodic functions (2.6) and positive α we put

$$(6.3)\qquad |a|_\alpha = \sup_{\tau\in\mathbb{Z}^m} |a^\tau| e^{\alpha\|\tau\|}.$$

We denote the space of functions with finite norm (6.3) by F_α.

LEMMA 4. *Let the set D be finite, $0 < \varkappa < 1$, $|\langle \tau , \omega\rangle| \geqslant 2\beta > \varepsilon^{\varkappa}$ for any $\tau \in D$, and the solutions $a_N^{\tau} = a_{N,\delta}^{\tau}$ of equations (6.2) satisfy the inequalities*

$$|a_N^{\tau}| < Ce^{-\beta\delta-\alpha'\|\tau\|} \quad \text{if } \tau \in D,$$
$$|a_N^{\tau}| < Ce^{-\alpha'\|\tau\|} \quad \text{if } \tau \in \mathbb{Z}^m \setminus D$$

with some positive constants C, β, α'. Then the functions

$$a_N = \sum_{\tau\in\mathbb{Z}^m} a_{N,\delta}^{\tau} e^{i\langle\tau , x\rangle} \tag{6.4}$$

converge as $N \to \infty$ in any norm $|\cdot|_{\alpha''}$, $0 < \alpha'' < \alpha'$, to a function $a = a_\delta$ that belongs to $F_{\alpha'}$ for any $0 \leqslant \delta \leqslant \infty$.

LEMMA 5. *Let $D = D_{2\Lambda'\varepsilon+\varepsilon^2}$, $0 \leqslant \varkappa \leqslant 1$, $|\langle \tau , \omega\rangle| \geqslant 2\Lambda'\varepsilon+\varepsilon^2$ for any $\tau \in D$, and for any natural N the solution $a_N^{\tau} = a_{N,\delta}^{\tau}$ of equations (6.2) satisfy the inequalities*

$$|a_N^{\tau}| \leqslant \varepsilon^3 e^{-\varepsilon^2\delta-\alpha'\|\tau\|} \quad \text{if } \tau \in D_{\varepsilon^{\varkappa}},$$
$$|a_N^{\tau}| \leqslant Ce^{-\varepsilon^2\delta-\alpha'\|\tau\|} \quad \text{if } \tau \in D \setminus D_{\varepsilon^{\varkappa}},$$
$$|a_N^{\tau}| \leqslant Ce^{-\alpha'\|\tau\|} \quad \text{if } \tau \in \mathbb{Z}^m \setminus D$$

with some positive constants α', C. Then the functions (6.4) converge, as $N \to \infty$, to the function $a = a_\delta \in F_{\alpha'}$, $0 \leqslant \delta \leqslant \infty$ in each norm $|\cdot|_{\alpha''}$, $0 < \alpha'' < \alpha'$.

Lemmas 1–3 (more precisely, their proofs in §§7–9) estimate solutions of equations (6.2). The remarkable fact is that these estimates can be chosen independently of N. Our plan is as follows. First, we prove the finite-dimensional versions of Lemmas 1–3 (§§7–9). Second, using them and Lemmas 4 and 5, we establish the existence of solutions of equations (6.1). Indeed, one can see that the estimates contained in Lemmas 1 and 2 can be regarded as the assumptions of Lemma 4 and the estimates contained in Lemma 3 can be regarded as the assumptions of Lemma 5. We note that in order to apply Lemmas 1 and 2 we can put $\alpha' = \alpha$ in Lemma 4 and to apply Lemma 3 we must put $\alpha' < \alpha$ in Lemma 5. After that the required infinite-dimensional versions of Lemmas 1–3 follow from the independence of the estimates in Lemmas 1–3 of N.

The proofs of Lemmas 1–4 use the auxiliary Proposition 4. This simple statement is a continuous analog of the usual method of mathematical induction.

PROPOSITION 4. *Consider the finite system of ordinary differential equations*

$$\begin{cases} u_j' = -q_j u_j + f_j(u , v , \delta), & j = 1, \dots , J, \\ v_l' = g_l(u , v , \delta), & l = 1, \dots , L, \end{cases} \tag{6.5}$$

with $u = (u_1 , \dots , u_J)$, $v = (v_1 , \dots , v_L)$. As usual, the primes denote derivatives with respect to δ. Also consider the inequalities

$$|u_j(\delta)| \leqslant U_j(\delta), \qquad |v_l(\delta) - v_l(0)| \leqslant V_l(\delta), \tag{6.6}$$

where $u(\delta)$, $v(\delta)$ are solutions of equations (6.5) *and U_j, V_l are smooth functions. Suppose that*

1. *Inequalities* (6.6) *hold for $\delta = 0$.*
2. *For any $\delta_0 \geqslant 0$ the statement "inequalities* (6.6) *hold for any $0 \leqslant \delta \leqslant \delta_0$" implies the estimates*

$$|f_j(u(\delta), v(\delta), \delta)| < e^{-q_j\delta}(e^{q_j\delta}U_j)', \qquad \left|\int_0^\delta g_l(u(\sigma), v(\sigma), \sigma)\,d\sigma\right| < V_l(\delta).$$

Then inequalities (6.6) *are true for all $\delta \geqslant 0$.*

PROOF. We transform the first part of equations (6.5) in the following way: $(e^{q_j\delta}u_j)' = f_j e^{q_j\delta}$ and put

$$\Delta = \max\{\delta_0 \geqslant 0 : \text{ all inequalities (6.6) hold for any } 0 \leqslant \delta < \delta_0\}.$$

This maximum evidently exists and is nonnegative. For $\delta = \Delta$ we get:

$$|e^{q_j\delta}u_j|' \leqslant |(e^{q_j\delta}u_j)'| = |f_j e^{q_j\delta}| < (e^{q_j\delta}U_j(\delta))',$$

$$|v_l(\delta) - v_l(0)| = \left|\int_0^\delta g_l(u(\sigma), v(\sigma), \sigma)\,d\sigma\right| < V_l(\delta);$$

thus inequalities (6.6) also hold on some interval $[\Delta, \Delta + \tilde{\delta}]$, $\tilde{\delta} > 0$. We have obtained a contradiction with the definition of Δ. Proposition 4 is proved.

We note that this proposition remains true for of vector-valued or matrix-valued variables u_j, v_l.

PROOF OF LEMMA 4. The spaces $(F_\alpha, |\cdot|_\alpha)$, $\alpha > 0$, are Banach and $F_\alpha \subset F_\beta$ for $\alpha > \beta$. Therefore, it is sufficient to show that a_N, $N \in \mathbb{N}$, is a Cauchy sequence with respect to any norm $|\cdot|_{\alpha''}$, $0 < \alpha'' < \alpha'$. Thus it is sufficient to prove the following inequalities:

$$|a_N^\tau - a_M^\tau| \leqslant \begin{cases} C'\beta^{-1}\varepsilon e^{-\beta\delta - \alpha' N} & \text{if } \tau \in D, \\ C'\beta^{-1}\varepsilon e^{-\alpha' N} & \text{if } \tau \in B(N) \setminus D \end{cases} \tag{6.7}$$

for large N and $M \geqslant N$, where $B(N) = \{\tau \in \mathbb{Z}^m : \|\tau\| \leqslant N\}$; C' is a positive constant.

Indeed, since N is large, we may assume that D is contained in the ball $B(N)$. If estimates (6.7) hold, then we have:

$$|a_N^\tau - a_M^\tau| \leqslant \begin{cases} C'\beta^{-1}\varepsilon e^{-(\alpha' - \alpha'')N} e^{-\alpha''\|\tau\|} & \text{if } \tau \in B(N), \\ 2Ce^{-(\alpha' - \alpha'')N} e^{-\alpha''\|\tau\|} & \text{if } \tau \in \mathbb{Z}^m \setminus B(N). \end{cases} \tag{6.8}$$

Here the first part of inequalities (6.8) follows from (6.7) and the second one is a consequence of conditions of Lemma 4. We see that

$$|a_N - a_M|_{\alpha''} \leqslant \max\{2C, C'\beta^{-1}\varepsilon\} e^{-(\alpha' - \alpha'')N},$$

thus a_N is a Cauchy sequence.

Let us verify inequalities (6.7). We put

$$z^\tau = a_N^\tau - a_M^\tau, \quad \zeta_1 = \sum_{\tau\in B(N)} z^\tau e^{i\langle\tau, x\rangle}, \quad \text{and} \quad \zeta_2 = \sum_{\tau\in\mathbb{Z}^m\setminus B(N)} z^\tau e^{i\langle\tau, x\rangle}.$$

The variables z^τ, $\|\tau\| \leqslant N$ satisfy the following equations:

$$(6.9) \qquad \begin{cases} (z^\tau)' = -|\langle\omega, \tau\rangle| z^\tau - G^\tau & \text{if } \tau \in D, \\ (z^\tau)' = -G^\tau & \text{if } \tau \in B(N)\setminus D, \\ z^\tau|_{\delta=0} = 0, \end{cases}$$
$$G^\tau = \varepsilon[\xi_D a_N, \zeta_1 + \zeta_2]_\tau + \varepsilon[\xi_D\zeta_1, a_M]_\tau.$$

These equations can be easy obtained using (6.2).

In accordance with Proposition 4, to prove the inequalities (6.7) it is sufficient to verify the following statements:

1. The inequalities (6.7) hold for $\delta = 0$.
2. If these inequalities hold for $0 \leqslant \delta \leqslant \delta_0$, then for $\delta = \delta_0$ we have:

$$(6.10) \qquad \begin{aligned} |G^\tau| &< C'(|\langle\omega, \tau\rangle| - \beta)\beta^{-1}\varepsilon e^{-\beta\delta - \alpha' N} \\ &\leqslant \frac{1}{2} C'\varepsilon e^{-\beta\delta-\alpha' N} \qquad \text{if } \tau \in D, \end{aligned}$$

$$(6.11) \qquad \begin{aligned} &\left| \int_0^\delta G^\tau(a_N(\sigma), a_M(\sigma), \zeta_1(\sigma), \zeta_2(\sigma))\, d\sigma \right| \\ &< C'\frac{\varepsilon}{\beta} e^{-\alpha' N} \qquad \text{if } \tau \in B(N)\setminus D. \end{aligned}$$

i) The first statement is obviously true. Let us verify inequality (6.10). We put $G^\tau = \Sigma_1 + \Sigma_2 + \Sigma_3$:

$$(6.12) \qquad \Sigma_1 = \varepsilon[\xi_D a_N, \zeta_2]_\tau, \quad \Sigma_2 = \varepsilon[\xi_D a_N, \zeta_1]_\tau, \quad \Sigma_3 = \varepsilon[\xi_D\zeta_1, a_M]_\tau.$$

We use Proposition 5 and obtain the estimate

$$(6.13) \qquad \begin{aligned} |\Sigma_1| &\leqslant \sum_{\mu\in D,\, \|\tau-\mu\|>N} 2n\varepsilon\, |a_N^\mu|\, |z^{\tau-\mu}| \\ &\leqslant 4\varepsilon n C^2 \sum_{\mu\in D} e^{-\beta\delta - \alpha'\|\mu\| - \alpha' N} \leqslant 4\varepsilon n C^2 K_1 e^{-\beta\delta - \alpha' N}, \end{aligned}$$

where the constant K_1 depends only on α' and m.

With the help of Proposition 5 we get the inequalities:

$$\begin{aligned} |\Sigma_2| &\leqslant \sum_{\mu\in D,\, \|\tau-\mu\|<N} 2n\varepsilon\, |a_N^\mu|\, |z^{\tau-\mu}| \\ &\leqslant 2\varepsilon n C C'\beta^{-1}\varepsilon \sum_{\mu\in D} e^{-\beta\delta - \alpha'\|\mu\| - \alpha' N} \leqslant 2\varepsilon^2\beta^{-1} n C C' K e^{-\beta\delta - \alpha' N}. \end{aligned}$$

The term Σ_3 is estimated in the same way. We have:

$$|G^\tau| \leqslant 4nCK\varepsilon(C + 2C'\varepsilon^{1-\varkappa})e^{-\beta\delta-\alpha' N}. \tag{6.14}$$

Thus inequality (6.10) holds for small ε and large C'.

ii) In the case $\tau \in B(N) \setminus D$, the function G^τ is estimated by (6.14) too. Thus

$$\left|\int_0^\delta G^\tau\, d\sigma\right| \leqslant 4nCK\varepsilon\beta^{-1}(C + 2C'\varepsilon^{1-\varkappa})e^{-\alpha' N}$$

and inequality (6.11) holds.

The limit function $a(x)$ belongs to $F_{\alpha'}$ because for any $\tau \in \mathbb{Z}^m$ the sequence a_N^τ converges as $N \to \infty$ to a limit a^τ such that $|a^\tau| \leqslant (C + C'\varepsilon\beta^{-1})e^{-\alpha'\|\tau\|}$. Lemma 4 is proved.

Proof of Lemma 5. Similarly to the proof of Lemma 4, we conclude that it is sufficient to verify the following inequalities:

$$|a_N^\tau - a_M^\tau| \leqslant \begin{cases} C'\varepsilon^3 e^{-\varepsilon^2\delta-\alpha' N} & \text{if } \tau \in D, \\ C'\varepsilon e^{-\alpha' N} & \text{if } \tau \in B(N) \setminus D. \end{cases} \tag{6.15}$$

The variables $z^\tau = a_N^\tau - a_M^\tau$ for large N and $\|\tau\| \leqslant N$ satisfy equations (6.9). In accordance with Proposition 4, to prove inequalities (6.15) it is sufficient to verify the following statements:

1. Inequalities (6.15) hold for $\delta = 0$.
2. If these inequalities hold for $0 \leqslant \delta \leqslant \delta_0$, then for $\delta = \delta_0$ we have:

$$|G^\tau| < \frac{1}{2}\, C'\varepsilon^3 e^{-\varepsilon^2\delta-\alpha' N} \quad \text{if } \tau \in D, \tag{6.16}$$

$$\left|\int_0^\delta G^\tau\, d\sigma\right| < C'\varepsilon e^{-\alpha' N} \quad \text{if } \tau \in B(N) \setminus D. \tag{6.17}$$

i) Statement 1 is evidently true. Let us verify inequality (6.16). Again we put $G^\tau = \Sigma_1 + \Sigma_2 + \Sigma_3$ (see (6.12)). The term Σ_1 satisfies the inequality

$$|\Sigma_1| \leqslant \sum_{\mu\in D_{\varepsilon^\varkappa},\ \|\tau-\mu\|>N} 2\varepsilon n\,|a_N^\mu|\,|z^{\tau-\mu}| + \sum_{\mu\in D\setminus D_{\varepsilon^\varkappa},\ \|\tau-\mu\|>N} 2\varepsilon n|a_N^\mu||z^{\tau-\mu}|.$$

We denote these sums by Σ_1' and Σ_1''. We have the estimate

$$\Sigma_1' \leqslant 2\varepsilon^5 nC' \sum_{\mu\in\mathbb{Z}^m} e^{-\varepsilon^2\delta-\alpha'\|\mu\|-\alpha' N} \leqslant 2\varepsilon^5 nC'Ke^{-\varepsilon^2\delta-\alpha' N}. \tag{6.18}$$

Using nonresonance conditions (2.5), we see that

$$\|\mu\| \geqslant \Gamma\varepsilon^{\frac{\varkappa}{m-1+\gamma}} \quad \text{for any } \mu \in D \setminus D_{\varepsilon^\varkappa}. \tag{6.19}$$

Thus for small ε,

$$
\begin{aligned}
|\Sigma_1''| &\leqslant 2\varepsilon^2 nC'C \exp(-\alpha'\Gamma\varepsilon^{\frac{\varkappa}{m-1+\gamma}}) \sum_{\|\nu\|>N} e^{-\varepsilon^2\delta-\alpha'\|\nu\|} \\
&\leqslant 2\varepsilon^5 nC'Ce^{-\varepsilon^2\delta-\alpha' N}.
\end{aligned}
\tag{6.20}
$$

Let us estimate the sum Σ_2. We write $|\Sigma_2| \leqslant \Sigma_2' + \Sigma_2''$, where

$$
\Sigma_2' = \sum_{\substack{\mu\in D_{\varepsilon^\varkappa} \\ \|\tau-\mu\|\leqslant N}} 2n\varepsilon\, |a_N^\mu||z^{\tau-\mu}|, \qquad \Sigma_2'' = \sum_{\substack{\mu\in D\setminus D_{\varepsilon^\varkappa} \\ \|\tau-\mu\|\leqslant N}} 2n\varepsilon\, |a_N^\mu||z^{\tau-\mu}|.
$$

The sum Σ_2' is estimated in the following way:

$$
\Sigma_2' \leqslant 2n\varepsilon^5 C' \sum_{\mu\in\mathbb{Z}^m} e^{-\varepsilon^2\delta-\alpha'\|\mu\|-\alpha' N} \leqslant 2nC'K_1\varepsilon^5 e^{-\varepsilon^2\delta-\alpha' N}.
\tag{6.21}
$$

Using inequality (6.19), we obtain the estimate

$$
\Sigma_2'' \leqslant 2n\varepsilon \sum_{\|\mu\|\geqslant\Gamma\varepsilon^{\frac{\varkappa}{m-1+\gamma}}} Ce^{-\varepsilon^2\delta-\alpha'\|\mu\|} C'\varepsilon e^{-\alpha' N} \leqslant 2nCC'\varepsilon^5 e^{-\varepsilon^2\delta-\alpha' N}
\tag{6.22}
$$

for small ε.

The sum Σ_3 is estimated in the same way as Σ_2. Thus we have

$$
|G^\tau| \leqslant 6nC'\varepsilon^5(K_1+C)e^{-\varepsilon^2\delta-\alpha' N}.
\tag{6.23}
$$

Inequality (6.16) is proved.

ii) The term G^τ in inequality (6.17) satisfies estimate (6.23). Therefore,

$$
\left|\int_0^\delta G^\tau\, d\sigma\right| \leqslant 6nC'\varepsilon^3(K_1+C)e^{-\alpha' N}.
$$

Thus inequality (6.17) holds and Lemma 5 is proved.

§7. Averaging procedure: the preliminary stage

In this section we prove the finite-dimensional version of Lemma 1. For brevity, we do not consider the equations $(a^\tau)' = 0$ in system (6.2). The author hopes that there will be no misunderstanding because of this.

In accordance with Proposition 4, to prove the finite-dimensional version of Lemma 1 it is sufficient to verify the following statements:

1. The estimates in Lemma 1 hold for $\delta = 0$.

2. Let these estimates be true for $0 \leqslant \delta \leqslant \delta_0$, $\delta_0 \geqslant 0$. Then for $\delta = \delta_0$ the following inequalities hold:

$$
\varepsilon\, |[\xi_{D_1} a\,,\, a]_\tau| < 2^{-1} c\, e^{-\delta/2-\alpha\|\tau\|} \qquad \text{if } \tau\in D_1\,,
\tag{7.1}
$$

$$
\int_0^\delta \varepsilon\, |[\xi_{D_1} a_\sigma\,,\, a_\sigma]_\tau|\, d\sigma < w\varepsilon^{\frac{\gamma}{m-1+\gamma}} \ln\varepsilon^{-1} e^{-\alpha\|\tau\|} \qquad \text{if } \tau\in B(N)\setminus D_1\,,
\tag{7.2}
$$

$$
\int_0^\delta \varepsilon\, |[\xi_{D_1} a_\sigma\,,\, a_\sigma]_0|\, d\sigma < w_0\varepsilon\,.
\tag{7.3}
$$

As usual, the index τ denotes the Fourier coefficient corresponding to τ.

The first statement is evidently true.

i) Let us verify inequality (7.1). The vector τ can be assumed to lie in $D_1^- = D_1 \cap D^-$ (see (4.4)) (the case $\tau \in D_1^+ = D_1 \cap D^+$ is studied in the same way). We use (4.5) and first estimate the term $\varepsilon 2i\,[a^-, a^+]_\tau$:

$$\varepsilon\,|2i\,[a^-, a^+]_\tau| \leqslant \sum_{\mu\in D_1^+,\ \tau-\mu\in D_1^-} 2\varepsilon\,|[a^{\tau-\mu}, a^\mu]|\,.$$

For $\delta = \delta_0$, the estimates contained in Lemma 1 are true thus for any $\psi \in D_1$ we have: $|a^\psi| \leqslant 2c\,e^{-\delta/2-\alpha\|\psi\|}$. Using this inequality and Proposition 5, we obtain the following estimate:

$$\varepsilon\,|2i\,[a^-, a^+]_\tau| \leqslant 16\varepsilon nc^2 e^{-\delta} \sum_{\mu\in D_1^+} e^{-\alpha(\|\mu\|+\|\tau-\mu\|)}.$$

The following inequalities are obtained with the help of Propositions 6 and 8.

$$\begin{aligned}
\sum_{\mu\in D_1^+} e^{-\alpha(\|\mu\|+\|\tau-\mu\|)} &\leqslant \sum_{\mu\in D_1^+} e^{-\alpha\langle\mu,\omega\rangle-\alpha\|\tau\|}\\
&\leqslant K(Q\varepsilon^{-\frac{1}{m-1+\gamma}})^{m-1} e^{-\alpha\|\tau\|}\sum_{j=0}^{\infty} e^{-\alpha j}\\
&\leqslant KQ^{m-1}(1-e^{-\alpha})^{-1}\varepsilon^{-\frac{m-1}{m-1+\gamma}} e^{-\alpha\|\tau\|}.
\end{aligned}\tag{7.4}$$

The second inequality in (7.4) is obtained by partitioning the set D_1^+ into subsets $D_1^+ \cap \{\mu \in \mathbb{Z}^m : j \leqslant \langle\mu, \omega\rangle \leqslant j+1\}$. After this we get the estimate

$$\varepsilon\,|2i\,[a^-, a^+]_\tau| \leqslant 16nc^2KQ^{m-1}(1-e^{-\alpha})^{-1}\varepsilon^{\frac{\gamma}{m-1+\gamma}} e^{-\delta-\alpha\|\tau\|}.\tag{7.5}$$

Let us analyze the term $\varepsilon\,[\xi_{D_1}a, a^*]_\tau$. We see that

$$\varepsilon\,|[\xi_{D_1}a, a^*]_\tau| \leqslant \sum_{\mu\in\mathbb{Z}^m\setminus D_1,\ \tau-\mu\in D_1} \varepsilon\,|[a^{\tau-\mu}, a^\mu]|\,.$$

For small ε, $\mu \in \mathbb{Z}^m \setminus D_1$, and $\delta \leqslant \delta_0$, we have the inequality $|a^\mu| \leqslant 2c\,e^{-\alpha\|\mu\|}$. Thus we obtain the following relation:

$$\varepsilon\,|[\xi_{D_1}a, a^*]_\tau| \leqslant 8\varepsilon nc^2 e^{-\delta/2} \sum_{\mu\in\mathbb{Z}^m\setminus D_1,\ \tau-\mu\in D_1} e^{-\alpha(\|\mu\|+\|\tau-\mu\|)}.$$

In accordance with Proposition 9, the right-hand side of this inequality does not exceed $8\varepsilon nc^2e^{-\delta/2}(KQ^{m-1}\varepsilon^{-\frac{m-1}{m-1+\gamma}} + K_1)e^{-\alpha\|\tau\|}$. For small ε we have the following inequality:

$$\varepsilon\,|[\xi_{D_1}a, a^*]_\tau| \leqslant 16nc^2KQ^{m-1}\varepsilon^{\frac{\gamma}{m-1+\gamma}} e^{-\delta/2-\alpha\|\tau\|}.\tag{7.6}$$

Inequality (7.1) for small ε follows from relations (4.5), (7.5), and (7.6).

ii) Let us verify inequality (7.2). For $\tau \in \mathbb{Z}^m \setminus D_1$ we have:

$$\varepsilon\,|[\xi_{D_1}a\,,\,a]_\tau| \leqslant 2\varepsilon n \sum_{\mu\in D_1}\left(\frac{c}{|\langle\mu\,,\,\omega\rangle|}\,e^{-\delta/2-\alpha\|\mu\|} + c\,e^{-|\langle\mu\,,\omega\rangle|\delta-\alpha\|\mu\|}\right)2c\,e^{-\alpha\|\tau-\mu\|}$$
$$\leqslant \Sigma_1 + \Sigma_2\,,$$

where

$$\Sigma_1 = \sum_{\mu\in D_1} 4\varepsilon nc^2 e^{-|\langle\mu\,,\omega\rangle|\delta-\alpha\|\tau\|}\,, \qquad \Sigma_2 = \sum_{\mu\in D_1}\frac{4\varepsilon nc^2}{|\langle\mu\,,\,\omega\rangle|}\,e^{-\delta/2-\alpha\|\tau\|}.$$

We subdivide the set D_1 into subsets $D_1\cap\{\mu\in\mathbb{Z}^m : j\leqslant|\langle\mu\,,\,\omega\rangle|\leqslant j+1\}$, $j\in\mathbb{N}$, and use Proposition 8. We see that the integrals $\int_0^\delta \Sigma_s(\sigma)\,d\sigma$, $s = 1\,,\,2$, do not exceed

$$16nc^2KQ^{m-1}\varepsilon^{\frac{\gamma}{m-1+\gamma}}e^{-\alpha\|\tau\|}\sum_{1\leqslant j\leqslant Q\varepsilon^{-\frac{1}{m-1+\gamma}}} j^{-1} \leqslant \operatorname{const}\varepsilon^{\frac{\gamma}{m-1+\gamma}}\ln\varepsilon^{-1}e^{-\alpha\|\tau\|}.$$

The constant at the right-hand side of this inequality does not depend on ε. Inequality (7.2) for small ε follows from this estimate.

iii) Inequality (7.3) follows immediately from the estimate

$$\varepsilon\,|[\xi_{D_1}a\,,\,a]_0| \leqslant 2\varepsilon n\sum_{\mu\in D_1}|a^\mu|\,|a^{-\mu}| \leqslant 8c^2\varepsilon n\sum_{\mu\in\mathbb{Z}^m}e^{-\delta-2\alpha\|\mu\|} \leqslant 8c^2\varepsilon nK_1e^{-\delta}$$

with a constant K_1 depending only on α and m.

§8. Averaging procedure: the inductive stage

In accordance with Proposition 4, to prove the finite-dimensional version of Lemma 2 it is sufficient to verify the following statements:

1. The estimates contained in Lemma 2 are true for $\delta = 0$.
2. Let these estimates be true for $0\leqslant\delta\leqslant\delta_0$, $\delta_0\geqslant 0$. Then for $\delta=\delta_0$ the following inequalities hold:

$$\varepsilon\,|[\xi_{D_\lambda}a\,,\,a]_\tau| < \frac{c_1p}{2}\,e^{-\lambda\delta/2-\alpha\|\tau\|} \quad \text{if } \tau\in D_\nu\,, \tag{8.1}$$

$$\varepsilon\,|[\xi_{D_\lambda}a\,,\,a]_\tau| < \frac{c_1\lambda}{2}\,e^{-\lambda\delta/2-\alpha\|\tau\|} \quad \text{if } \tau\in D_\lambda\setminus D_\nu\,, \tag{8.2}$$

$$\int_0^\delta \varepsilon\,|[\xi_{D_\lambda}a_\sigma\,,\,a_\sigma]_\tau|\,d\sigma < w'\varepsilon^\vartheta e^{-\alpha\|\tau\|} \quad \text{if } \tau\in B(N)\setminus D_\lambda\,, \tag{8.3}$$

$$\int_0^\delta \varepsilon\,|[\xi_{D_\lambda}a_\sigma\,,\,a_\sigma]_0|\,d\sigma < w_0'\varepsilon\lambda^{-1}(\exp(-\alpha\Gamma\nu^{-\frac{1}{m-1+\gamma}}) + p^2)\,. \tag{8.4}$$

The first statement is evidently true.

i) Let us verify the inequality (8.1). We consider τ to be in $D_\nu^- = D_\nu \cap D^-$, see (4.4) (the case $\tau \in D_\nu^+$ is considered in a similar way). We write the equality

$$2\varepsilon [a^-, a^+]_\tau = \sum_{\substack{\mu\in D_\lambda^+\setminus D_\nu^+ \\ \tau-\mu\in D_\nu^-}} 2\varepsilon [a^{\tau-\mu}, a^\mu] + \sum_{\substack{\mu\in D_\nu^+ \\ \tau-\mu\in D_\nu^-}} 2\varepsilon [a^{\tau-\mu}, a^\mu]. \tag{8.5}$$

Note that $\tau-\mu$ always belongs to D_ν^- because $\mu \in D^+$ and $\tau \in D_\nu^-$.

Let us estimate the first sum Σ_1 at the right-hand side of (8.5). Using Propositions 5 and 8 and the inequalities

$$|a^\mu| \leqslant 2c_1 e^{-\lambda\delta/2-\alpha\|\mu\|}, \qquad |a^{\tau-\mu}| \leqslant \frac{c_1 p}{\nu} e^{-\lambda\delta/2-\alpha\|\tau-\mu\|},$$

we obtain:

$$\begin{aligned} |\Sigma_1| &\leqslant \sum_{\mu\in D_\lambda^+\setminus D_\nu^+} 8nc_1^2 p\varepsilon\nu^{-1} e^{-\lambda\delta-\alpha\|\tau\|} \\ &\leqslant 8nc_1^2 pKQ^{m-1}\varepsilon^{\frac{\gamma}{m-1+\gamma}}\nu^{-\frac{\gamma}{m-1+\gamma}} e^{-\lambda\delta-\alpha\|\tau\|}. \end{aligned} \tag{8.6}$$

The second sum Σ_2 at the right-hand side of (8.5) is estimated with the help of Propositions 5 and 6 in the following way:

$$|\Sigma_2| \leqslant \sum_{\mu\in D_\nu^+} \frac{4nc_1^2p^2\varepsilon}{|\langle\mu, \omega\rangle|\nu} e^{-\lambda\delta-\alpha\|\tau\|-\alpha\langle\mu,\omega\rangle} \leqslant 4nc_1^2p^2e^{-\lambda\delta-\alpha\|\tau\|}(\Sigma_3+\Sigma_4), \tag{8.7}$$

$$\Sigma_3 = \sum_{\mu\in D_\nu^+,\ \nu\leqslant\langle\mu,\omega\rangle\leqslant 1} \frac{\varepsilon}{\langle\mu, \omega\rangle\nu} \quad\text{and}\quad \Sigma_4 = \sum_{\mu\in D_\nu^+,\ \langle\mu,\omega\rangle\geqslant 1} \frac{\varepsilon}{\nu} e^{-\alpha\langle\mu,\omega\rangle}.$$

Let us consider the sets

$$Y_j = \{\mu \in D_\nu^+ : 2^j\nu \leqslant |\langle\mu, \omega\rangle| \leqslant 2^{j+1}\nu\}, \qquad j = 0, 1, \ldots. \tag{8.8}$$

We have the following inequalities:

$$\begin{aligned} \Sigma_3 &\leqslant \sum_{0\leqslant j\leqslant 1+\log_2\nu^{-1}} \sum_{\mu\in Y_j} \frac{\varepsilon}{2^j\nu^2} \leqslant \sum_{j\geqslant 0} \frac{1}{2^j\nu^2} K(2^j\nu)^{\frac{m-1}{m-1+\gamma}} Q^{m-1}\varepsilon^{\frac{\gamma}{m-1+\gamma}} \\ &\leqslant KQ^{m-1}(1-2^{-\frac{\gamma}{m-1+\gamma}})^{-1}\varepsilon^{\frac{\gamma}{m-1+\gamma}}\nu^{-1-\frac{\gamma}{m-1+\gamma}}, \end{aligned} \tag{8.9}$$

$$\Sigma_4 \leqslant \nu^{-1}KQ^{m-1}\varepsilon^{\frac{\gamma}{m-1+\gamma}}\sum_{j\geqslant 0} e^{-\alpha j} \leqslant \nu^{-1}(1-e^{-\alpha})^{-1}KQ^{m-1}\varepsilon^{\frac{\gamma}{m-1+\gamma}}. \tag{8.10}$$

Let us combine these estimates with inequalities (8.6), (8.7). Using the simple relations $0 \leqslant \nu \leqslant 1$, $p/\nu \leqslant 2$ we get:

$$2\varepsilon\,|[a^-, a^+]_\tau| \leqslant \text{const}\, p\,(\varepsilon/\nu)^{\frac{\gamma}{m-1+\gamma}} e^{-\lambda\delta-\alpha\|\tau\|}. \tag{8.11}$$

To use equality (4.5), we estimate the term $\varepsilon\,[\xi_{D_\lambda}a\,,\,a^*]_\tau$. We have the relation

$$\varepsilon\,|[\xi_{D_\lambda}a\,,\,a^*]_\tau| \leqslant \sum_{\mu\in\mathbb{Z}^m\setminus D_\lambda\,,\;\tau-\mu\in D_\lambda} 8nc_1^2\varepsilon e^{-\lambda\delta/2-\alpha(\|\mu\|+\|\tau-\mu\|)}.$$

In accordance with Proposition 9, this sum does not exceed

$$8nc_1^2(K\lambda^{\frac{m-1}{m-1+\gamma}}Q^{m-1}\varepsilon^{\frac{\gamma}{m-1+\gamma}}+\varepsilon K_1)\,e^{-\lambda\delta/2-\alpha\|\tau\|}$$
$$\leqslant \text{const}\,\lambda^{\frac{m-1}{m-1+\gamma}}\varepsilon^{\frac{\gamma}{m-1+\gamma}}e^{-\lambda\delta/2-\alpha\|\tau\|},$$

because $\lambda>\varepsilon$. As a result we obtain the estimate

$$\varepsilon\,|[\xi_{D_\lambda}a\,,\,a]_\tau| \leqslant \text{const}\,(p\,(\varepsilon/\nu)^{\frac{\gamma}{m-1+\gamma}}\lambda^{\frac{m-1}{m-1+\gamma}}\varepsilon^{\frac{\gamma}{m-1+\gamma}})\,e^{-\lambda\delta/2-\alpha\|\tau\|}$$
$$\leqslant (c_1p/2)\,e^{-\lambda\delta/2-\alpha\|\tau\|}.$$

Inequality (8.1) is proved.

ii) Let us verify inequality (8.2). We assume τ to be in $D_\lambda^-\setminus D_\nu^-$ and have the relation

$$2\varepsilon\,|[a^-\,,\,a^+]_\tau| = \sum_{\substack{\mu\in D_\lambda^+\setminus D_\nu^+\\ \tau-\mu\in D_\lambda^-}} 2\varepsilon\,[a^{\tau-\mu}\,,\,a^\mu] + \sum_{\substack{\mu\in D_\nu^+\\ \tau-\mu\in D_\nu^-}} 2\varepsilon\,[a^{\tau-\mu}\,,\,a^\mu]. \tag{8.12}$$

The first sum Σ_1' is estimated in the same way as the sum Σ_1 in part i):

$$|\Sigma_1'| \leqslant \sum_{\mu\in D_\lambda^+\setminus D_\nu^+} 16nc_1^2e^{-\lambda\delta-\alpha\|\tau\|} \leqslant 16nc_1^2K\nu^{\frac{m-1}{m-1+\gamma}}Q^{m-1}\varepsilon^{\frac{\gamma}{m-1+\gamma}}e^{-\lambda\delta-\alpha\|\tau\|}.$$

The second sum at the right-hand side of (8.12) has the same estimate as the sum Σ_2 in part i). Thus we have:

$$2\varepsilon\,|[a^-\,,\,a^+]_\tau| \leqslant \text{const}\,(\nu^{\frac{m-1}{m-1+\gamma}}\varepsilon^{\frac{\gamma}{m-1+\gamma}}+p\,(\varepsilon/\nu)^{\frac{\gamma}{m-1+\gamma}})\,e^{-\lambda\delta-\alpha\|\tau\|}. \tag{8.13}$$

The term $\varepsilon\,[\xi_{D_\lambda}a\,,\,a^*]_\tau$ has the same estimate as the corresponding term in part i). Using the conditions $p\leqslant 2\nu$, $\lambda<\nu$, $\nu^{m-1}\varepsilon^\gamma\lambda^{-m+1-\gamma}\leqslant\varepsilon^{\vartheta(m-1+\gamma)}$ of Lemma 2, we obtain the inequality

$$\varepsilon\,|[\xi_{D_\lambda}a\,,\,a]_\tau| \leqslant \text{const}\,\nu^{\frac{m-1}{m-1+\gamma}}\varepsilon^{\frac{\gamma}{m-1+\gamma}}e^{-\lambda\delta/2-\alpha\|\tau\|} \leqslant \text{const}\,\varepsilon^{\frac{\vartheta}{m-1+\gamma}}\lambda e^{-\lambda\delta/2-\alpha\|\tau\|}.$$

Inequality (8.2) is proved for small ε.

iii) Let us verify inequality (8.3). If $\tau\in\mathbb{Z}^m\setminus D_\lambda$, then

$$\varepsilon\,|[\xi_{D_\lambda}a\,,\,a]_\tau| \leqslant \sum_{\mu\in D_\nu}\frac{4nc_1^2p\varepsilon}{|\langle\mu\,,\,\omega\rangle|}e^{-\lambda\delta/2-\alpha\|\tau\|} + \sum_{\mu\in D_\lambda\setminus D_\nu} 8nc_1^2\varepsilon e^{-\lambda\delta/2-\alpha\|\tau\|}. \tag{8.14}$$

Here we have used the inequality $|a^\psi|\leqslant 2c_1e^{-\alpha\|\psi\|}$ for any $\psi\in\mathbb{Z}^m\setminus D_\lambda$.

The first sum in (8.14) can be divided into two parts:

$$\Sigma_1'' = \sum_{\mu \in D_\nu,\ |\langle \mu, \omega\rangle| \leqslant 1} \frac{4nc_1^2 p\varepsilon}{|\langle \mu, \omega\rangle|} e^{-\lambda\delta/2 - \alpha\|\tau\|},$$

$$\Sigma_2'' = \sum_{\mu \in D_\nu,\ |\langle \mu, \omega\rangle| > 1} \frac{4nc_1^2 p\varepsilon}{|\langle \mu, \omega\rangle|} e^{-\lambda\delta/2 - \alpha\|\tau\|}.$$

We estimate the integral of Σ_1'' (Y_j is defined by (8.8)):

$$\begin{aligned}\int_0^\delta \Sigma_1''(\sigma)\, d\sigma &\leqslant \frac{2}{\lambda} 4nc_1^2 p\varepsilon e^{-\alpha\|\tau\|} \sum_{0 \leqslant j \leqslant 1 + \log_2 \nu^{-1}} \sum_{\nu \in Y_j} \frac{1}{2^j \nu} \\ &\leqslant \frac{16nc_1^2 p}{\lambda\nu} e^{-\alpha\|\tau\|} \sum_{j \geqslant 0} \frac{K(2^j\nu)^{\frac{m-1}{m-1+\gamma}}}{2^j} Q^{m-1} \varepsilon^{\frac{\gamma}{m-1+\gamma}} \\ &\leqslant \frac{16nc_1^2 K Q^{m-1}}{1 - 2^{-\frac{\gamma}{m-1+\gamma}}} \frac{p}{\lambda} \left(\frac{\varepsilon}{\nu}\right)^{\frac{\gamma}{m-1+\gamma}} e^{-\alpha\|\tau\|}.\end{aligned}$$

The integral of Σ_2'' is estimated similarly to the integral of Σ_2 in part ii) of §7:

$$\int_0^\delta \Sigma_2''(\sigma)\, d\sigma \leqslant 16nc_1^2 K Q^{m-1} \frac{p}{\lambda} \varepsilon^{\frac{\gamma}{m-1+\gamma}} \ln\left(2Q\varepsilon^{-\frac{1}{m-1+\gamma}}\right) e^{-\alpha\|\tau\|}.$$

The integral of the second sum at the right-hand side of (8.14) does not exceed the magnitude $16nc_1^2 K Q^{m-1} \lambda^{-1} \nu^{\frac{m-1}{m-1+\gamma}} \varepsilon^{\frac{\gamma}{m-1+\gamma}} e^{-\alpha\|\tau\|}$. Thus we have:

$$\int_0^\delta \varepsilon\, |[\xi_{D_\lambda} a_\sigma, a_\sigma]_\tau|\, d\sigma \leqslant \text{const}\, \varepsilon^{\frac{\gamma}{m-1+\gamma}} \lambda^{-1} \left(p\nu^{-\frac{\gamma}{m-1+\gamma}} + p \ln \varepsilon^{-1} + \nu^{\frac{m-1}{m-1+\gamma}}\right).$$

To complete the verification of inequality (8.3) it is sufficient to use the relations

$$\nu^{m-1} \varepsilon^\gamma \lambda^{-m+1-\gamma} \leqslant \varepsilon^{\vartheta(m-1+\gamma)}, \qquad p = L\varepsilon^{\frac{\gamma}{m-1+\gamma}} \lambda^{\frac{m-1}{m-1+\gamma}} \leqslant 2\nu.$$

iv) Let us verify inequality (8.4). We have:

$$\begin{aligned}\varepsilon\, |[\xi_{D_\lambda} a, a]_0| \leqslant &\sum_{\mu \in D_\lambda \setminus D_\nu} 8\varepsilon nc_1^2 e^{-\lambda\delta - 2\alpha\|\mu\|} \\ &+ \sum_{\mu \in D_\nu \setminus D_1} \frac{2\varepsilon nc_1^2 p^2}{\langle \mu, \omega\rangle^2} e^{-\lambda\delta - 2\alpha\|\mu\|} + \sum_{\mu \in D_1} \frac{2\varepsilon nc_1^2 p^2}{\langle \mu, \omega\rangle^2} e^{-\lambda\delta - 2\alpha\|\mu\|}.\end{aligned}$$

Let us denote these sums by Σ_1, Σ_2, and Σ_3 respectively. In accordance with the nonresonance conditions (2.5), every vector $\mu \in D_\lambda \setminus D_\nu$ satisfies the inequality $\|\mu\| \geqslant M = \Gamma\nu^{-\frac{1}{m-1+\gamma}}$. Thus

$$\Sigma_1 < \sum_{\|\mu\| \geqslant M} 8\varepsilon nc_1^2 e^{-\lambda\delta - 2\alpha\|\mu\|} < \text{const}\, \varepsilon \exp\left(-\lambda\delta - \alpha\Gamma\nu^{-\frac{1}{m-1+\gamma}}\right).$$

Using similar arguments, we see that (j is integer):

$$\Sigma_2 \leqslant \sum_{0\leqslant j\leqslant \log_2 \nu^{-1}} \sum_{\mu\in D_{2^{-j-1}}\setminus D_{2^{-j}}} 2\varepsilon n c_1^2 p^2 4^{j+1} e^{-\lambda\delta-2\alpha\|\mu\|}$$
$$\leqslant \sum_{0\leqslant j\leqslant \log_2 \nu^{-1}} \sum_{\|\mu\|\geqslant \Gamma 2^{-\frac{j}{m-1+\gamma}}} 8\varepsilon n c_1^2 p^2 4^{j} e^{-\lambda\delta-2\alpha\|\mu\|} \leqslant \operatorname{const} \varepsilon p^2 e^{-\lambda\delta}.$$

The estimate of Σ_3 is simple:

$$\Sigma_3 \leqslant \sum_{\mu\in\mathbb{Z}^m} 2\varepsilon n c_1^2 p^2 e^{-\lambda\delta-2\alpha\|\mu\|} \leqslant \operatorname{const} \varepsilon p^2 e^{-\lambda\delta}.$$

As a result we have:

$$\int_0^\delta (\Sigma_1(\sigma)+\Sigma_2(\sigma)+\Sigma_3(\sigma))\, d\sigma \leqslant \operatorname{const} \varepsilon \lambda^{-1}(\exp(-\alpha\Gamma\nu^{-\frac{1}{m-1+\gamma}})+p^2).$$

Lemma 2 is proved.

§9. Averaging procedure: the final stage

In this section we prove the finite-dimensional version of Lemma 3. To use Proposition 4, we need some additional arguments. For simplicity, we assume that the matrix a_0^0 is diagonalizable. Let $\lambda_1, \dots, \lambda_n$ be its eigenvalues, $v_1, \dots, v_n$ be the corresponding right eigenvectors (i.e., v_s are the column vectors and $a_0^0 v_s = \lambda_s v_s$) and $u_1, \dots, u_n$ be the corresponding left eigenvectors (i.e., u_s are the row vectors and $u_s a_0^0 = \lambda_s u_s$). Then the matrices $e_{ks} = v_k u_s$, $k, s = 1, \dots, n$ form an eigenbasis of the operator $[\cdot, a_0^0]$ in the linear space $L(n, \mathbb{C})$. Indeed, we have:

$$[e_{ks}, a_0^0] = v_k u_s a_0^0 - a_0^0 v_k u_s = (\lambda_s - \lambda_k) e_{ks}$$

and e_{ks} are evidently linearly independent.

Let us consider equations (6.2) in the basis e_{ks} for every $\tau \in \mathbb{Z}^m$ (as usual, we ignore the equations $(a^\tau)' = 0$ for brevity):

$$\begin{cases} (a_{ks}^\tau)' = -(|\langle\omega, \tau\rangle| - i\varepsilon(\lambda_s-\lambda_k)\operatorname{sgn}\langle\omega,\tau\rangle)a_{ks} & \\ \qquad\qquad -\varepsilon[\xi_D a, a - a_0^0]_{ks,\tau} & \text{if } \tau\in\widehat{D}, \\ (a_{ks}^\tau)' = -\varepsilon[\xi_D a, a-a_0^0]_{ks,\tau} & \text{if } \tau\in B(N)\setminus\widehat{D}. \end{cases}$$

We consider the following inequalities

$$(9.1)\qquad \begin{cases} |a_{ks,\delta}^\tau| \leqslant c_2 p'(c_*|\langle\tau,\omega\rangle|)^{-1} e^{-\varepsilon^2\delta-\alpha\|\tau\|} & \text{if } \tau\in D_{\varepsilon^\varkappa}, \\ |a_{ks,\delta}^\tau| \leqslant c_2 c_*^{-1} e^{-\varepsilon^2\delta-\alpha\|\tau\|} & \text{if } \tau\in\widehat{D}\setminus D_{\varepsilon^\varkappa}, \\ |a_\delta^\tau - a_0^\tau| \leqslant \varepsilon e^{-\alpha\|\tau\|} & \text{if } \tau\in B_0\setminus\widehat{D}, \\ |a_\delta^\tau| \leqslant \varepsilon^{j(\varkappa-2)} e^{-\alpha\|\tau\|} & \text{if } \tau\in B_j,\ j\in\mathbb{N}, \\ |a_\delta^0 - a_0^0| \leqslant \varepsilon^3. & \end{cases}$$

Lemma 3 follows from these inequalities for a sufficiently large constant c_* because of the relations

$$c_*^{-1}|f| \leqslant \max_{1\leqslant k, s\leqslant n} |f_{ks}| \leqslant c_*|f| \quad \text{for any } f = \sum_{k,s=1}^{n} f_{ks}e_{ks}. \tag{9.2}$$

In accordance with Proposition 4 to verify inequalities (9.1) it is sufficient to prove the following two statements:

1. Inequalities (9.1) are true for $\delta = 0$.
2. If these inequalities hold for $0 \leqslant \delta \leqslant \delta_0$, $\delta_0 \geqslant 0$, then for $\delta = \delta_0$ the following relations are true:

$$\begin{cases} \varepsilon\,|[\xi_{\widehat{D}}a\,,\, a - a_0^0]_{ks\,,\,\tau}| < 2c_2p'c_*^{-1}e^{-\varepsilon^2\delta-\alpha\|\tau\|} & \text{if } \tau \in D_{\varepsilon^{\varkappa}}\,, \\ \varepsilon\,|[\xi_{\widehat{D}}a\,,\, a - a_0^0]_{ks\,,\,\tau}| < \varepsilon^2 c_2c_*^{-1}e^{-\varepsilon^2\delta-\alpha\|\tau\|} & \text{if } \tau \in \widehat{D}\setminus D_{\varepsilon^{\varkappa}}\,, \\ \int_0^\delta \varepsilon\,|[\xi_{\widehat{D}}a_\sigma\,,\, a_\sigma - a_0^0]_\tau|\,d\sigma < \varepsilon e^{-\alpha\|\tau\|} & \text{if } \tau \in B_0\setminus\widehat{D}\,, \\ \int_0^\delta \varepsilon\,|[\xi_{\widehat{D}}a_\sigma\,,\, a_\sigma - a_0^0]_\tau|\,d\sigma < \frac{1}{2}\varepsilon^{j(\varkappa-2)}e^{-\alpha\|\tau\|} & \text{if } \tau \in B_j\,, \\ \int_0^\delta \varepsilon\,|[\xi_{\widehat{D}}a_\sigma\,,\, a_\sigma - a_0^0]_0|\,d\sigma < \varepsilon^3. \end{cases} \tag{9.3}$$

The first statement is obviously valid. Before verifying inequalities (9.3), let us transform them. Due to the conditions of statement 2, we have $|a_\delta^0 - a_0^0| \leqslant \varepsilon^3$, so that $\varepsilon\,|[\xi_{\widehat{D}}a_\delta\,,\, a_0^0 - a_\delta^0]_\tau| \leqslant 2\varepsilon^4 n|a^\tau|$. Using this estimate and inequalities (9.2), we conclude that for small ε inequalities (9.3) are consequences of the following five inequalities:

$$\varepsilon\,|[\xi_{\widehat{D}}a_\delta\,,\, a_\delta - a_\delta^0]_\tau| < c_2p'c_*^{-2}e^{-\varepsilon^2\delta-\alpha\|\tau\|} \quad \text{if } \tau \in D_{\varepsilon^{\varkappa}}\,, \tag{9.4}$$

$$\varepsilon\,|[\xi_{\widehat{D}}a_\delta\,,\, a_\delta - a_\delta^0]_\tau| < \frac{1}{2}\varepsilon^2c_2c_*^{-2}e^{-\varepsilon^2\delta-\alpha\|\tau\|} \quad \text{if } \tau \in \widehat{D}\setminus D_{\varepsilon^{\varkappa}}\,, \tag{9.5}$$

$$\int_0^\delta \varepsilon\,|[\xi_{\widehat{D}}a_\sigma\,,\, a_\sigma - a_\sigma^0]_\tau|\,d\sigma < \frac{1}{2}\varepsilon e^{-\alpha\|\tau\|} \quad \text{if } \tau \in B_0\setminus\widehat{D}\,, \tag{9.6}$$

$$\int_0^\delta \varepsilon\,|[\xi_{\widehat{D}}a_\sigma\,,\, a_\sigma - a_\sigma^0]_\tau|\,d\sigma < \frac{1}{4}\varepsilon^{j(\varkappa-2)}e^{-\alpha\|\tau\|} \quad \text{if } \tau \in B_j\,, \tag{9.7}$$

$$\int_0^\delta \varepsilon\,|[\xi_{\widehat{D}}a_\sigma\,,\, a_\sigma - a_\sigma^0]_0|\,d\sigma < \frac{1}{2}\varepsilon^3. \tag{9.8}$$

i) Let us verify inequality (9.4). The vector τ is assumed to lie in $D_{\varepsilon^{\varkappa}}^- = D_{\varepsilon^{\varkappa}} \cap D^-$. We use formula (4.5) again. The term $2\varepsilon\,|[a^-\,,\, a^+]_\tau|$ is estimated in the same way as in part i) of §8. We put $c_1 = c_2$, $p = p'$, $v = \varepsilon^{\varkappa}$. In accordance with (8.11) we have:

$$2\varepsilon\,|[a^-\,,\, a^+]_\tau| \leqslant \text{const}\, p'\varepsilon^{\frac{(1-\varkappa)\gamma}{m-1+\gamma}}e^{-\varepsilon^2\delta-\alpha\|\tau\|}.$$

Let us estimate the term $\varepsilon\,|[\xi_{\widehat{D}}a\,,\,a^*-a^0]_\tau|$:

$$(9.9)\quad \varepsilon\,|[\xi_{\widehat{D}}a\,,\,a^*-a^0]_\tau| \leqslant \Bigg(\sum_{\substack{\tau-\mu\in B_0\setminus(\widehat{D}\cup\{0\})\\ \mu\in D_{\varepsilon^\varkappa}}} + \sum_{\substack{\tau-\mu\in B_N\setminus B_0\\ \mu\in D_{\varepsilon^\varkappa}}} + \sum_{\substack{\tau-\mu\in D_0\setminus(\widehat{D}\cup\{0\})\\ \mu\in\widehat{D}\setminus D_{\varepsilon^\varkappa}}} + \sum_{\substack{\tau-\mu\in B_N\setminus D_0\\ \mu\in\widehat{D}\setminus D_{\varepsilon^\varkappa}}} \Bigg) 2n\varepsilon\,|a^\mu|\,|a^{\tau-\mu}|\,,$$

where $D_0=\{\tau\in\mathbb{Z}^m : \|\tau\|\leqslant Q\varepsilon^{-\frac{1}{m-1+\gamma}}\}$ and N is the integer part of $6\varepsilon^{-\frac{1-\varkappa}{m-1+\gamma}}$. The vectors $\tau-\mu$ belong to B_N since μ and τ belong to $\widehat{D}$. We denote the sums (9.9) by Σ_1, Σ_2, Σ_3, and Σ_4.

The sum Σ_1 is estimated using Proposition 9:

$$\begin{aligned}\Sigma_1 &\leqslant \sum_{\mu\in D_{\varepsilon^\varkappa}\,,\ \tau-\mu\in B_0\setminus(\widehat{D}\cup\{0\})} 4n\varepsilon^{1-\varkappa}c_2^2p'e^{-\varepsilon^2\delta-\alpha(\|\mu\|+\|\tau-\mu\|)}\\ &\leqslant 4n\varepsilon^{1-\varkappa}c_2^2p'\big(K\varepsilon^{\frac{\varkappa(m-1)}{m-1+\gamma}}Q^{m-1}\varepsilon^{-\frac{m-1}{m-1+\gamma}}+K_1\big)e^{-\varepsilon^2\delta-\alpha\|\tau\|}\\ &\leqslant \mathrm{const}\,\varepsilon^{\frac{(1-\varkappa)\gamma}{m-1+\gamma}}p'e^{-\varepsilon^2\delta-\alpha\|\tau\|}.\end{aligned}$$

We have the following inequalities for Σ_2:

$$\begin{aligned}\Sigma_2 &\leqslant \sum_{\mu\in D_{\varepsilon^\varkappa}\,,\ \tau-\mu\in B_N\setminus B_0} 2n\varepsilon^{1-\varkappa+N(\varkappa-2)}c_2p'e^{-\varepsilon^2\delta-\alpha(\|\mu\|+\|\tau-\mu\|)}\\ &\leqslant \sum_{\mu\in D_{\varepsilon^\varkappa}} 2n\varepsilon^{(N-1)(\varkappa-2)-1}c_2p'\exp\big(-\varepsilon^2\delta-\alpha(\|\mu\|+\|\tau\|+(\Gamma/2)\varepsilon^{-\frac{\varkappa}{m-1+\gamma}})\big)\\ &\leqslant \mathrm{const}\,\varepsilon^{(N-1)(\varkappa-2)-1}p'\exp\big(-\varepsilon^2\delta-\alpha(\|\tau\|+(\Gamma/2)\varepsilon^{-\frac{\varkappa}{m-1+\gamma}})\big).\end{aligned}$$

The vectors μ and $\tau-\mu$ in the sum Σ_3 are linearly independent. Indeed, let us assume the contrary. Since $|\langle\mu\,,\,\omega\rangle|>|\langle\tau-\mu\,,\,\omega\rangle|$, we have $\mu=s(\tau-\mu)$, $|s|>1$. Moreover, $|s|\geqslant 2$ since s is integer. Using inequality $|\langle\tau-\mu\,,\,\omega\rangle|\leqslant\varepsilon\,(2\Lambda'+2\varepsilon)$ and the nonresonance conditions (2.5), we see that

$$\|\tau-\mu\|\geqslant\Gamma(\varepsilon\,(2\Lambda'+2\varepsilon))^{-\frac{1}{m-1+\gamma}}>Q\varepsilon^{-\frac{1}{m-1+\gamma}}/2\,.$$

Thus we obtain the inequality $\|\mu\|>Q\varepsilon^{-\frac{1}{m-1+\gamma}}$, which contradicts to the condition $\mu\in\widehat{D}$.

For linearly independent μ and $\tau-\mu$, we can apply Proposition 10. We have:

$$\begin{aligned}\Sigma_3 &\leqslant \sum_{\mu\in\widehat{D}\setminus D_{\varepsilon^\varkappa}\,,\ \tau-\mu\in D_0\setminus(\widehat{D}\cup\{0\})} 8nc_2^2\varepsilon e^{-\varepsilon^2\delta-\alpha(\|\mu\|+\|\tau-\mu\|)}\\ &\leqslant \sum_{\mu\in\widehat{D}\setminus D_{\varepsilon^\varkappa}} 8nc_2^2\varepsilon\exp(-\varepsilon^2\delta-\alpha\|\tau\|-\alpha\varepsilon^{-\beta})\\ &\leqslant 8nc_2^2KQ^{m-1}\varepsilon^{\frac{\varkappa(m-1)+\gamma}{m-1+\gamma}}\exp(-\varepsilon^2\delta-\alpha\|\tau\|-\alpha\varepsilon^{-\beta})\,.\end{aligned}$$

Here we also used Proposition 8.

The sum Σ_4 is estimated by using the inequalities $\|\tau-\mu\| \geqslant \|\tau\|$, $\|\mu\| \geqslant M = \gamma\varepsilon^{-\frac{\varkappa}{m-1+\gamma}}$:

$$\begin{aligned}\Sigma_4 &\leqslant \sum_{\mu\in\widehat{D}\setminus D_{\varepsilon^\varkappa},\ \tau-\mu\in B_N\setminus D_0} 8nc_2\varepsilon^{1+N(\varkappa-2)}e^{-\varepsilon^2\delta-\alpha(\|\mu\|+\|\tau-\mu\|)} \\ &\leqslant \sum_{\|\mu\|\geqslant M} 8nc_2\varepsilon^{1+N(\varkappa-2)}e^{-\varepsilon^2\delta-\alpha(\|\mu\|+\|\tau\|)} \\ &\leqslant \operatorname{const}\varepsilon^{1+N(\varkappa-2)}\exp(-\varepsilon^2\delta-\alpha\|\tau\|-(\alpha/2)\Gamma\varepsilon^{-\frac{\varkappa}{m-1+\gamma}}).\end{aligned}$$

Thus we obtain the estimate

$$\begin{aligned}|\varepsilon[\xi_{\widehat{D}}a\,,\ a-a^0]_\tau| \leqslant \operatorname{const}(p'\varepsilon^{\frac{(1-\varkappa)\gamma}{m-1+\gamma}} &+ \exp(-\alpha\varepsilon^{-\beta}) \\ &+\varepsilon^{1+N(\varkappa-2)}\exp(-\alpha\Gamma\varepsilon^{-\frac{\varkappa}{m-1+\gamma}}/2))e^{-\varepsilon^2\delta-\alpha\|\tau\|}.\end{aligned}$$

Inequality (9.4) is proved for small ε.

ii) Let us verify inequality (9.5). We consider τ to lie in $\widehat{D}^-\setminus D^-_{\varepsilon^\varkappa}$. We have the relation

$$2\varepsilon[a^-\,,\ a^+]_\tau = \Bigg(\sum_{\mu\in\widehat{D}^+\setminus D^+_{\varepsilon^\varkappa},\ \tau-\mu\in\widehat{D}^-} + \sum_{\mu\in D^+_{\varepsilon^\varkappa},\ \tau-\mu\in\widehat{D}^-_{\varepsilon^\varkappa}}\Bigg)2\varepsilon[a^{\tau-\mu}\,,\ a^\mu].$$

The second sum Σ_2' in this equality is estimated in the same way as the sum Σ_2 in part i) of §8. We put $c'=c''$, $p=p'$, $v=\varepsilon^\varkappa$ and, in accordance with (8.9), (8.10), we obtain:

$$\Sigma_2' \leqslant \operatorname{const}(p')^2\varepsilon^{\frac{\gamma(1-\varkappa)}{m-1+\gamma}-\varkappa}e^{-\varepsilon^2\delta-\alpha\|\tau\|}.$$

The first sum Σ_1' is estimated in the following way:

$$\Sigma_1' \leqslant \sum_{\mu\in\widehat{D}^+\setminus D^+_{\varepsilon^\varkappa},\ \tau-\mu\in\widehat{D}^-} 16nc_2^2\varepsilon e^{-\varepsilon^2\delta-\alpha(\|\mu\|+\|\tau-\mu\|)}.$$

Since in this sum $\mu\in\widehat{D}^+\setminus D^+_{\varepsilon^\varkappa}$ and $\tau\in\widehat{D}^-\setminus D^-_{\varepsilon^\varkappa}$ then $\tau-\mu\in\widehat{D}^-\setminus D^-_{2\varepsilon^\varkappa}$. We apply Propositions 10 and 8:

$$\begin{aligned}\Sigma_1' &\leqslant \sum_{\mu\in\widehat{D}^+\setminus D^+_{\varepsilon^\varkappa}} 16nc_2^2\varepsilon\exp(-\varepsilon^2\delta-\alpha\|\tau\|-\alpha\varepsilon^{-\beta}) \\ &\leqslant 16nc_2^2K(3\Gamma/2)^{m-1}\varepsilon^{\frac{\varkappa(m-1)+\gamma}{m-1+\gamma}}\exp(-\varepsilon^2\delta-\alpha\|\tau\|-\alpha\varepsilon^{-\beta}).\end{aligned}$$

The term $2\varepsilon[\xi_{\widehat{D}}a\,,\ a^*-a^0]_\tau$ is estimated in the same way as in part i) of this section. Thus

$$\begin{aligned}\varepsilon|[\xi_{\widehat{D}}a\,,\ a-a^0]_\tau| \leqslant \operatorname{const}((p')^2\varepsilon^{\frac{\gamma(1-\varkappa)}{m-1+\gamma}-\varkappa} &+ \exp(-\alpha\varepsilon^{-\beta}) \\ &+\varepsilon^{1-N(\varkappa-2)}\exp(-\alpha\Gamma\varepsilon^{-\frac{\varkappa}{m-1+\gamma}}/2))e^{-\varepsilon^2\delta-\alpha\|\tau\|}\end{aligned}$$

and inequality (9.5) holds for small ε.

iii) Let us verify inequality (9.6). We assume τ to be in $B_0 \setminus \widehat{D}$ and have the following estimate:

$$\varepsilon \left|[\xi_{\widehat{D}} a\,,\, a - a^0]_\tau\right| \leqslant \sum_{\substack{\tau-\mu\in B_{N+1}\setminus\widehat{D}\\ \mu\in D_{\varepsilon^\varkappa}}} \frac{4nc_2 p' \varepsilon^{1-(N+1)(\varkappa-2)}}{|\langle \mu\,,\,\omega\rangle|} e^{-\varepsilon^2\delta - \alpha(\|\mu\|+\|\tau-\mu\|)}$$
$$+ \Bigg(\sum_{\substack{\tau-\mu\in D_0\setminus\widehat{D}\\ \mu\in\widehat{D}\setminus D_{\varepsilon^\varkappa}}} + \sum_{\substack{\tau-\mu\in B_{N+1}\setminus D_0\\ \mu\in\widehat{D}\setminus D_{\varepsilon^\varkappa}}} \Bigg) 2n\varepsilon\, |a^\mu|\, |a^{\tau-\mu}|\,.$$

We denote these sums by Σ_1, Σ_2, and Σ_3.

The sum Σ_1 satisfies the inequality

$$\Sigma_1 \leqslant \Bigg(\sum_{\substack{\mu\in D_{\varepsilon^\varkappa}\\ \varepsilon^\varkappa\leqslant|\langle\mu\,,\,\omega\rangle|\leqslant 1}} + \sum_{\substack{\mu\in D_{\varepsilon^\varkappa}\\ |\langle\mu\,,\,\omega\rangle|\geqslant 1}} \Bigg) \frac{4nc_2 p' \varepsilon^{1-(N+1)(\varkappa-2)}}{|\langle \mu\,,\,\omega\rangle|} e^{-\varepsilon^2\delta - \alpha\|\tau\|}.$$

The first sum Σ_1' in this relation is estimated in the same way as the sum Σ_3 in part i) of §8:

$$\Sigma_1' \leqslant \text{const}\, p' \varepsilon^{\frac{\gamma(1-\varkappa)}{m-1+\gamma}+(N+1)(\varkappa-2)} e^{-\varepsilon^2\delta-\alpha\|\tau\|}.$$

The second sum Σ_1'' is similar to the sum Σ_2 in part ii) of §7. Consequently,

$$\Sigma_1'' \leqslant \text{const}\, p' \varepsilon^{\frac{\gamma}{m-1+\gamma}+(N+1)(\varkappa-2)} \ln \varepsilon^{-1} e^{-\varepsilon^2\delta-\alpha\|\tau\|}.$$

The sum Σ_2 is estimated similarly to the sum Σ_3 in part i) of this section:

$$\Sigma_2 \leqslant \text{const}\exp(-\alpha\varepsilon^{-\beta} - \varepsilon^2\delta - \alpha\|\tau\|)\,.$$

The vectors μ and $\tau - \mu$ in the sum Σ_3 satisfy the inequalities

$$\|\mu\| + \|\tau-\mu\| \geqslant \Gamma\varepsilon^{-\frac{\varkappa}{m-1+\gamma}} + Q\varepsilon^{-\frac{1}{m-1+\gamma}} \geqslant \Gamma\varepsilon^{-\frac{\varkappa}{m-1+\gamma}}/2 + \|\tau\|\,.$$

Consequently,

$$\Sigma_3 \leqslant \sum_{\mu\in\widehat{D}\setminus D_{\varepsilon^\varkappa}\,,\ \tau-\mu\in B_{N+1}\setminus D_0} 8nc_2\varepsilon^{1+(N+1)(\varkappa-2)} e^{-\varepsilon^2\delta-\alpha(\|\mu\|+\|\tau-\mu\|)}$$
$$\leqslant \sum_{\mu\in\widehat{D}\setminus D_{\varepsilon^\varkappa}} 8nc_2\varepsilon^{1+(N+1)(\varkappa-2)} \exp(-(\alpha/2)\Gamma\varepsilon^{-\frac{\varkappa}{m-1+\gamma}} - \varepsilon^2\delta - \alpha\|\tau\|)$$
$$\leqslant \text{const}\,\varepsilon^{\frac{\gamma+\varkappa(m-1)}{m-1+\gamma}+(N+1)(\varkappa-2)} \exp(-(\alpha/2)\Gamma\varepsilon^{-\frac{\varkappa}{m-1+\gamma}} - \varepsilon^2\delta - \alpha\|\tau\|)\,.$$

Thus

$$\varepsilon \left|[\xi_{\widehat{D}} a\,,\, a - a^0]_\tau\right| \leqslant \text{const}\,\varepsilon^2 \exp(-l\varepsilon^{-q} - \varepsilon^2\delta - \alpha\|\tau\|)$$

with some positive constants l and q. Inequality (9.6) follows immediately from this relation.

iv) Let us verify inequality (9.7). We assume that $\tau \in B_j$. The term $[\xi_{\widehat{D}} a, a - a^0]_\tau$ satisfies the inequality

$$
\begin{aligned}
|[\xi_{\widehat{D}} a, a - a^0]_\tau| \leqslant & \sum_{\substack{\tau-\mu \in B_{N+j+1} \\ \mu \in D_{\varepsilon^\varkappa}}} \frac{4nc_2 p' \varepsilon^{1+(N+j+1)(\varkappa-2)}}{|\langle \mu, \omega \rangle|} e^{-\varepsilon^2\delta - \alpha(\|\mu\|+\|\tau-\mu\|)} \\
& + \sum_{\substack{\tau-\mu \in B_{j-1} \\ \mu \in \widehat{D} \setminus D_{\varepsilon^\varkappa}}} 4nc_2 \varepsilon^{1+(j-1)(\varkappa-2)} e^{-\varepsilon^2\delta - \alpha(\|\mu\|+\|\tau-\mu\|)} \\
& + \sum_{\substack{\tau-\mu \in B_{N+j+1} \setminus B_{j-1} \\ \mu \in \widehat{D} \setminus D_{\varepsilon^\varkappa}}} 4nc_2 \varepsilon^{1+(N+j+1)(\varkappa-2)} e^{-\varepsilon^2\delta - \alpha(\|\mu\|+\|\tau-\mu\|)}.
\end{aligned}
$$

We denote these sums by $\widehat{\Sigma}_1$, $\widehat{\Sigma}_2$, and $\widehat{\Sigma}_3$.

The sum $\widehat{\Sigma}_1$ is estimated similarly the sum Σ_1 in part (iii):

$$
\widehat{\Sigma}_1 \leqslant \operatorname{const} p' \varepsilon^{(N+j+1)(\varkappa-2)} e^{-\varepsilon^2\delta - \alpha\|\tau\|} < \varepsilon^{2+j(\varkappa-2)} e^{-\varepsilon^2\delta - \alpha\|\tau\|}/12.
$$

The sum $\widehat{\Sigma}_2$ is estimated by using Proposition 8:

$$
\begin{aligned}
\widehat{\Sigma}_2 \leqslant & \sum_{\mu \in \widehat{D} \setminus D_{\varepsilon^\varkappa}} 4nc_2 \varepsilon^{1+(j-1)(\varkappa-2)} e^{-\varepsilon^2\delta - \alpha\|\tau\|} \\
& \leqslant 4nc_2 K Q^{m-1} \varepsilon^{\frac{\varkappa(m-1)+\gamma}{m-1+\gamma} + (j-1)(\varkappa-2)} e^{-\varepsilon^2\delta - \alpha\|\tau\|} < \varepsilon^{2+j(\varkappa-2)} e^{-\varepsilon^2\delta - \alpha\|\tau\|}/12.
\end{aligned}
$$

The vectors μ and $\tau - \mu$ in the sum $\widehat{\Sigma}_3$ satisfy the inequality

$$
\|\mu\| + \|\tau - \mu\| \geqslant \Gamma \varepsilon^{-\frac{\varkappa}{m-1+\gamma}} + Q \varepsilon^{-\frac{1}{m-1+\gamma}} + \Gamma j \varepsilon^{-\frac{\varkappa}{m-1+\gamma}}/2 \geqslant \Gamma \varepsilon^{-\frac{\varkappa}{m-1+\gamma}}/2 + \|\tau\|,
$$

so that

$$
\begin{aligned}
\widehat{\Sigma}_3 \leqslant & \sum_{\mu \in \widehat{D} \setminus D_{\varepsilon^\varkappa}} 4nc_2 \varepsilon^{1+(N+j+1)(\varkappa-2)} \exp\left(-(\alpha/2)\Gamma \varepsilon^{-\frac{\varkappa}{m-1+\gamma}} - \varepsilon^2\delta - \alpha\|\tau\|\right) \\
& < \varepsilon^{2+j(\varkappa-2)} e^{-\varepsilon^2\delta - \alpha\|\tau\|}/12.
\end{aligned}
$$

Thus we obtain the estimate

$$
|[\xi_{\widehat{D}} a, a - a_0]_\tau| < \varepsilon^{2+j(\varkappa-2)} e^{-\varepsilon^2\delta - \alpha\|\tau\|}/4.
$$

Inequality (9.7) follows immediately.

v) Let us verify inequality (9.8). We use the inequality

$$
\varepsilon\,|[\xi_{\widehat{D}} a, a - a^0]_0| \leqslant \sum_{\mu \in D_{\varepsilon^\varkappa}} 2n\varepsilon\,|a^\mu|\,|a^{-\mu}| + \sum_{\mu \in \widehat{D} \setminus D_{\varepsilon^\varkappa}} 2n\varepsilon\,|a^\mu|\,|a^{-\mu}|.
$$

The first sum in this relation is less then $\varepsilon^3/2$ because each term in it is exponentially small. The second sum is also less than $\varepsilon^3/2$, since for each $\mu \in \widehat{D} \setminus D_{\varepsilon^\varkappa}$ we have the inequality $\|\mu\| \geqslant \Gamma \varepsilon^{-\frac{\varkappa}{m-1+\gamma}}$ and thus

$$
|a^{\pm\mu}| \leqslant c_2 \exp\left(-\varepsilon^2\delta - \alpha\Gamma \varepsilon^{-\frac{\varkappa}{m-1+\gamma}}\right).
$$

Inequality (9.8) is proved.

§10. Discussion

The averaging method developed in §4 admits far reaching generalizations. Indeed, let us try to transform the system of ordinary differential equations

$$\dot{z} = v(z), \tag{10.1}$$

using the change of the variables

$$z \to z_\Delta. \tag{10.2}$$

Here z is a point of a manifold M, v is a smooth vector field on M, Δ is a nonnegative parameter, and the change (10.2) is defined as the shift along the trajectories of the equation

$$z'_\delta = f(z_\delta, \delta), \qquad z_0 = z, \tag{10.3}$$

where the prime denotes the derivative with respect to δ and the trajectories are considered on the closed interval $[0, \Delta]$.

Suppose that the change $z \to z_\delta$ transforms system (10.1) to

$$\dot{z}_\delta = v_\delta(z_\delta). \tag{10.4}$$

Let us differentiate (10.4) with respect to δ. We have:

$$(f(z_\delta, \delta))^{\cdot} = v'_\delta(z_\delta) + \partial_f v_\delta(z_\delta) \quad \text{or} \quad v'_\delta = [v_\delta, f].$$

Here ∂_f is the differential operator on M corresponding to the vector field f and $[\cdot, \cdot]$ is the vector commutator: $[u_1, u_2] = \partial_{u_1} u_2 - \partial_{u_2} u_1$. If we set $f = \xi v$, where ξ is some linear operator, we obtain the equation

$$v' = -[\xi v, v] \tag{10.5}$$

(cf. (4.3)). Recall that for brevity we often omit the subscript δ.

If the system (10.1) is Hamiltonian with the Hamiltonian function $H = H(z)$, then it is natural to assume that the change (10.2) is symplectic and equation (10.3) is Hamiltonian with some Hamiltonian function $F(z, \delta)$. If these conditions hold, then systems (10.4) are also Hamiltonian. Their Hamiltonian functions H_δ satisfy the relation $H_\delta(z_\delta) = H(z)$.

Let us differentiate this relation with respect to δ. We get:

$$H'_\delta(z_\delta) + \partial_{f(z_\delta, \delta)} H_\delta(z_\delta) = 0 \quad \text{or} \quad H'_\delta = \{F, H_\delta\}$$

since $\partial_f H_\delta$ is equal to the ordinary Poisson bracket of H_δ and F. We set $F = \xi H$ with some linear operator ξ and obtain the equation

$$H' = \{\xi H, H\} \tag{10.6}$$

(cf. (4.3), (10.5)). We note that equation (10.6) is close to the relations of the well-known Deprit–Hori method.

The author believes that after a proper choice of the operators ξ, equations (4.3), (10.5), (10.6) become a powerful device for the investigation of exponentially small effects in the theory of dynamical systems. Let us present an example of a possible application of equation (10.6).

We consider the Hamiltonian system with one and half degrees of freedom

$$(10.7) \qquad \dot{y} = -\varepsilon \partial H/\partial x\,, \quad \dot{x} = \varepsilon \partial H/\partial y\,, \qquad H = H(x\,,\,y\,,\,t)\,.$$

Here $y \in \mathbb{R}$, $x \in \mathbb{T} = \mathbb{R}/2\pi\mathbb{Z}$; the function H is real-analytic in x, y and 2π-periodic in t. We try to weaken the rate of the dependence of H on t with the help of a symplectic change of variables $(x\,,\,y) \to (x_\Delta(x\,,\,y\,,\,t)\,,\,y_\Delta(x\,,\,y\,,\,t))$, $\Delta > 0$, where

$$(10.8) \qquad y'_\delta = -\partial F/\partial x_\delta\,, \quad x'_\delta = \partial F/\partial y_\delta\,, \qquad F = F_\delta(x_\delta\,,\,y_\delta\,,\,t)\,.$$

To use equality (10.6), we represent equations (10.7), (10.8) in the following form:

$$\frac{dy}{ds} = -\frac{\partial \mathcal{H}}{\partial x}\,, \quad \frac{dx}{ds} = \frac{\partial \mathcal{H}}{\partial y}\,, \quad \frac{dE}{ds} = -\frac{\partial \mathcal{H}}{\partial t}\,, \quad \frac{dt}{ds} = \frac{\partial \mathcal{H}}{\partial E} = 1\,, \quad \mathcal{H} = \varepsilon H + E\,,$$

$$y' = -\frac{\partial F}{\partial x}\,, \quad x' = \frac{\partial F}{\partial y}\,, \quad E' = -\frac{\partial F}{\partial t}\,, \quad t' = \frac{\partial F}{\partial E} = 0\,.$$

We put $F = \xi H$ and obtain, in accordance with (10.6),

$$(10.9) \qquad \mathcal{H}' = \{\xi H\,,\,\mathcal{H}\}_*\,,$$

where the Poisson bracket $\{\cdot\,,\,\cdot\}_*$ is determined by the symplectic structure $dx \wedge dy + dt \wedge dE$. If we set $\mathcal{H}_\delta = H_\delta + E$, then equation (10.9) is transformed in the following way:

$$(10.10) \qquad H' = -\partial(\xi H)/\partial t + \varepsilon\{\xi H\,,\,H\}\,,$$

where the Poisson bracket $\{\cdot\,,\,\cdot\}$ is defined by the symplectic structure $dx \wedge dy$ on the cylinder $(\mathbb{R}/2\pi\mathbb{Z}) \times \mathbb{R}$.

We set

$$\xi H_\delta = -\sum_{k\in\mathbb{Z}} i \operatorname{sgn} k H_{\delta\,,\,k}(x\,,\,y)e^{ikt}\,,$$

where $H_{\delta\,,\,k}$ are the Fourier coefficients in the expansion

$$H_\delta = \sum_{k\in\mathbb{Z}} H_{\delta\,,\,k}(x\,,\,y)\,e^{ikt}.$$

We obtain the following system:

$$(10.11) \qquad H'_k = -|k|\,H_k + \varepsilon\{\xi H\,,\,H\}_k\,, \qquad k \in \mathbb{Z}\,.$$

The symbol $\{\ ,\ \}_k$ denotes the Fourier coefficient corresponding to k.

The terms $\varepsilon\{\xi H, H\}_k$ in (10.11) are small, at least for values of δ that are not too large. If we neglect these terms, then we get the equations

$$H_k' = -|k|\,H_k\,. \tag{10.12}$$

For $k \neq 0$ the solutions of (10.12) rapidly tend to zero as $\delta \to \infty$.

A more precise approximation of (10.11) is obtained if we take into account the term $\varepsilon\{\xi H, H_0\}_k = -i\varepsilon \operatorname{sgn} k\{H_k, H_0\}$. We get:

$$H_k' = -|k|\,H_k - i\varepsilon \operatorname{sgn} k\,\{H_k, H_0\}\,. \tag{10.13}$$

The solutions of these equations have the form

$$H_{\delta, k} = H_{0, k} \circ g^{i\varepsilon\delta \operatorname{sgn} k} e^{-|k|\delta}, \tag{10.14}$$

where g^q is the shift $(x, y)|_{t=0} \to (x, y)|_{t=q}$ along the trajectories of the Hamiltonian system

$$\dot{x} = \partial H_0/\partial y\,, \qquad \dot{y} = -\partial H_0/\partial x\,.$$

We see that complex singularities of the functions $\beta_k(q) = H_{0,k} \circ g^q$, $q \in \mathbb{C}$ hinder us from the unrestricted prolongation of solution (10.14) to the set of positive values of δ. Nevertheless, the functions (10.14) can be made exponentially small in ε since we can choose $\delta = \text{const}/\varepsilon$. Apparently for δ not too close to a singular value, the functions (10.14) are close to the actual solutions (10.11).

It is well known [8] that the Hamiltonian (10.7) can be transformed to one that contains time t only in exponentially small terms. An analysis of equation (10.10) should make this result much more precise.

§11. Appendix 1

PROPOSITION 1. *Let $\|\cdot\|$ be a norm in $\mathbb{R}^k$ such that the sphere $S = \{\mu \in \mathbb{R}^k : \|\mu\| = 1\}$ is a smooth surface. Then the norm $\|\cdot\|$ is strictly convex if and only if the normal curvature of S at any point along any direction in the standard metric $\langle\cdot,\cdot\rangle$ exceeds some positive constant depending only on the norm $\|\cdot\|$.*

PROOF. a) (Sufficiency). Let us assume that the normal curvatures of S are not bounded away from zero. The sphere S being compact, there exists a curve $\mu(s) \subset S$ whose curvature at the point $\mu(0)$ is zero. The curve $\mu(s)$ has the form $\mu(s) = \mu(0) + vs + O(s^3)$; $v \in \mathbb{R}^k$ is not equal to 0. Using Proposition 2 we see that

$$\lim_{s\to 0} \frac{\|\mu(s)\| + \|\mu(0)\| - \|\mu(s) + \mu(0)\|}{\|\mu(s) - \mu(0)\|^2} = 0$$

and the norm $\|\cdot\|$ is not strictly convex.

b) (Necessity). Let us assume that the norm $\|\cdot\|$ is not strictly convex. Then there exists a sequence M of pairs $(\mu_1^{(j)}, \mu_2^{(j)})$, $j \in \mathbb{N}$, such that $\|\mu_1^{(j)}\| = \|\mu_2^{(j)}\| = 1$ and

$$\lim_{j\to\infty} \frac{\|\mu_1^{(j)}\| + \|\mu_2^{(j)}\| - \|\mu_1^{(j)} + \mu_2^{(j)}\|}{\|\mu_1^{(j)} - \mu_2^{(j)}\|^2} = 0\,.$$

The sequence M on the compact set $S \times S$ contains a converging subsequence. To simplify the notation, we assume that M converges itself. Let μ_1^∞ and μ_2^∞ be the limit of M. We consider two cases: $\mu_1^\infty \neq \mu_2^\infty$ and $\mu_1^\infty = \mu_2^\infty$.

In the first case, the three points μ_1^∞, μ_2^∞, and $(\mu_1^\infty + \mu_2^\infty)/2$ are situated on the sphere S. The sphere S is the boundary of the convex ball $\{\mu \in \mathbb{R}^k : \|\mu\| \leqslant 1\}$, thus the whole line segment $I \subset \mathbb{R}^k$ with end points μ_1^∞ and μ_2^∞ belongs to S. The curvature of S at any point of I along I equals zero.

The second case is more complex. We consider the function

$$f(\mu_1, \mu_2) = \frac{\|\mu_1\| + \|\mu_2\| - \|\mu_1 + \mu_2\|}{\|\mu_1 - \mu_2\|^2}, \qquad \mu_1, \mu_2 \in S.$$

This function cannot be extended to a continuous function on $S \times S$. But it can be extended to a continuous function on some other manifold N. This manifold can be regarded as a subbundle of the tangent bundle TS:

$$N = \{v_\mu \in T_\mu S : \mu \in S, \ \|v_\mu\| \leqslant 1\}.$$

We have the natural projection $\pi\colon N \to S$, $\pi(v_\mu) = \mu$. Let us denote by d_μ the $(k-1)$-dimensional disk $\pi^{-1}(\mu)$.

Now we define maps $p_\mu\colon d_\mu \to S$ satisfying the following cinditions:

1) every p_μ is continuous,

2) $p_\mu(0) = \mu$,

3) p_μ maps the boundary ∂d_μ to $-\mu$,

4) p_μ maps every diameter q of d_μ onto a circle $c_q \subset S$ such that $\mu \in c_q$, $-\mu \in c_q$ and any vector $v \in q$ is tangent to c_q,

5) p_μ maps $d_\mu \setminus \partial d_\mu$ diffeomorphically onto $S \setminus \{-\mu\}$.

Consider the continuous map $p\colon N \to S \times S$ given by $p(v_\mu) = (\mu, p_\mu(v_\mu))$. It transforms $N \setminus \partial N$ diffeomorphically to $(S \times S) \setminus D$, where D is the diagonal:

$$D = \{(\mu_1, \mu_2) \in S \times S : \mu_1 = \mu_2\}.$$

Thus N can be regarded as $(S \times S) \setminus D$ with $S \times S^{k-2}$ attached to it instead of the diagonal D. Here S^{k-2} is the $(k-2)$-dimensional sphere and an element $(\mu, \nu) \in S \times S^{k-2}$ from the geometrical point of view represents a point $\mu \in S$ and a direction ν from μ to some point of S infinitely close to μ.

The function f and the map p generate the continuous function $f \circ p\colon N \setminus \partial N \to \mathbb{R}$. Due to Proposition 2, this function can be prolonged to some continuous function $F\colon N \to \mathbb{R}$. For any point $v_\mu \in \partial N$ we have

$$F(v_\mu) = \frac{|v_\mu|^2 \operatorname{curv}}{4\langle \mathbf{n}, \mu\rangle},$$

where curv is the normal curvature of S at the point μ in the direction $v_\mu \in T_\mu S$.

The sequence M' of points $p^{-1}(\mu_1^{(j)}, \mu_2^{(j)})$ tends to the boundary ∂N. Since ∂N is compact, M' has a limit point $u \in \partial N$. The function F vanishes at the point u, thus the corresponding curvature equals zero. $\square$

Proposition 2. *Suppose that the conditions of Proposition* 1 *hold. Consider a curve* $\mu(s) \subset S$ *such that* $\mu(s) = \mu_0 + \dot{v}s + O(s^2)$. *Then*

$$\lim_{s \to 0} \frac{\|\mu(s)\| + \|\mu_0\| - \|\mu(s) + \mu_0\|}{\|\mu(s) - \mu_0\|^2} = \frac{|v_\mu|^2 \operatorname{curv}}{4 \langle \mathbf{n}, \mu \rangle},$$

where $\mathbf{n}$ *is the unit normal to* S *at the point* μ_0 *and* curv *is the normal curvature of* S *at the point* μ_0 *along the direction* $v \in T_{\mu_0} S$. *Both* $\mathbf{n}$ *and* curv *are considered with respect to the standard metric* $\langle \cdot , \cdot \rangle$ *of* $\mathbb{R}^k$.

The proof of Proposition 2 is based on a direct computations. We omit it.

Proposition 3. *Let the norm* $\| \cdot \|$ *in* $\mathbb{R}^k$ *be strictly convex. Then for any two vectors* $\mu_1, \mu_2 \in \mathbb{R}^k$ *such that* $\|\mu_1\| \geqslant \|\mu_2\|$ *and* $\|\mu_1\| + \|\mu_2\| - \|\mu_1 + \mu_2\| \leqslant b$, *one can find* $s \in \mathbb{R}$ *and* $w \in \mathbb{R}^k$ *such that*

$$\mu_1 = s\mu_2 + w \quad \text{and} \quad \|w\|^2 \leqslant P^{-1} b \, \|\mu_2\|, \tag{11.1}$$

where the constant P *is the same as in inequality* (2.2).

Proof. i). Let us assume that $\|\mu_1\| = \|\mu_2\| = c$. We put $s = 1$, $w = \mu_1 - \mu_2$. In accordance with (2.2) we have:

$$P\|w/c\|^2 \leqslant \|\mu_1/c\| + \|\mu_2/c\| - \|(\mu_1 + \mu_2)/c\| \leqslant b/c.$$

Thus $\|w\|^2 \leqslant P^{-1}bc$ and inequality (11.1) holds.

ii). In the general case, we put $\mu_1 = \mu' + \mu''$, where the directions of the three vectors μ_1, μ', μ'' coincide and $\|\mu'\| = \|\mu_2\|$. Conditions of Proposition 3 and the triangle inequality imply the following estimate:

$$\|\mu'\| + \|\mu_2\| - \|\mu' + \mu_2\| = \|\mu_1\| + \|\mu_2\| - \|\mu_1 + \mu_2\| - (\|\mu''\| + \|\mu' + \mu_2\| - \|\mu_1 + \mu_2\|) \leqslant b.$$

We put $w = \mu' - \mu_2$. Thus we have reduced the general case to the case i).

§12. Appendix 2

Proposition 5. *For any two square matrices* a, b *of order* n *the inequality* $\|[a, b]\| \leqslant 2n\,|a|\,|b|$ *holds.*

The proof follows immediately from the definition of the norm $| \cdot |$ (see §1) and the simple inequality $|ab| \leqslant n\,|a|\,|b|$.

Proposition 6. *Let the norm* $\| \cdot \|$ *be acceptable and* τ, μ *be two vectors in* $\mathbb{R}^m$ *such that* $\langle \tau, \omega \rangle < 0$ *and* $\langle \mu, \omega \rangle > 0$. *Then* $\|\mu\| + \|\tau - \mu\| \geqslant \|\tau\| + \langle \mu, \omega \rangle$.

Proof. By the definition of acceptable norm, we have

$$\begin{aligned} \|\mu\| + \|\tau - \mu\| &= \langle \mu, \omega \rangle/2 + \|\mu - \langle \mu, \omega \rangle \omega\|_\pi \\ &\quad - \langle \tau - \mu, \omega \rangle/2 + \|\tau - \mu - \langle \tau - \mu, \omega \rangle \omega\|_\pi \\ &\leqslant \langle \mu, \omega \rangle + |\langle \tau, \omega \rangle|/2 + \|\tau - \langle \tau, \omega \rangle \omega\|_\pi = \langle \mu, \omega \rangle + \|\tau\|. \end{aligned}$$

Here we used the triangle inequality for the norm $\| \cdot \|_\pi$. Proposition 6 is proved.

PROPOSITION 7. *Let $\|\cdot\|$ be a norm in $\mathbb{R}^p$. Then for any $N \in \mathbb{R}$, $k \in \mathbb{N}$, the ball $B_N^0 = \{\tau \in \mathbb{R}^p : \|\tau\| \leqslant N\}$ can be covered by k balls*

$$B_M^{\tau^{(j)}} = \{\tau \in \mathbb{R}^p : \|\tau - \tau^{(j)}\| \leqslant M\,,\ \tau^{(j)} \in \mathbb{R}^p\}, \qquad j = 1, \dots, k,$$

with $M \leqslant K'Nk^{-1/p}$. The constant K' depends only on the norm $\|\cdot\|$.

PROOF. Let C_r^μ be a cube $\{\tau = (\tau_1, \dots, \tau_p) \in \mathbb{R}^p : |\tau_l - \mu_l| \leqslant r,\ \mu \in \mathbb{R}^p\}$. There exist positive numbers q and Q such that $C_{qr}^\tau \subset B_r^\tau \subset C_{Qr}^\tau$ for any $\tau \in \mathbb{R}^p$, $r > 0$.

The ball B_N^0 belongs to the cube C_{NQ}^0. For any integer l this cube can be covered by l^p cubes $C_{NQ/l}^{\tau^{(j)}}$, $j = 1, \dots, l^p$, where $\tau^{(j)}$ are some points in $\mathbb{R}^p$. We put l = integer part of $k^{1/p}$. We have the following relations:

$$l^p \leqslant k\,, \qquad b_N^0 \subset C_{NQ}^0 \subset \bigcup_{j=1}^{l^p} C_{NQ/l}^{\tau^{(j)}} \subset \bigcup_{j=1}^{l^p} B_{NQ/(ql)}^{\tau^{(j)}}\,.$$

Thus we obtain the estimate $M \leqslant QN(ql)^{-1} \leqslant K'Nk^{-1/p}$. □

PROPOSITION 8. *Let the norm $\|\cdot\|$ be acceptable and ν_1, ν_2 be two real numbers such that $\widehat{K}N^{-m+1-\gamma} \leqslant \nu_2 - \nu_1 \leqslant 1$, $\widehat{K} > 0$. Then the set*

$$S_N = \{\tau \in \mathbb{Z}^m : \nu_1 \leqslant \langle \tau\,, \omega\rangle \leqslant \nu_2\,,\ \|\tau\| \leqslant N\}$$

consists of no more than $K(\nu_2 - \nu_1)^{\frac{m-1}{m-1+\gamma}} N^{m-1}$ elements, where $K = K(\widehat{K})$ is some positive constant.

PROOF. We put

$$S_M^\mu = \{\tau \in \mathbb{Z}^m : \nu_1 \leqslant \langle \tau\,, \omega\rangle \leqslant \nu_2\,,\ \|\tau - \mu\| \leqslant M\,,\ \mu \in \mathbb{R}^m\,,\ \langle \mu\,, \omega\rangle = 0\}$$

and assume $l = l_M^\mu$ to be the number of elements in S_M^μ. We can find $\tau_1\,, \tau_2 \in S_M^\mu$ such that

$$|\langle \tau_1 - \tau_2\,, \omega\rangle| \leqslant (\nu_2 - \nu_1)/(l - 1)\,. \tag{12.1}$$

The inequality $\|\tau_1 - \tau_2\| \leqslant 2M$ holds. Consequently, using the nonresonance conditions (2.5), we obtain the estimate

$$(2M/\Gamma)^{-m+1-\gamma} \leqslant |\langle \tau_1 - \tau_2\,, \omega\rangle|\,. \tag{12.2}$$

Bringing together inequalities (12.1) and (12.2), we get the following estimate:

$$l \leqslant 1 + (\nu_2 - \nu_1)(2M/\Gamma)^{m-1+\gamma}. \tag{12.3}$$

Let us cover the $(m-1)$-dimensional ball $B_N^0 = \{\tau \in \mathbb{R}^m : \langle \tau\,, \omega\rangle = 0\,,\ \|\tau\| \leqslant N\}$ by k balls

$$B_{M'}^{\tau^{(j)}} = \{\tau \in \mathbb{R}^m : \langle \tau\,, \omega\rangle = 0\,,\ \|\tau - \tau^{(j)}\| \leqslant M',\ \langle \tau^{(j)}\,, \omega\rangle = 0\}, \qquad j = 1, \dots, k.$$

By Proposition 7, we can take M' to be at most $K'Nk^{-\frac{1}{m-1}}/4$. Let

$$k = 1 + \text{integer part of } (v_2 - v_1)^{\frac{m-1}{m-1+\gamma}} (K'N/\Gamma)^{m-1}, \tag{12.4}$$
$$M' = K'Nk^{-\frac{1}{m-1}} = \Gamma(v_2 - v_1)^{-\frac{1}{m-1+\gamma}}, \qquad M = M' + 1.$$

Due to the conditions of Proposition 8, we have $k \leqslant \text{const}(v_2 - v_1)^{\frac{m-1}{m-1+\gamma}} N^{m-1}$. Using the definition of acceptable norm, we get the inclusion: $S_N \subset \bigcup_{j=1}^k S_M^{\tau^{(j)}}$. Thus the number of elements in the set S_N, in accordance with relations (12.3), (12.4), does not exceed

$$k(1 + (v_2 - v_1)(2M/\Gamma)^{m-1+\gamma})$$
$$\leqslant k(1 + (v_2 - v_1)(2(v_2 - v_1)^{-\frac{1}{m-1+\gamma}} + 2/\Gamma)^{m-1+\gamma}) \leqslant \text{const}\, k.$$

The last inequality follows from the condition $0 \leqslant v_2 - v_1 \leqslant 1$. Now Proposition 8 follows from the definition of k (12.4).

PROPOSITION 9. *Let $v \in (0, 1)$ and $\tau \in D_v$, where D_v is defined by (5.4) with $Q\varepsilon^{-\frac{1}{m-1+\gamma}} = N$. Then*

$$\sum_{\mu \in \mathbb{Z}^m \setminus D_v,\ \tau - \mu \in D_v} e^{-\alpha(\|\mu\| + \|\tau - \mu\|)} \leqslant (Kv^{\frac{m-1}{m-1+\gamma}} N^{m-1} + K_1) e^{-\alpha\|\tau\|}.$$

Here K, K_1 are positive constants.

PROOF. Let us divide the sum in question into two parts:

$$\Sigma_1 = \sum_{\substack{\mu \in D_0 \setminus D_v \\ \tau - \mu \in D_v}} e^{-\alpha(\|\mu\| + \|\tau - \mu\|)}, \qquad \Sigma_2 = \sum_{\substack{\|\mu\| > N \\ \tau - \mu \in D_v}} e^{-\alpha(\|\mu\| + \|\tau - \mu\|)}.$$

The sum Σ_1 is estimated by using Proposition 8:

$$\Sigma_1 \leqslant \sum_{\|\mu\| \leqslant N,\ |\langle \mu, \omega \rangle| < v} e^{-\alpha\|\tau\|} \leqslant Kv^{\frac{m-1}{m-1+\gamma}} N^{m-1} e^{-\alpha\|\tau\|}.$$

Let us estimate the sum Σ_2. We have

$$\Sigma_2 \leqslant \sum_{\psi \in D_v} e^{-\alpha(N + \|\psi\|)} \leqslant K_1 e^{-\alpha N} \leqslant K_1 e^{-\alpha\|\tau\|},$$

where the constant K_1 depends only on α and m. □

PROPOSITION 10. *Let us assume the following conditions to be true:*
1) *The vector ω satisfies nonresonance conditions* (2.5).
2) *The norm $\|\cdot\|$ is acceptable.*
3) *The vectors τ_1 and τ_2 belong to $\mathbb{R}^m$ and are linearly independent.*
4) $\|\tau_j\| \leqslant 3\Gamma\varepsilon^{-\frac{1}{m-1+\gamma}}/2$, $j = 1, 2$.
5) $|\langle \tau_j, \omega \rangle| \leqslant 2\varepsilon^{\varkappa}$, $j = 1, 2$, *where $\varkappa$ satisfies inequalities* (5.2).

Then for small ε *and* β *satisfying* (5.3), *the following inequality holds*:

$$\|\tau_1\| + \|\tau_2\| > \|\tau_1 + \tau_2\| + \varepsilon^{-\beta}.$$

REMARK. Proposition 10 remains valid if we replace condition 3) by the following condition

3′). The vectors τ_1 and τ_2 belong to $\mathbb{R}^m$ and $\langle \tau_1, \omega\rangle\langle\tau_2, \omega\rangle < 0$.

PROOF OF THE REMARK. If the vectors τ_1 and τ_2 are linearly independent then condition 3) holds. If the vectors τ_1 and τ_2 are parallel, then, by 3′), their directions are opposite. Thus

$$\|\tau_1\| + \|\tau_2\| - \|\tau_1 + \tau_2\| = 2\min\{\|\tau_1\|, \|\tau_2\|\}$$

and Proposition 10 holds because the nonresonance conditions (2.5) and condition 5) of Proposition 10 imply that

$$\|\tau_j\| \geqslant \Gamma(2\varepsilon^{\varkappa})^{-\frac{1}{m-1+\gamma}}.$$

PROOF OF PROPOSITION 10. Let us assume that

$$\|\tau_1\| + \|\tau_2\| \leqslant \|\tau_1 + \tau_2\| + \varepsilon^{-\beta}.$$

One can assume that $\|\tau_1\| \geqslant \|\tau_2\|$. Thus, in accordance with Proposition 3,

$$\tau_1 = s\tau_2 + w, \quad s \geqslant 1, \qquad \|w\|^2 \leqslant P\|\tau_2\|\varepsilon^{-\beta}.$$

A standard lemma from number theory [6] states that for any natural M there exist integers l_1 and l_2 such that $|l_1| \leqslant M$, $|l_2| \leqslant M$, and $|l_1 + l_2/s| < 1/M$.

Let $\tau = l_1\tau_1 + l_2\tau_2$ and M be the integer part of $\varepsilon^{\frac{\beta}{4} - \frac{1}{4(m-1+\gamma)}} + 1$. The vector τ is nonzero by the linear independence of τ_1 and τ_2. We see that

$$\begin{aligned}
\|\tau\| &= \|s\tau_2(l_1 + l_2/s) + l_1 w\| \leqslant \|s\tau_2\|/M + M\|w\| \\
&\leqslant 3\Gamma\varepsilon^{-\frac{1}{m-1+\gamma}} M^{-1} + M(3\Gamma P/2)^{\frac{1}{2}}\varepsilon^{-\frac{1}{2(m-1+\gamma)} - \frac{\beta}{2}} \\
&\leqslant (3\Gamma + (3\Gamma P)^{\frac{1}{2}})\varepsilon^{-\frac{3}{4(m-1+\gamma)} - \frac{\beta}{4}}.
\end{aligned}$$

We have the estimate

$$|\langle\tau, \omega\rangle| \leqslant |l_1\langle\tau_1, \omega\rangle| + |l_2\langle\tau_2, \omega\rangle| \leqslant 2M\varepsilon^{\varkappa} < 4\varepsilon^{\frac{\beta}{4} - \frac{1}{4(m-1+\gamma)} + \varkappa}.$$

Thus

$$|\langle\tau, \omega\rangle|(\Gamma/\|\tau\|)^{-\frac{1}{m-1+\gamma}} < \operatorname{const}\varepsilon^{q}, \tag{12.5}$$

where $q = \varkappa - (m-1+\gamma)^{-1}/4 - 3/4 - \beta(m-2+\gamma)/4 > 0$. For small ε inequality (12.5) contradicts the nonresonance conditions (2.5). Proposition 10 is proved.

References

1. Ph. Hartman, *Ordinary differential equations*, Wiley, New York, 1964.
2. A. G. Baskakov, *A theorem on the reducibility of linear differential equations with quasi-periodic coefficients*, Ukrain. Mat. Zh. **35** (1983), no. 4, 416–421; English transl., Ukranian Math. J. **35** (1983), no. 4, 357–361.
3. A. R. Johnson and R. G. Sell, *Smoothness of spectral subbundles and reducibility of quasi-periodic linear differential systems*, J. Differential Equations **41** (1981), 262–288.
4. A. Jorba and C. Simo, *On the reducibility of linear differential equations with quasiperiodic coefficients*, J. Differential Equations **98** (1992), 111–124.
5. N. N. Bogolyubov, Yu. A. Mitropolskiĭ, and A. M. Samoĭlenko, *Methods of accelerated convergence in nonlinear mechanics*, Springer-Verlag, New York, 1976.
6. W. M. Schmidt, *Diophantine approximation*, Lectures Notes in Math., vol. 785, Springer-Verlag, Heidelberg, 1980.
7. D. V. Treshchev, *A mechanism for the destruction of resonance tori in Hamiltonian systems*, Mat. Sb. **180** (1989), no. 10, 1325–1346; English transl., Math. USSR-Sb. **68** (1991), no. 1, 181–203.
8. A. I. Neishtadt, *The separation of motions in systems with rapidly rotating phase*, Prikl. Mat. Mekh. **48** (1984), no. 2, 197–204; English transl., J. Appl. Math. Mech. **48** (1984), no. 2, 133–139.

Translated by THE AUTHOR

DEPARTMENT OF MATHEMATICS, MOSCOW STATE UNIVERSITY, MOSCOW 119899, RUSSIA

Amer. Math. Soc. Transl.
(2) Vol. **168**, 1995

On the Reducibility of the One-Dimensional Schrödinger Equation with Quasi-Periodic Potential

DMITRY V. TRESHCHEV

§1. Introduction

Let us consider the Schrödinger equation

$$-\frac{d^2}{dx^2}\psi + u(\omega x)\psi = E\psi, \tag{1.1}$$

where $x \in \mathbb{R}$ is the independent variable, E is the energy constant, $\omega \in \mathbb{R}^m$ is the constant frequency vector, and $u\colon \mathbb{T}^m \to \mathbb{R}$ is the real-analytic potential. As usual, $\mathbb{T}^m$ is the m-dimensional torus $\mathbb{R}^m/2\pi\mathbb{Z}^m$. Such an equation was studied by Dinaburg and Sinai [1]; some generalizations appear in [2–4]; the results of [1] are extended to the case of the difference Schrödinger operator in [5].

The following problem connected with equation (1.1) arises. What values of the energy E admit quasi-periodic solutions $\psi_E(x)$? Dinaburg and Sinai noticed [1] that if for some E equation (1.1) is reducible in a certain sense, then a pair of quasi-periodic solutions ψ_E exists.

We set

$$x = t, \quad \psi = y_1, \quad \frac{d\psi}{dx} = \sqrt{E}\, y_2, \quad y = \begin{pmatrix} y_1 \\ y_2 \end{pmatrix}, \quad \varepsilon = E^{-1/2}.$$

Equation (1.1) is transformed to the form

$$\begin{cases} \dot{y} = (\varepsilon^{-1} e_* + \varepsilon U(\vartheta))\, y, \\ \dot{\vartheta} = \omega, \end{cases} \tag{1.2}$$

where the dot denotes the derivative with respect to t, $\vartheta \in \mathbb{T}^m$, $\vartheta(0) = 0$, and

$$e_* = \begin{pmatrix} 0 & 1 \\ -1 & 0 \end{pmatrix}, \qquad U = \begin{pmatrix} 0 & 0 \\ -u & 0 \end{pmatrix}. \tag{1.3}$$

We refer to equations (1.2) as *reducible* if the change of variables

$$y = B(\vartheta, \varepsilon)\, z \tag{1.4}$$

1991 *Mathematics Subject Classification*. Primary 34C27; 34L40.

with a real nondegenerate square matrix B of order 2 transforms them to the form

$$\begin{cases} \dot{z} = \widehat{A} z\,, \\ \dot{\vartheta} = \omega\,, \end{cases} \tag{1.5}$$

with a real constant matrix $\widehat{A} = \widehat{A}(\varepsilon)$.

It is easy to show that for large values of the energy (or equivalently, for small values of ε) the reducibility of system (1.2), (1.3) implies the existence of two linearly independent quasi-periodic solutions $\psi_E(x)$ of equation (1.1).

Indeed, if equations (1.2), (1.3) are reducible, then their solution with the initial conditions $\vartheta(0) = 0$, $y(0) = y_0$ has the form $y(t\,,\, y_0) = B(\omega t) e^{\widehat{A}t} y_0$. The functions $\psi_E(x\,,\, y_0) = y_1(x\,,\, y_0)$ form a two-parameter family of solutions of (1.1). If the eigenvalues λ_1, λ_2 of the matrix $\widehat{A}$ are pure imaginary and different, then the functions $\psi_E(x\,,\, y_0)$ are quasi-periodic. For small ε we have $\lambda_1 \neq \lambda_2$, since the eigenvalues of $\widehat{A}$ are close to $\pm\varepsilon^{-1} i$ (those of $\varepsilon^{-1} e_*$). Here $i = \sqrt{-1}$. To show that λ_1 and λ_2 are pure imaginary, it is sufficient to prove that $\operatorname{tr} \widehat{A} = \lambda_1 + \lambda_2 = 0$. We set

$$\partial = \sum_{s=1}^{m} \omega_s \frac{\partial}{\partial \vartheta_s} \tag{1.6}$$

and obtain the following relations:

$$\widehat{A} = \partial B B^{-1} + B(\varepsilon^{-1} e_* + \varepsilon U) B^{-1}\,,$$
$$\operatorname{tr} B(\varepsilon^{-1} e_* + \varepsilon U) B^{-1} = \operatorname{tr}(\varepsilon^{-1} e_* + \varepsilon U) = 0\,,$$

which imply $\operatorname{tr} \widehat{A} = \operatorname{tr}(\partial B B^{-1})$. Let us use the matrix relation

$$\operatorname{tr}(\partial B B^{-1}) = \partial \ln(\det B)\,.$$

We get $\operatorname{tr} \widehat{A} = \partial \ln(\det B) = 0$ since any constant function of the form ∂f on $\mathbb{T}^m$ equals zero.

Eliasson [4] showed that if the potential u is real-analytic, then for almost all large values of the energy E the system (1.2), (1.3) is reducible in a certain weaker sense. It can be noted that in [4] the corresponding changes (1.4) do not depend continuously on E. Apparently, irreducibility does not occur for all large values of E.

In this paper we try to weaken the rate of the dependence of the right-hand side of (1.2) on ϑ for all small ε by using the change of variables (1.4). We obtain an asymptotics of quasi-periodic terms, which are irremovable in the framework of our averaging method. These terms turn out to be exponentially small and their asymptotics depends on the complex singularities of the potential u. We hope that these asymptotics will help to investigate the spectrum of the Schrödinger operator corresponding to equation (1.1). Unlike papers [1–4], we do not completely eliminate the dependence of the matrix U on ϑ. But our change (1.4) is valid for all large values of energy. Moreover, it is continuous (and even smooth) with respect to E. We note also that the Diophantine conditions for the frequencies ω in our paper are much weaker than the standard ones (see e.g. [4]).

§2. The main result

We suppose that the potential $u(\vartheta)$ satisfies the following two conditions.

A. For any real ϑ the function $f_\vartheta(\zeta) = u(\vartheta + \omega\zeta)$ is real-analytic in ζ in the strip

$$S_\alpha = \{\zeta \in \mathbb{C} : |\operatorname{Im}\zeta| < \alpha\}, \qquad \alpha > 0.$$

Moreover, all the singularities of the functions f_ϑ for $|\operatorname{Im}\zeta| = \alpha$ are poles of the first order.

B. For some positive constant β and any $s \in (-\alpha, \alpha)$, the function $g_s(\vartheta) = u(\vartheta + i\omega s)$ is analytic in the closed complex domain $D_{\alpha-|s|}$, where

$$D_\nu = \{\vartheta \in \mathbb{C}^m : |\operatorname{Im}\vartheta_j| \leqslant \beta\nu, \ j = 1, \dots, m\}. \tag{2.1}$$

We call the vector $\omega \in \mathbb{R}^m$ *admissible* if for any nonzero vector $\tau \in \mathbb{Z}^m$ the following inequalities hold:

$$|\langle\omega, \tau\rangle| \geqslant \Gamma \exp(-\lambda \|\tau\|^{1/\sigma}), \qquad \sigma > 2. \tag{2.2}$$

Here Γ and λ are some positive constants, $\langle \cdot , \cdot \rangle$ is the standard scalar product, and

$$\|\tau\| = |\tau_1| + \cdots + |\tau_m|. \tag{2.3}$$

Evidently the measure of the unadmissible vectors $\omega \in \mathbb{R}^m$ is zero.

Let us introduce two matrices

$$e_+ = \begin{pmatrix} i & 1 \\ 1 & -i \end{pmatrix}, \qquad e_- = \begin{pmatrix} -i & 1 \\ 1 & i \end{pmatrix}. \tag{2.4}$$

We see that $e_+ = \overline{e}_-$, where the bar denotes complex conjugation. We set

$$u_0 = (2\pi)^{-m} \int_{\mathbb{T}^m} u(\vartheta)\, d\vartheta.$$

THEOREM. *Suppose that the potential u satisfies conditions* A *and* B, *and the frequency vector ω is admissible. Then for any positive constant c, $\varkappa$, and small values of ε there exists a real-analytic change of variables* (1.4) *which transforms equations* (1.2), (1.3) *to the following form*:

$$\begin{cases} \dot z = (\varepsilon^{-1} e_* + \varepsilon \widehat{U}(\vartheta))\, z, \\ \dot\vartheta = \omega, \end{cases} \tag{1.2}$$

where

$$\begin{aligned} \widehat{U} = -(u_0/2 + \varepsilon\widetilde{u})\, e_* &+ e^{-G_+(\vartheta) - 2(\alpha/\varepsilon - c)} u(\vartheta + i\omega(\alpha - \varepsilon c))\, e_+/4 \\ &+ e^{-G_-(\vartheta) - 2(\alpha/\varepsilon - c)} u(\vartheta - i\omega(\alpha - \varepsilon c))\, e_-/4, \end{aligned}$$

$\widetilde{u} = \widetilde{u}(c, \varepsilon, \varkappa)$ *does not depend on* ϑ *and* $|\widetilde{u}| < a_1$,

$$|G_\pm| < a_2 \varepsilon^{1-\varkappa}, \qquad G_+(\overline{\vartheta}) = \overline{G}_-(\vartheta)$$

for $\vartheta \in D_{\varepsilon c/2}$. *Here* a_1 *and* a_2 *are positive constants not depending on* ε.

REMARK 1. An unusual fact is that the estimates in the theorem do not depend on the arithmetic properties of the admissible frequency vector ω.

REMARK 2. One can see from the proof of the theorem that the matrix $B(\vartheta)$ in (1.4) has the form $I + O(\varepsilon)$ and depends smoothly on ε.

§3. Averaging method

The proof of the theorem is based on a certain averaging method which is described in this section. We construct the change of variables (1.4) as a solution of the differential equation

$$B'_\delta = B_\delta b_\delta , \qquad B_0 = I .$$

Here the prime denotes the derivative with respect to a real parameter $\delta \geqslant 0$; $b_\delta = b_\delta(\vartheta , \varepsilon)$ is some square matrix. In accordance with the theory of linear differential equations, the matrix B_δ is nondegenerate.

Let the change $y = B_\delta y_\delta$ transform system (1.2), (1.3) to the system

$$\begin{cases} \dot{y}_\delta = (\varepsilon^{-1} e_* + \varepsilon U_\delta(\vartheta))\, y_\delta , \\ \dot{\vartheta} = \omega . \end{cases}$$

Then the following equation holds:

$$\partial B_\delta = (\varepsilon^{-1} e_* + \varepsilon U)\, B_\delta - B_\delta(\varepsilon^{-1} e_* + \varepsilon U_\delta) , \tag{3.1}$$

where the differential operator ∂ is defined by (1.6). Let us differentiate relation (3.1) with respect to δ. We obtain the equation

$$\varepsilon U'_\delta = -\partial b_\delta - [b_\delta , \varepsilon^{-1} e_* + \varepsilon U_\delta] . \tag{3.2}$$

Here $[\cdot , \cdot]$ is the matrix commutator: $[A_1 , A_2] = A_1 A_2 - A_2 A_1$. We set $b_\delta = \varepsilon \xi U_\delta$, where ξ is some linear operator. Relation (3.2) takes the following form:

$$U' = -\partial \xi U - [\xi U , \varepsilon^{-1} e_* + \varepsilon U] . \tag{3.3}$$

Here and further we usually omit the subindices δ for the sake of brevity.

Let us define the operator ξ. First, we note that due to our constructions, tr U_δ is identically zero. The square complex matrices of order 2 with zero trace constitute the Lie algebra $sl(2, \mathbb{C})$. The matrices e_*, e_+, e_- (see (1.3), (2.1)) form a basis in it. We have the following relations:

$$[e_\pm , e_*] = \pm 2i e_\pm , \qquad [e_+ , e_-] = 4 e_* . \tag{3.4}$$

For each $\delta \geqslant 0$ we write

$$U = v_+ e_+ + v_- e_- + v_* e_* . \tag{3.5}$$

In particular, for $\delta = 0$,

$$v_+ = v_- = u/4 , \qquad v_* = -u/2 . \tag{3.6}$$

Due to the fact that the frequency vector ω is nonresonant, the periodic function

$$v_* = \sum_{\tau \in \mathbb{Z}^m} v_*^\tau e^{i\langle \tau , \vartheta \rangle}$$

is the sum of three terms:

$$v_* = v_0 + v_\oplus + v_\ominus \,, \tag{3.7}$$

where

$$v_0 = v_*^0 \,, \qquad v_\oplus = \sum_{\langle \tau , \omega \rangle > 0} v_*^\tau e^{i\langle \tau , \vartheta \rangle} \,, \qquad v_\ominus = \sum_{\langle \tau , \omega \rangle < 0} v_*^\tau e^{i\langle \tau , \vartheta \rangle} . \tag{3.8}$$

We combine relations (3.5) and (3.7) and get:

$$U = v_+ e_+ + v_- e_- + (v_0 e_0 + v_\oplus + v_\ominus) e_* \,. \tag{3.9}$$

Now we set $\Delta_* = \alpha/\varepsilon - c$ and define the operator ξ in the following way:

$$\begin{aligned} &\xi U = \xi_1 U \equiv -i\varepsilon (v_+ e_+ - v_- e_-) && \text{if } 0 \leqslant \delta \leqslant \Delta_* \,, \\ &\xi U = \xi_2 U \equiv -i(v_\oplus - v_\ominus) e_* && \text{if } \delta > \Delta_* \,. \end{aligned} \tag{3.10}$$

Let us explain the reasons of for a choice of ξ. Using the equalities (3.9), (3.10), and (3.4), we get:

$$\begin{aligned} -[\xi_1 U \,, \varepsilon^{-1} e_* + \varepsilon U] &= -(2 + 2\varepsilon^2 v_*)(v_+ e_+ + v_- e_-) - 8i\varepsilon^2 v_+ v_- e_* \,, \\ -[\xi_2 U \,, \varepsilon^{-1} e_* + \varepsilon U] &= 2\varepsilon (v_\oplus - v_\ominus)(v_+ e_+ - v_- e_-) \,. \end{aligned}$$

Thus equation (3.3) is equivalent to the system of the two following systems of first order partial differential equations

$$\begin{cases} v_+' = -(2 - i\varepsilon\partial - 2\varepsilon^2 v_*) v_+ \,, \\ v_-' = -(2 + i\varepsilon\partial + 2\varepsilon^2 v_*) v_- \,, \\ v_*' = 8i\varepsilon^2 v_+ v_- \,, \end{cases} \tag{3.11}$$

$$\begin{cases} v_+' = 2\varepsilon (v_\oplus - v_\ominus) v_+ \,, & v_-' = -2\varepsilon (v_\oplus - v_\ominus) v_- \,, \\ v_\oplus' = i\partial v_\oplus \,, & v_\ominus' = -i\partial v_\ominus \,, \\ v_0' = 0 \,, \end{cases} \tag{3.12}$$

where system (3.11) must be used for $0 \leqslant \delta \leqslant \Delta_*$ and system (3.12) must be used for $\delta > \Delta_*$.

Let us carry out an heuristic analysis of system (3.11)–(3.12). We keep the "main" terms in this system and omit the other ones. We get the following systems:

$$\begin{cases} v_+' = -(2 - i\varepsilon\partial) v_+ \,, \\ v_-' = -(2 + i\varepsilon\partial) v_- \,, \\ v_*' = 0 \,, \end{cases} \tag{3.11$'$}$$

$$\begin{cases} v_+' = 0 \,, & v_-' = 0 \,, \\ v_\oplus' = i\partial v_\oplus \,, & v_\ominus' = -i\partial v_\ominus \,, \\ v_0' = 0 \,. \end{cases} \tag{3.12$'$}$$

The solution of these equations with initial data (3.6) is

$$\begin{cases} v_\pm(\vartheta, \delta) = e^{-2\delta} u(\vartheta \pm i\varepsilon\omega\delta)/4, \\ \qquad v_* = -u/2, \end{cases} \qquad \text{if } 0 \leqslant \delta \leqslant \Delta_*,$$

$$\begin{cases} v_\pm(\vartheta, \delta) = e^{-2\Delta_*} u(\vartheta \pm i\omega\Delta_*), \\ v_\oplus(\vartheta, \delta) = -u_\oplus(\vartheta + i\omega(\delta - \Delta_*))/2, \\ v_\ominus(\vartheta, \delta) = -u_\ominus(\vartheta - i\omega(\delta - \Delta_*))/2, \\ \qquad v_0 = -u_0/2, \end{cases} \qquad \text{if } \delta > \Delta_*,$$

where u_0, $u_\oplus$, $u_\ominus$ are defined in the same way as v_0, $v_\oplus$, $v_\ominus$ in (3.8). We see that the functions $v_\oplus$ and $v_\ominus$ rapidly tend to zero as $\delta \to +\infty$, since all their Fourier coefficients vanish as $\delta \to +\infty$. The absolute values of the functions $e^{-2\delta}u(\vartheta \pm i\varepsilon\omega\delta)/4$ decrease while $\vartheta \pm i\varepsilon\omega\delta$ is not too close to a singularity of the potential u, therefore we use the operator ξ_1 on some subinterval of $[0, \alpha/\varepsilon]$. In spite of the fact that equations (3.11′)–(3.12′) have only heuristic value, their properties turn out to be close to those of the exact equations (3.11)–(3.12). In the following two sections we analyze system (3.11)–(3.12) and prove the theorem.

§4. Properties of the averaging procedure

The theorem is a direct consequence of the following three lemmas, which are proved in this section.

LEMMA 1. *If the solution $U_\delta(\vartheta)$ of equation* (3.3) *with the operator ξ defined by* (3.10) *and with the initial condition $U_0 = U$ (see* (1.3)*) exists in a vicinity V of $\mathbb{R}^m \subset \mathbb{C}^m = \{\vartheta\}$, then it is real analytic in V.*

PROOF. Recall that a function f of a complex variable ζ is called *real analytic* if it is holomorphic and $f(\overline{\zeta}) = \overline{f}(\zeta)$.

The function U_δ is holomorphic since it is a solution of the system of analytic differential equations (3.11). The relation $U_\delta(\overline{\vartheta}) = \overline{U}_\delta(\vartheta)$ is a consequence of the following two statements:

1. $U_0(\vartheta)$ is real analytic;
2. if $U_\delta(\vartheta)$ is real analytic, then so is $U'_\delta(\vartheta)$.

The first statement is an immediate consequence of our assumptions. Let us verify the second one. Since $U_\delta(\overline{\vartheta}) = \overline{U}_\delta(\vartheta)$, with the help of relations (3.7), (3.9) we get

$$v_\pm(\overline{\vartheta}) = \overline{v}_\mp(\vartheta), \quad v_*(\overline{\vartheta}) = \overline{v}_*(\vartheta), \quad v_\oplus(\overline{\vartheta}) = \overline{v}_\ominus(\vartheta), \quad v_\ominus(\overline{\vartheta}) = \overline{v}_\oplus(\vartheta).$$

Using the definition of the operators ξ_1 and ξ_2, we obtain

$$\xi_1 U(\overline{\vartheta}) = -i\varepsilon(\overline{v}_-(\vartheta)e_+ - \overline{v}_+(\vartheta)e_-) = \overline{\xi_1 U}(\vartheta),$$
$$\xi_2 U(\overline{\vartheta}) = -i(\overline{v}_\ominus(\vartheta) - \overline{v}_\oplus(\vartheta))e_* = \overline{\xi_2 U}(\vartheta).$$

Thus the right-hand side of (3.3) is real-analytic. □

LEMMA 2. *For any positive constant c the solution of equations* (3.11) *with initial conditions* (3.6) *for* $0 \leqslant \delta \leqslant \Delta_* - |s|$, $s \in \mathbb{R}$, $\vartheta \in D_{\alpha-\varepsilon\delta}$ *satisfies the following relations*

$$v_\pm(\vartheta, \delta) = e^{H_\pm(\vartheta, \delta) - 2\delta} u(\vartheta \pm i\omega\varepsilon\delta)/4, \tag{4.1}$$

$$|v_*(\vartheta + is, \delta)| \leqslant a_0(\alpha - |s|)^{-1-2\varkappa}, \tag{4.2}$$

$$|v_*(\vartheta, \delta) + u(\vartheta, \delta)/2| \leqslant a_1\varepsilon, \tag{4.3}$$

$$|H_\pm(\vartheta + is, \delta)| \leqslant a_2\varepsilon^{1-\varkappa}. \tag{4.4}$$

Moreover, the Fourier coefficients $v_*^\tau(\Delta_*)$, $\tau \in \mathbb{Z}^n$ *of the function* $v_*(\vartheta, \Delta_*)$ *satisfy the estimates*

$$|v_*^\tau(\Delta_*)| \leqslant \begin{cases} a_3 e^{-\beta\alpha\|\tau\|} & \text{if } \varepsilon\beta\|\tau\| \leqslant 1, \\ a_3\varepsilon^{-1} e^{-\varepsilon\beta c\|\tau\| - 2\alpha/\varepsilon} & \text{if } \varepsilon\beta\|\tau\| > 1. \end{cases} \tag{4.5}$$

PROOF. The estimates (4.2)–(4.4) are evidently true for $\delta = 0$. We set

$$\Delta = \sup\{\delta_0 \in [0, \alpha/\varepsilon - c] : \text{ inequalities (4.2)–(4.4) hold for any } 0 \leqslant \delta \leqslant \delta_0\}$$

and show that $\Delta = \alpha/\varepsilon - c$.

Indeed, suppose that $\Delta < \alpha/\varepsilon - c$. Then for $\delta = \Delta$ at least one of inequalities (4.2)–(4.4) becomes an equality. Now we prove that this conversion is impossible if the constants a_0, a_1, a_2 are chosen in an appropriate way.

i) We begin with relations (4.2) and (4.3). We have for $\delta = \Delta$:

$$v_*(\vartheta + is, \Delta) = -\frac{1}{2} u(\vartheta + is) + \int_0^\Delta 8i\varepsilon^2 (v_+ v_-)(\vartheta + is, \delta)\, d\delta. \tag{4.6}$$

The inequality $|u(\vartheta + is)/2| \leqslant C(\alpha - |s|)^{-1-\varkappa}$ follows from Lemma 5. Lemma 2 is true on the half-open interval $0 \leqslant \delta < \Delta$, consequently for $0 \leqslant \delta \leqslant \Delta - |s|$, $\vartheta \in D_{(\alpha - \varepsilon\Delta)/2}$, and small ε we have

$$|(v_+ v_-)(\vartheta + is, \delta)| \leqslant e^{-4\delta} C^2 (\alpha - |s| - \varepsilon\delta)^{-2-2\varkappa}/8.$$

Here we used relations (4.1), (4.4), and Lemma 5. Thus

$$\left| \int_0^\Delta 8i\varepsilon^2 (v_+ v_-)(\vartheta + is, \delta)\, d\delta \right| \leqslant \int_0^{\Delta_*} e^{-4\delta} \varepsilon^2 C^2 (\alpha - |s| - \varepsilon\delta)^{-2-2\varkappa}\, d\delta.$$

The last integral is less than $\mathrm{const}\, \varepsilon^2 (\alpha - |s|)^{-2-2\varkappa} < \mathrm{const}\, \varepsilon(\alpha - |s|)^{-1-2\varkappa}$ due to Lemma 6. Inequalities (4.2) and (4.3) are proved.

ii) Let us verify inequalities (4.4). Using the first and the second equations (3.11), we get

$$H_\pm(\vartheta + is, \Delta) = \int_0^\Delta 2\varepsilon^2 v_*(\vartheta + is\omega \pm i\varepsilon\omega(\Delta - \delta), \delta)\, d\delta.$$

Thus, using estimate (4.2) for $0 \leqslant \delta < \Delta$ and $\vartheta \in D_{\alpha-\varepsilon\delta}$, we obtain

$$|H_\pm(\vartheta + is\,,\,\Delta)| \leqslant \int_0^\Delta 2\varepsilon^2 a_0(\alpha - |s| - \varepsilon(\Delta-\delta))^{-1-2\varkappa}\,d\delta$$
$$\leqslant \varepsilon\varkappa^{-1}a_0(\alpha - |s| - \varepsilon\Delta)^{-2\varkappa} \leqslant \varepsilon^{1-2\varkappa}\varkappa^{-1}a_0 c^{-\varkappa}.$$

iii) In order to verify inequalities (4.5), we set $s = 0$ in (4.6). Then

$$v_*^\tau(\Delta) = -\frac{1}{2}\,u(\vartheta) + \int_0^\Delta 8i\varepsilon^2(v_+v_-)^\tau(\delta)\,d\delta\,, \qquad \tau \in \mathbb{Z}^n\,,$$

where u^τ and $(v_+v_-)^\tau(\delta)$ are the Fourier coefficients of the functions $u(\vartheta)$ and $(v_+v_-)(\vartheta\,,\,\delta)$. Since the function $u(\vartheta)$ is analytic in the closed domain D_α and $|u(\vartheta)| \leqslant C\alpha^{-1-\varkappa}$ in D_α (see Lemma 5), we have the estimate

$$|u^\tau| \leqslant C\alpha^{-1-\varkappa}e^{-\beta\alpha\|\tau\|}. \tag{4.7}$$

In a similar way, with the help of relations (4.1), (4.4), and Lemma 5, we get

$$|(v_+v_-)\tau(\delta)| \leqslant C(\alpha - \varepsilon\delta)^{-1-\varkappa}e^{-\beta(\alpha-\varepsilon\delta)\|\tau\|-2\delta}.$$

We see that

(4.8)
$$\left|\int_0^{\Delta_*} 8i\varepsilon^2(v_+v_-)^\tau(\delta)\,d\delta\right| \leqslant \int_0^{\Delta_*} 8i\varepsilon^2 C(\alpha-\varepsilon\delta)^{-1-\varkappa}e^{-\beta(\alpha-\varepsilon\delta)\|\tau\|-2\delta}\,d\delta$$
$$\leqslant \text{const}\, e^{-\alpha\beta\|\tau\|}\big(e^{(-2+\varepsilon\beta\|\tau\|)\Delta_*} - 1\big)(-2+\beta\varepsilon\|\tau\|)^{-1}.$$

In the case $-2 + \beta\varepsilon\|\tau\| = 0$, the right-hand side of this inequality has the form $\text{const}\, e^{-\alpha\beta\|\tau\|}\Delta_*$.

Inequalities (4.7)–(4.8) give estimate (4.5) in the case $\varepsilon\beta\,\|\tau\| \leqslant 1$. If $\varepsilon\beta\,\|\tau\| \geqslant 1$, then one can use the simple inequality

$$(e^{q\Delta} - 1)q^{-1} < \Delta e^{q\Delta} \text{ for all real } q \text{ and } \Delta,\ q \neq 0.$$

We set $q = -2 + \varepsilon\beta\,\|\tau\|$, $\Delta = \Delta_*$ and see that the right-hand side of inequality (4.7) does not exceed $\text{const}\,\alpha\varepsilon^{-1}e^{-2\alpha/\varepsilon-\varepsilon\beta c\|\tau\|}$. In the case $-2+\varepsilon\beta\|\tau\| = 0$, we have the same estimate. □

LEMMA 3. *For any $\vartheta \in D_{\varepsilon c/2}$ the solution of equation* (3.3), (3.10) *with the initial conditions* (3.6) *satisfies for $\delta > \Delta_*$ the following estimates*:

$$v_\pm(\vartheta\,,\,\delta) = e^{h_\pm(\vartheta\,,\,\delta)}v_\pm(\vartheta\,,\,\Delta_*)\,,$$
$$|v_\oplus(\vartheta\,,\,\delta)| \leqslant a_4\exp(-a_5\ln^\sigma(s+1)) + a_4'\varepsilon^{-m-1}\exp(-2\alpha/\varepsilon - \varepsilon a_5'\ln^\sigma(s+1))\,, \tag{4.9}$$
$$|v_\ominus(\vartheta\,,\,\delta)| \leqslant a_4\exp(-a_5\ln^\sigma(s+1)) + a_4'\varepsilon^{-m-1}\exp(-2\alpha/\varepsilon - \varepsilon a_5'\ln^\sigma(s+1))\,, \tag{4.10}$$
$$|h_\pm(\vartheta\,,\,\delta)| \leqslant a_6\varepsilon\,. \tag{4.11}$$

PROOF. i) Estimate (4.9) is derived from the formula

$$v_{\oplus}(\vartheta\,,\delta)=v_{\oplus}(\vartheta+i\omega(\delta-\Delta_*)\,,\Delta_*)\,,\qquad \delta\geqslant\Delta_*\,.$$

We use estimates (2.2), (4.5), and Lemma 4, and obtain the following inequality for $\vartheta\in D_{\varepsilon c/2}$:

$$|v_{\oplus}(\vartheta\,,\delta)|\leqslant c_0a_3(\beta\alpha)^{-m}\exp(-\beta\alpha(3\lambda^{\sigma})^{-1}\ln^{\sigma}(s+1)) \\ +c_0a_3\varepsilon^{-1}(\varepsilon\beta c)^{-m}\exp(-2\alpha/\varepsilon-\varepsilon\beta c\,(3\lambda^{\sigma})^{-1}\ln^{\sigma}(s+1))\,.$$

Estimate (4.10) is verified in the same way.

ii) Let us verify inequality (4.11). For $\Delta>\Delta_*$ and $\vartheta\in D_{\varepsilon c/2}$ we have

$$h_+(\vartheta\,,\Delta)=\int_{\Delta_*}^{\Delta}2\varepsilon(v_{\oplus}-v_{\ominus})(\vartheta\,,\delta)\,d\delta\leqslant I_1+I_2\,,$$

where

$$I_1=\int_{\Delta_*}^{\Delta}4\varepsilon a_4\exp(-a_5\ln^{\sigma}(\delta-\Delta_*+1))\,d\delta\,,$$

$$I_2=\int_{\Delta_*}^{\Delta}4a_4'\varepsilon^{-m-2}\exp(-\varepsilon a_5'\ln^{\sigma}(\delta-\Delta_*+1))\,d\delta\,.$$

The integral I_1 is less than const $\cdot\,\varepsilon$ since $\sigma>2$. Let us estimate I_2. We have

$$I_2\leqslant 4a_4'\varepsilon^{-m-2}e^{-2\alpha/\varepsilon}(I_2'+I_2'')\,,$$

where

$$I_2'=\int_0^M\exp(-\varepsilon a_5'\ln^{\sigma}(q+1))\,dq\quad\text{and}\quad I_2''=\int_M^{\infty}\exp(-\varepsilon a_5'\ln^{\sigma}(q+1))\,dq\,.$$

We set $M=\exp((a_5'\varepsilon/2)^{-1/(\sigma-1)})$ obtaining $I_1'\leqslant M$, and also

$$I_2'\leqslant\int_0^M(1+q)^{-\varepsilon a_5\ln^{\sigma-1}q}\,dq\leqslant\int_0^M(1+q)^{-2}\,dq<1\,.$$

Thus

$$I_2\leqslant 4a_4'\varepsilon^{-m-2}e^{-2\alpha/\varepsilon}(M+1)<\text{const}\cdot\varepsilon$$

for small ε since $\sigma>2$. □

§5. Three lemmas

This auxiliary section contains formulations and proofs of three lemmas used in the proof of the theorem.

LEMMA 4. *Let the vector ω be admissible, the function $f = f(\vartheta)$ be periodic in $\vartheta = (\vartheta_1, \dots, \vartheta_m)$, and the Fourier coefficients f_τ satisfy the inequalities*

$$|f_\tau| \leqslant c_f e^{-\beta\nu\|\tau\|}. \tag{5.1}$$

Then for any $\vartheta \in D_{\nu/2}$ the following estimates hold:

$$|f_\oplus(\vartheta + i\omega s)| \leqslant c_0 c_f (\beta\nu)^{-m} \exp(-\beta\nu(3\lambda^\sigma)^{-1}\ln^\sigma(s+1)), \tag{5.2}$$

$$|f_\ominus(\vartheta - i\omega s)| \leqslant c_0 c_f (\beta\nu)^{-m} \exp(-\beta\nu(3\lambda^\sigma)^{-1}\ln^\sigma(s+1)). \tag{5.3}$$

PROOF. We verify inequality (5.2) only (inequality (5.3) is verified in the same way). We have the relation

$$f_\oplus(\vartheta + i\omega s) = \sum_{\langle\tau,\omega\rangle>0} f_\tau e^{-\langle\tau,\omega\rangle s + i\langle\tau,\vartheta\rangle}. \tag{5.4}$$

For any $\vartheta \in D_{\nu/2}$ the inequality $\operatorname{Re}(i\langle\tau, \vartheta\rangle) \leqslant \beta\nu/2$ holds; consequently, using relations (5.3), (5.4), we obtain the estimate $|f_\oplus(\vartheta + i\omega s)| \leqslant \Sigma_1 + \Sigma_2$ for $\vartheta \in D_{\nu/2}$, where

$$\Sigma_1 = \sum_{\langle\tau,\omega\rangle\geqslant q} c_f e^{-\langle\tau,\omega\rangle s - \beta\nu\|\tau\|/2}, \qquad \Sigma_2 = \sum_{0<\langle\tau,\omega\rangle\leqslant q} c_f e^{-\langle\tau,\omega\rangle s - \beta\nu\|\tau\|/2}.$$

Here q is any positive constant.

Let us estimate the sum Σ_1:

$$\Sigma_1 \leqslant c_f e^{-qs} \sum_{\tau\in\mathbb{Z}^m} e^{-\beta\nu\|\tau\|/2} \leqslant c_f c_1 (\beta\nu)^{-m} e^{-qs},$$

where the constant c_1 depends only on m. The last inequality can be obtained if we estimate the sum with the help of the corresponding integral.

The sum Σ_2 is estimated by using the inequality

$$\|\tau\| > N = (-\lambda^{-1}\ln(q/\Gamma))^\sigma \text{ for any } \tau \in \mathbb{Z}^m \text{ such that } |\langle\tau, \omega\rangle| < q.$$

We have

$$\Sigma_2 \leqslant c_f \sum_{\|\tau\|>N} e^{-\beta\nu\|\tau\|/2} \leqslant c_f c_2 (\beta\nu)^{-m}(1 + (\beta\nu N)^{m-1}) e^{-\beta\nu N/2},$$

where the constant c_2 depends only on m. The last inequality is also obtained by estimating the sum with the help of an integral.

We set $q = \beta\nu(3\lambda^\sigma(s+1))^{-1}\ln^\sigma(s+1)$ and obtain the relation

$$\Sigma_1 + \Sigma_2 \leqslant c_0 c_f (\beta\nu)^{-m} \exp(-\beta\nu(3\lambda^\sigma)^{-1}\ln^\sigma(s+1))$$

with some positive constant c_0. $\square$

LEMMA 5. *Suppose the potential u satisfies conditions* A *and* B *from* §2. *Then for any positive* $\varkappa$, *any* $0 < \nu \leqslant \alpha$, $\vartheta \in D_\nu$, *and* $|s| \leqslant \alpha - \nu$ *there exists a positive constant* $C(\varkappa)$ *such that the following estimate holds*:

$$|u(\vartheta + i\omega s)| \leqslant C(\alpha - |s|)^{-1-\varkappa}.$$

PROOF. Let us consider the function

$$F(\vartheta\,,\,s) = \begin{cases} |(\alpha - |s|)^{1+\varkappa} u(\vartheta + i\omega s)| & \text{if } |s| < \alpha\,, \\ 0 & \text{if } |s| = \alpha\,. \end{cases}$$

It is well defined and continuous on the compact set

$$P = \{(\vartheta\,,\,s) : |s| \leqslant \alpha\,,\ \vartheta \in D_{\alpha - |s|}\}.$$

Indeed, it is defined and continuous on the set $P \setminus \{|s| = \alpha\}$ due to condition B. Due to condition A, the function $F(\vartheta\,,\,s)$ vanishes as $|s| \to \alpha$, $(s\,,\,\vartheta) \in P \setminus \{|s| = \alpha\}$. Lemma 4 is proved because any function continuous on a compact set is bounded.

LEMMA 6. *For any* $\Delta > 0$, $u > \varepsilon(c + \Delta)$, *and* $m > 0$ *there exists a constant* c'' *depending only on* c *and* m *such that the following inequality*

$$\int_0^\Delta e^{-4\delta}(u - \varepsilon\delta)^{-m}\, d\delta < c'' u^{-m}$$

holds.

PROOF. We set $f(\delta) = e^{-2\delta}(u - \varepsilon\delta)^{-m}$. Since

$$df/d\delta = (-2(u - \varepsilon\delta) + m\varepsilon)e^{-2\delta}(u - \varepsilon\delta)^{-m-1}\,,$$

the function $f(\delta)$ decreases for $\delta < \Delta_0 = u/\varepsilon - m/2$. Consider two cases:

a) $\Delta_0 \geqslant \Delta$. Then

$$\int_0^\Delta e^{-2\delta} f(\delta)\, d\delta \leqslant \int_0^{\Delta_0} e^{-2\delta} f(0)\, d\delta \leqslant \int_0^\infty e^{-2\delta} u^{-m}\, d\delta = u^{-m}/2\,.$$

b) $\Delta_0 < \Delta$. In this case

$$\int_0^\Delta e^{-2\delta} f(\delta)\, d\delta \leqslant \int_0^{\Delta_0} e^{-2\delta} f(\delta)\, d\delta + \int_{\Delta_0}^\Delta e^{-2\delta} f(\delta)\, d\delta\,.$$

The first integral in the right-hand side of this inequality was estimated in case a). The second integral I_2 is estimated with the help of the following inequalities:

$$\Delta - \Delta_0 = -u/\varepsilon + m/2 \leqslant c + m/2\,.$$

For $\delta \in [\Delta_0\,,\,\Delta]$ we have:

$$\begin{aligned} f(\delta) \leqslant e^{-2\Delta_0}(\varepsilon c)^{-m} &= (2c/m)^{-m} e^{-2\Delta_0}(\varepsilon m/2)^{-m} \\ &= (2c/m)^{-m} f(\Delta_0) < (2c/m)^{-m} f(0)\,. \end{aligned}$$

Thus $I_2 \leqslant (c + m/2)(2c/m)^{-m} u^{-m}$. □

This work was partially supported by the Russian Foundation of Basic Research and the International Science Foundation.

References

1. E. Dinaburg and Ya. Sinai, *The one-dimensional Schrödinger equation with a quasi-periodic potential*, Funktsional. Anal. i Prilozhen. **9** (1975), no. 4, 8–21; English transl., Funct. Anal. Appl. **9** (1975), no. 4, 279–289.
2. H. Rüssmann, *On the one-dimensional Schrödinger equation with a quasi-periodic potential*, Ann. New York Acad. Sci. **357** (1980), 90–107.
3. J. Moser and J. Pöschel, *An extension of a result by Dinaburg and Sinai on quasi-periodic potentials*, Comment. Math. Helv. **59** (1984), 39–85.
4. L. H. Eliasson, *Floquet solutions for the 1-dimensional quasi-periodic Schrödinger equation*, Comm. Math. Phys. **146** (1992), 447–482.
5. Ya. Sinai, *Structure of spectrum of a Schrödinger difference operator with almost periodic potential near the left boundary*, Funktsional. Anal. i Prilozhen. **19** (1985), no. 1, 42–48; English transl., Funct. Anal. Appl. **19** (1985), no. 1, 34–39.

Translated by THE AUTHOR

Department of Mathematics, Moscow State University, Moscow 119899, Russia

Amer. Math. Soc. Transl.
(2) Vol. 168, 1995

Various Aspects of n-Dimensional Rigid Body Dynamics

YU. N. FEDOROV AND V. V. KOZLOV

Before we start to consider the motion of a "multi-dimensional body," it is useful to answer a question usually ignored by mathematicians dealing with this subject. Namely, is there any sense in such an occupation? Concerning the physical aspect, the answer seems to be negative. It does not even matter that the space in which we live is definitely three-dimensional. The main reason is deeper. As early as at the end of the last century, Paul Erenfest, the prominent physicist of the time, noted that in the n-dimensional space rigid bodies and even matter itself, as we understand it, could not exist. Such an unexpected conclusion is based on the natural assumption of universality of the energy conservation law and its mathematical expression, the Gauss–Ostrogradskiĭ principle. It follows from the latter that in the case $n \neq 3$ any central force must vary with the distance from an attracting center according to a law which is different from the Newtonian inverse square law. By the known theorems of dynamics, this implies that stable stationary orbits in the classical and quantum two body problems—the simplest models of the atom—cannot exist.

However, one should not hurry with the conclusion that the n-dimensional top and similar objects are no more than toys for mathematicians. There are several examples when equations of multi-dimensional dynamics can describe real mechanical systems.

Besides, while various Lie group constructions allow one to "see" multi-dimensional systems, studying such systems gives a remarkable opportunity to understand the properties of abstract groups better. In the present article, we consider multi-dimensional systems as the main source of new interesting integrable problems.

§1. Momentum theorem

In classical mechanics the term "momentum" is usually associated with a linear momentum (impulse), or an angular momentum, or a momentum of a force (torque) relative to an axis or a point. At the same time, this term has a universal interpretation and can be related to various geometric objects playing a major role in the description of dynamical systems.

Let M^n be the configuration space of a mechanical system with n degrees of freedom. Suppose an arbitrary Lie group $\mathfrak{G}$ acts on M^n. Let g be the Lie algebra of $\mathfrak{G}$ and g^* the space of linear functions on g (the dual space). To each vector

1991 *Mathematics Subject Classification*. Primary 70E99.

$Y \in g$ there corresponds a one-parameter subgroup $\mathfrak{G}_Y^\alpha \subset \mathfrak{G}$, $\alpha \in \mathbb{R}$, whose action on M^n determines the tangent vector field

$$v_Y(x) = \frac{d}{d\alpha} \mathfrak{G}_Y^\alpha(x)\bigg|_{\alpha=0} . \tag{1.1}$$

The mapping $Y \to v_Y$ is a homomorphism of the Lie algebra g to the Lie algebra of all vector fields on M^n (if $Y \to v_Y$ and $Z \to v_Z$, then $v_{[Y,Z]} = [v_Y, v_Z]$).

Let $T: TM^n \to \mathbb{R}$ be the kinetic energy of the mechanical system, which defines the inertial properties of the system. Here TM^n is the tangent bundle of M^n. In the applications, T is usually a positive definite quadratic form. We may think of it as a Riemannian metric on M^n. Define the function

$$\mathcal{I}_{\mathfrak{G}}(x, \dot{x} \mid Y) = \left(\frac{\partial T}{\partial \dot{x}}, v_Y\right) = \sum_{i=1}^{n} \frac{\partial T}{\partial \dot{x}_i} (v_Y)_i . \tag{1.2}$$

Under changes of the local (generalized) coordinates $(x_1, \dots, x_n)$, the collection of the derivatives $\partial T/\partial \dot{x}_i$ transforms according to the covariant law. Therefore, the function $\mathcal{I}_{\mathfrak{G}}(x, \dot{x} \mid Y)$ does not depend on the choice of coordinates. Besides, it is a linear function in Y, so $\mathcal{I}_{\mathfrak{G}} \in g^*$.

The *kinetic momentum* of the mechanical system *relative to the Lie group* $\mathfrak{G}$ is, by definition, the mapping $\mathcal{I}_{\mathfrak{G}}: TM^n \to g^*$ that assigns to each point $(x, \dot{x})$ of the phase space TM^n the linear function $\mathcal{I}_{\mathfrak{G}}(Y)$ on the algebra g. Let $y_1, \dots, y_n$ be a basis in g and $f^1, \dots, f^n$ be the dual basis in g^* with respect to the bilinear form $\langle \cdot, \cdot \rangle$ on $g \times g^*$ such that $\langle y_i, f^j \rangle = \delta_i^j$. Then the function $\mathcal{I}_{\mathfrak{G}}(Y)$ is uniquely defined by the vector $K = K_1 f^1 + \dots + K_n f^n \in g^*$ such that

$$\mathcal{I}_{\mathfrak{G}}(Y) = \langle Y, K \rangle_g . \tag{1.3}$$

The vector K is also called the kinetic momentum[1].

EXAMPLE 1. Let the mechanical system be a set of mass points $m_1, \dots, m_N$ with the position vectors $r_1, \dots, r_N \in \mathbb{R}^3$. Consider the action of the Lie group $E(3)$ on the configuration space $M^n = \{r_1, \dots, r_N\}$, $n = 3N$. Recall that $E(3)$ is the group of all rigid motions of three-dimensional Euclidean space $\mathbb{R}^3$, i.e., compositions of rotations and translations of this space:

$$r_k \to R r_k + s, \qquad k = 1, \dots, N.$$

Here $R: \mathbb{R}^3 \to \mathbb{R}^3$ is an orthogonal rotation matrix, $s \in \mathbb{R}^3$ is a translation vector.

The action of a general one-parameter subgroup $\mathfrak{R}(\alpha)$ of $E(3)$ is realized as a helical motion of the space $\mathbb{R}^3$, i.e., it is the rotation about a fixed axis l through the angle αw ($w = \text{const}$) followed by the translation $s = \alpha v_l$ along the same axis:

$$r_k \to R_l(\alpha)(r_k - d) + d + \alpha v_l, \qquad k = 1, \dots, N.$$

[1] In the sequel, instead of saying "the kinetic momentum relative to the group $SO(3)$ (or even the group $SO(n)$)", we shall use the classical term "angular momentum".

Here $l = \{e_l t + d \mid t \in \mathbb{R}\}$, where e_l is the unit vector along this axis, $d \in \mathbb{R}^3$ is a constant translation vector, v_l is a velocity vector, α is the parameter of the subgroup. Using homogeneous coordinates in $\mathbb{R}^3$, we can write

$$r_k \to \mathfrak{R}(\alpha) r_k ,$$

$$\mathfrak{R}_l(\alpha) = \begin{pmatrix} R_l(\alpha) & d - R_l(\alpha) d + \alpha v_l \\ 0 & 1 \end{pmatrix}, \qquad r_k = (r_{k1}, r_{k2}, r_{k3}, 1)^T .$$

The element of the Lie algebra $e(3)$ generating the given one-parameter subgroup is represented by the *screw velocity* matrix

$$\mathfrak{W} = \frac{\partial}{\partial \alpha} \mathfrak{R}\mathfrak{R}^{-1} = \begin{pmatrix} \Omega & v_l - \Omega d \\ 0 & 0 \end{pmatrix}, \qquad \Omega = \frac{\partial}{\partial \alpha} R(\alpha) R^{-1}(\alpha), \tag{1.4}$$

where Ω is a skew-symmetric 3×3 matrix (an element of the Lie algebra $so(3)$): $\Omega_{\alpha\beta} = \varepsilon_{\alpha\beta\gamma}\omega_\gamma$, $\alpha, \beta, \gamma = 1, 2, 3$, and $\omega = w e_l = (\omega_1, \omega_2, \omega_3)^T$ is the angular velocity vector directed along the axis l.

According to (1.1), on $M^n = \{r_1, \dots, r_N\}$ this matrix generates the vector field

$$v_{\mathfrak{W}} = \{\omega \times (r_1 - d) + v_l, \dots, \omega \times (r_N - d) + v_l\} = \{\mathfrak{W} r_1, \dots, \mathfrak{W} r_N\} .$$

Using the expression for the kinetic energy

$$T = \frac{1}{2} \sum_{k=1}^{N} m_k (\dot r_k, \dot r_k)$$

and the definition (1.2), we obtain the following linear function on $e(3)$:

(1.5)

$$\mathcal{I}_{E(3)}(r, \dot r \mid \mathfrak{W}) = \sum_{k=1}^{N} m_k (\dot r_k, \omega \times (r_k - d) + v_l) = (\omega, M_d) + (v_l, p),$$

$$M_d = \sum_{k=1}^{N} (r_k - d) \times m_k \dot r_k , \qquad p = \sum_{k=1}^{N} m_k \dot r_k .$$

We see that M_d is the angular momentum of the system relative to the origin displaced at the vector d, whereas p is the total momentum of this system.

Now for any two matrices

$$\mathfrak{W}_1 = \begin{pmatrix} \Omega_1 & V_1 \\ 0 & 0 \end{pmatrix}, \quad \mathfrak{W}_2 = \begin{pmatrix} \Omega_2 & V_2 \\ 0 & 0 \end{pmatrix}, \qquad \Omega_1, \Omega_2 \in so(3), \ V_1, V_2 \in \mathbb{R}^3,$$

we define the Euclidean bilinear form $\langle \cdot , \cdot \rangle$ on $e(3) \times e^*(3)$ as follows

$$\langle \mathfrak{W}_1, \mathfrak{W}_2 \rangle = -\frac{1}{2} \operatorname{tr}(\Omega_1 \Omega_2) + (V_1, V_2) . \tag{1.6}$$

The first term at the right-hand side is the Killing form of the Lie algebra $so(3)$, i.e., the scalar product invariant with respect to the adjoint action of the rotation group $SO(3)$. Comparing the expressions (1.6) and (1.5), one can rewrite the latter in the form

$$\mathcal{I}_{E(3)}(\mathfrak{W}) = \langle \mathfrak{W}, K \rangle,$$

(1.7)
$$K = \begin{pmatrix} \widehat{M}_d & p \\ 0 & 0 \end{pmatrix} \in e^*(3), \qquad (\widehat{M}_d)_{\alpha\beta} = \varepsilon_{\alpha\beta\gamma}(M_d)_\gamma .$$

According to definition (1.3), the matrix K represents the kinetic momentum of the system relative to the group $E(3)$.

In the special case $s = 0$, the group $E(3)$ is reduced to the group $SO(3)$, the corresponding one-parameter subgroup $R(\alpha)$ is generated by the angular velocity matrix Ω, and from (1.5) we obtain

$$\mathcal{I}_{SO(3)}(\Omega) = (\omega, M_0) = -\frac{1}{2}\operatorname{tr}(\Omega\widehat{M}_0), \qquad (\widehat{M}_0)_{\alpha\beta} = \varepsilon_{\alpha\beta\gamma}(M_0)_\gamma ,$$

i.e., the kinetic momentum relative to the group $SO(3)$ is represented by the 3×3 matrix $\widehat{M}_0 \in so^*(3)$, which is isomorphic to the vector M_0 of the classical angular momentum relative to the origin O.

The action of the group $\mathfrak{G}$ on M^n may depend on time t. An example is the group of rotations of Euclidean space about a moving axis. In this case the vector field (1.1) depends on t, which should be regarded as a parameter.

Note that the action of $\mathfrak{G}$ on M^n extends naturally to an action on TM^n. Therefore, one can speak about invariants $f : TM^n \to \mathbb{R}$ of the action of the group $\mathfrak{G}$: these are the functions that are constant on the orbits of the action of this group on TM^n. The condition for a function $f(x, \dot{x})$ to be invariant has the form

(1.8)
$$\frac{d}{d\alpha} f(\mathfrak{G}_Y^\alpha(x, \dot{x}))\Big|_{\alpha=0} = \left(\frac{\partial f}{\partial \dot{x}}, \dot{v}_Y\right) + \left(\frac{\partial f}{\partial x}, v_Y\right) \equiv 0 \quad \text{for all } Y \in g.$$

If $f = T$ is the kinetic energy of a mechanical system and $\partial T/\partial x = 0$, then, in view of (1.2), (1.3), condition (1.8) can be represented in the form

(1.9)
$$\langle \dot{Y}, K \rangle = 0 \quad \text{for all } Y \in g.$$

For example, let $\mathfrak{G}$ be the group S_l of translations of the space $\mathbb{R}^3$ along the moving axis l with unit vector $e(t)$. The group S_l can be regarded as a one-parameter subgroup of the translation group $\mathbb{R}^3$. Put $g = \mathbb{R}^3$ in (1.9). Then the invariance condition for the function $T(\dot{x})$ takes the simple form

(1.10)
$$(p, \dot{e}) = 0 .$$

A less trivial example arises when $\mathfrak{G}$ is the group $R_l = SO(2)$ of rotations of $\mathbb{R}^3$ about the same moving axis that passes through a moving point with the position vector $s(t)$. Similarly, the group R_l can be considered as a one-parameter subgroup of the group $E(3)$ generated by the element $\mathfrak{W}_l$ of the algebra $e(3)$ (compare with (1.4)):

$$\mathfrak{W}_l = \begin{pmatrix} \Omega_l & -\Omega_l s \\ 0 & 0 \end{pmatrix}, \qquad \Omega_{\alpha\beta} = \varepsilon_{\alpha\beta\gamma}\omega_\gamma . \tag{1.11}$$

Here $\omega_l = we(t)$ is the angular velocity of rotation about the axis l. According to (1.9), the invariance condition for the kinetic energy is $\langle\dot{\mathfrak{W}}_l, K\rangle_{e(3)} = 0$, where K is the kinetic momentum relative to the group $E(3)$ defined in (1.7). In the vector form, this condition transforms to

$$(p, (s \times e)^{\cdot}) + (M_O, \dot{e}) = 0. \tag{1.12}$$

Here M_O is the angular momentum of the system relative to the origin O. Note that, as one may expect, the obtained condition depends only on the Plücker coordinates $(s \times e, e)$ of the axis l, but does not depend on the coordinates of the position vector s itself. In particular, if the axis l does not change its direction in space, then (1.12) takes the form

$$(e, \dot{s} \times \dot{r}_C) = 0, \tag{1.13}$$

where $r_C = \Sigma m_k r_k / \Sigma m_k$ is the position vector of the mass center C. This condition was obtained by Chaplygin in connection with a generalization of the angular momentum theorem (the area theorem) in certain mechanical problems (1897, [7]). It is obviously fulfilled provided the axis l always passes through the mass center.

Now consider the general situation, when the following constraints are imposed on a mechanical system:

$$f_1(x, \dot{x}) = \cdots = f_m(x, \dot{x}) = 0. \tag{1.14}$$

In applications, the functions f_i are usually linear in the velocities $\dot{x}$. Assume that the gradients $\partial f_1/\partial \dot{x}, \ldots, \partial f_m/\partial \dot{x}$ are linearly independent. If equations (1.14) can be reduced to the form

$$\dot{h}_1(x) = \cdots = \dot{h}_m(x) = 0, \tag{1.15}$$

then the constraints and the mechanical system itself are called *holonomic* (following H. Herz).

A simple example of a nonholonomic system is a ball rolling without sliding on a horizontal plane (the velocity of the contact point equals zero): without violating the constraints, the ball can be transformed from any configuration to any other one.

The action of the group $\mathfrak{G}$ is said to be *consistent with the constraints* (1.14) if

$$\left(\frac{\partial f_i}{\partial \dot{x}}, v_Y\right) = 0, \qquad \text{for all } Y \in g, i = 1, \ldots, m. \tag{1.16}$$

In the case of integrable constraints (1.15), this is equivalent to the invariance of the functions $h_1, \ldots, h_m$ with respect to the action of $\mathfrak{G}$ on M^n. The tangent vectors v_Y satisfying condition (1.14) are called *virtual velocities*.

The equations of motion of nonholonomic systems can be written in the form of Lagrange equations with *multipliers*

$$\left(\frac{\partial T}{\partial \dot{x}}\right)^{\cdot} - \frac{\partial T}{\partial x} = F + \sum_{i=1}^{m} \lambda_i \frac{\partial f_i}{\partial \dot{x}_i}, \tag{1.17}$$

where $F = (F_1, \dots, F_n)$ are *generalized forces*. These equations, together with the constraint equations (1.14), constitute a closed system for determining the generalized coordinates x and the multipliers λ_i as functions of time.

As well as the gradients $\partial T/\partial \dot{x}$, the generalized force F is a covector. Hence, by analogy with (1.2) and (1.3), it is natural to define the mapping

$$\Phi_{\mathfrak{G}}(x, \dot{x} \mid Y) = \sum_{i=1}^{n} F_i (v_Y)_i = \langle \mathfrak{F}, Y \rangle.$$

This is the *torque relative to the group* $\mathfrak{G}$. The same name is given to the vector $\mathfrak{F} \in g^*$ that uniquely defines the linear function $\Phi_{\mathfrak{G}}(x, \dot{x} \mid Y)$.

THEOREM 1. *Suppose that the kinetic energy of the mechanical system is an invariant of the group* $\mathfrak{G}$, *and the action of this group is consistent with the constraints. Then for any* $Y \in g$ *the following relation holds*

$$\frac{d}{dt} \mathcal{I}_{\mathfrak{G}}(x, \dot{x} \mid Y) = \Phi_{\mathfrak{G}}(x, \dot{x} \mid Y), \tag{1.18}$$

which, in view of (1.16), (1.18), *implies the equation in* g^*

$$\frac{d}{dt} K = \mathfrak{F}. \tag{1.19}$$

This theorem goes back to Lagrange and Jacobi, who noticed the connection between conservation of momentum and angular momentum, and the groups of translation and rotation respectively. If the forces are potential ($F = -\partial V/\partial x$) and the potential energy $V\colon M^n \to \mathbb{R}$ is invariant with respect to the action of the group $\mathfrak{G}$, then (1.18) yields the first integral $\mathcal{I}_{\mathfrak{G}}(x, \dot{x}) = \text{const}$. This is a classical result of E. Noether (1918).

PROOF OF THEOREM 1. According to (1.16) and (1.17), we have

$$\left(\frac{\partial T}{\partial \dot{x}}\right)^{\cdot} v_Y - \frac{\partial T}{\partial x} v_Y = F v_Y,$$

or, in another form,

$$\left(\frac{\partial T}{\partial \dot{x}} v_Y\right)^{\cdot} - \left[\frac{\partial T}{\partial \dot{x}} \dot{v}_Y + \frac{\partial T}{\partial x} v_Y\right] = F v_Y.$$

The expression in the square brackets vanishes by the invariance property of the function T.

COROLLARY. *Suppose that condition* (1.10) *is fulfilled and the vectors* $v_e = \{e, \dots, e\}$ *are virtual velocities on the configuration space* $M^n = \{r_1, \dots, r_N\}$, *i.e., the constraints admit infinitesimal translations of the point mass system as a rigid body along the axis* l. *Then, by* (1.18),

$$(p, e)^{\cdot} = (F, e).$$

In a similar way, if condition (1.12) *is fulfilled and the constraints admit infinitesimal rotations of the system as a rigid body about the (moving) axis* l, *then the following angular momentum theorem holds*

$$\dot{M}_l = (\mathcal{M}_F, e),$$

$$M_l = \left(\sum_{k=1}^{N}(r_k - d) \times m_k \dot{r}_k, e\right), \qquad \mathcal{M}_F = \sum_{k=1}^{N}(r_k - d) \times F_k. \tag{1.20}$$

EXAMPLE 2. As a remarkable application of Theorem 1, we consider Chaplygin's problem on a dynamically nonsymmetric ball rolling without sliding on a horizontal plane H (1903, [7]). The mass center C of the ball is assumed to coincide with its geometric center. Here the configuration space is the group $E(3)$, i.e., the set of the matrices

$$R = \begin{pmatrix} R & r_C \\ 0 & 1 \end{pmatrix},$$

where $R \in SO(3)$ is the orthogonal rotation matrix of the ball, r_C is the position vector of the mass center C with the initial point at the origin O of a fixed frame (we assume that O belongs to the plane H). Let P be the point of contact of the ball with the plane H. Then the condition for rolling without sliding has the form

$$v_P \equiv \dot{r}_C - \omega \times \rho\gamma = 0, \tag{1.21}$$

where ω, ρ are the angular velocity and the radius of the ball, γ is the vertical unit vector.

Converted to scalar form, condition (1.21) gives three nonholonomic constraints (in addition to one holonomic constraint $(r_C, \gamma) = \rho$).

PROPOSITION 2. *The constraints* (1.21) *admit infinitesimal rotations of the ball around the contact point* P.

The proof is obvious: under any such rotations one has $v_P = 0$, which coincides with the condition (1.21).

However, it is useful to consider also the following formal proof relying directly on the invariance condition (1.28). Since the constraints (1.21) depend neither on the orientation of the ball nor on its position on the plane H, it is sufficient to verify this invariance condition for only one configuration (i.e., at a certain point on the group $E(3)$). Assume that this configuration (point) is the identity of the group. Picking the expressions for the screw velocity (1.4) and the bilinear form (1.6), then taking into account the relation $r_P = r_C - \rho\gamma$, one can represent the constraint equations (1.21) in the following way

$$\langle N_\alpha, \mathfrak{W}\rangle_{e(3)} = 0, \qquad \alpha = 1, 2, 3,$$

$$\mathfrak{W} = \dot{R}R^{-1} = \begin{pmatrix} \Omega & \dot{r}_c - \Omega r_c \\ 0 & 0 \end{pmatrix} \in e(3), \qquad N_\alpha \in e^*(3),$$

$$N_1 = \begin{pmatrix} 0 & R_2 & R_3 & 1 \\ -R_2 & 0 & 0 & 0 \\ -R_3 & 0 & 0 & 0 \\ 0 & 0 & 0 & 0 \end{pmatrix}, \quad N_2 = \begin{pmatrix} 0 & -R_1 & 0 & 0 \\ R_1 & 0 & R_3 & 1 \\ 0 & -R_3 & 0 & 0 \\ 0 & 0 & 0 & 0 \end{pmatrix},$$

$$N_3 = \begin{pmatrix} 0 & 0 & -R_1 & 0 \\ 0 & 0 & -R_2 & 0 \\ R_1 & R_2 & 0 & 1 \\ 0 & 0 & 0 & 0 \end{pmatrix},$$

where R_1, R_2, R_3 are the projections of the position vector r_P to the fixed axes, and the form $\langle \cdot , \cdot \rangle_{e(3)}$ is specified in (1.6).

Using the terminology introduced above, one may call the matrices N_α the momenta of constraints (1.21) relative to the action of the group $E(3)$. These stand for the covectors $\partial f_i/\partial \dot{x}$ in (1.16). According to (1.11), the velocity vector $v_{\mathfrak{W}}$ in $e(3) = T_e E(3)$ induced by infinitesimal rotations around the point P is represented by matrices of the form

$$\mathfrak{W}_l = \begin{pmatrix} \Omega_l & -\Omega_l r_P \\ 0 & 0 \end{pmatrix} \in e(3)$$

(the axis l passes through P). It follows that

$$\langle N_\alpha, \mathfrak{W}_l\rangle_{e(3)} = 0 \qquad \text{for any } l,$$

i.e., condition (1.16) is satisfied on $T_e E(3)$ and, as a consequence, on the whole tangent bundle $TE(3)$.

Besides, for any axes l passing through the contact point P, Chaplygin's condition (1.13) is also satisfied. Since the torque of the constraint reaction at P relative to P is zero, Theorem 1 and equations (1.20) imply that the angular momentum of the ball K_P relative to the same point turns out to be a fixed vector in space.

Then, according to a well-known theorem of mechanics, we may write

$$K_P = K_C + \rho\gamma \times p, \qquad p = mv_C, \tag{1.22}$$

where m is the mass of the ball, K_C is its angular momentum relative to the mass center C.

In the moving axes attached to the ball, the momentum K_P satisfies the equations

$$\dot{K}_P + \omega \times K_P = 0, \tag{1.23}$$

where, in view of (1.21), (1.22), one can put

$$K_P = I\omega + D\gamma \times (\omega \times \gamma), \qquad D = m\rho^2,$$

I being the inertia tensor of the ball relative to the C. Then equations (1.23), together with the Poisson equations, form a closed system for determining $\omega(t)$ and $\gamma(t)$:

$$\begin{gathered}\Lambda\dot{\omega} = \Lambda\omega \times \omega + (D/\varphi)((\Lambda\omega \times \omega)\Lambda^{-1})\gamma\,, \\ \dot{\gamma} = \gamma \times \omega\,, \qquad \varphi = 1 - D(\gamma\,,\,\Lambda^{-1}\gamma)\,.\end{gathered} \tag{1.24}$$

From (1.23) and (1.24) it follows immediately that the system has three independent geometric integrals

$$\begin{gathered}(K\,,\,K) = (\Lambda\omega\,,\,\Lambda\omega) - 2D(\Lambda\omega\,,\,\gamma) + D^2(\omega\,,\,\gamma)^2\,, \\ (K\,,\,\gamma) = (I\omega\,,\,\gamma)\,, \qquad (\gamma\,,\,\gamma) = 1\,,\end{gathered} \tag{1.25}$$

as well as the kinetic energy integral

$$(K\,,\,\omega) = (\Lambda\omega\,,\,\omega) - D^2(\omega\,,\,\gamma)^2. \tag{1.26}$$

Besides, as shown by Chaplygin [7], in the phase space $(\omega\,,\,\gamma)$ the system possesses an integral invariant with density

$$\mu = \sqrt{\varphi}\,. \tag{1.27}$$

Therefore, by the well-known Jacobi theorem on the last multiplier, equations (1.24) are integrable by quadratures.

After Chaplygin, various integrable mechanical generalizations of the problem were found in [8, 13, 17] (see also Example 3 for a multi-dimensional generalization).

§2. Multi-dimensional dynamics

Now we proceed to the generalized Euler problem concerning the free motion of an n-dimensional rigid body around the fixed point in $\mathbb{R}^n$. This problem can be regarded as a classical one, since the idea of such a generalization was stated by A. Cayley as early as in the middle of the 19th century [6]. At present there exists a great number of publications devoted to this problem. Nevertheless, using the momentum theorem, here we give a detailed derivation of the equations of motion in order to set the background for constructing more complicated multi-dimensional systems.

What can be taken for the configuration manifold of the n-dimensional rigid body? The answer to this question is not unique. One may think of this manifold as a set of all positions of the body. It is known to be the group $SO(n)$, i.e., the group of orthogonal rotation matrices $R(t)$ such that

$$r_t = R(t)\rho\,, \qquad r_t\,,\,\rho \in \mathbb{R}^n\,, \tag{2.1}$$

where ρ is the position vector of some fixed point of the body relative to some frame $\mathcal{B}$ attached to the body (i.e., $\rho = \text{const}$), while r_t represents the same vector relative to some fixed frame. The components of the matrix R stand for the redundant generalized coordinates on $SO(n)$. In this case there are no constraints imposed.

At the same time, the configuration manifold can be defined as a set $\mathcal{R}$ of the position vectors of all points of the body

$$\mathcal{R} = \{r_1, r_2, \dots, r_N\} = \mathbb{R}^n \times \mathbb{R}^n \times \cdots \times \mathbb{R}^n.$$

It is obvious that this manifold is actually infinite-dimensional and the generalized coordinates $r_1, r_2, \dots, r_N$ are strongly redundant in view of a great number of constraints fixing distances between all points of the body, as well as their distances from a fixed point O. These constraints are obviously invariant with respect to the action of the group $SO(n)$.

In the sequel we shall operate with both configuration varieties.

Differentiating (2.1), we can write

$$\dot{r}_t = \Omega_s r_t, \qquad \Omega_s = \dot{R}R^{-1}.$$

As in the three-dimensional case, the matrix Ω_s is called the angular velocity of the body relative to the fixed frame (the spatial angular velocity). This matrix turns out to be skew symmetric and, thereby, it is a vector in the Lie algebra $so(n)$.

Now let v be the velocity vector of a fixed point taken in the moving frame $\mathcal{B}$: $v = R^{-1}\dot{r}_t$. Then

$$v = \Omega_c \rho, \qquad \Omega_c = R^{-1}\dot{R}, \tag{2.2}$$

where $\Omega_c \in so(n)$ is the angular velocity matrix in the frame attached to the body (the body angular velocity). It is said that Ω_c and Ω_s arise as a result of left and right displacements (multiplications by R^{-1}) of the tangent vector $\dot{R} \in T_R SO(3)$ to the algebra $so(n) = T_E SO(n)$ regarded as the tangent linear space at the identity E of the group. Conversely, on the group regarded as a manifold, any matrix Ω generates the left-invariant (right-invariant) tangent vector field $v(R) = R\Omega \in T_R SO(n)$ ($v(R) = \Omega R \in T_R SO(n)$), where R runs through the group $SO(n)$.

From (2.1) and (2.2) we find that the angular velocities in the space and in the body are related as follows

$$\Omega_s = R\Omega_c R^{-1}, \tag{2.3}$$

or, in the vector form,

$$\Omega_s = \mathrm{Ad}_R\, \Omega_c, \qquad \Omega_s, \Omega_c \in \mathbb{R}^m,$$
$$\mathrm{Ad}_R \colon \mathbb{R}^m \to \mathbb{R}^m, \qquad m = \dim(so(n)) = n(n-1)/2,$$

where Ad_R is the operator of the adjoint action of the group $SO(n)$ on the algebra $so(n)$.

Using (2.2), we write the kinetic energy of the n-dimensional body in the form

$$T = \frac{1}{2}\sum_{i=1}^{N} m_i(\dot{r}_i, \dot{r}_i) = \frac{1}{2}\sum_{i=1}^{N} m_i(\Omega_c\rho_i, \Omega_c\rho_i), \qquad \rho = \mathrm{const}, \tag{2.4}$$

where $(\cdot\,,\,\cdot)$ is the Euclidean scalar product in $\mathbb{R}^n$.

Now consider the *right* action of the group $SO(n)$ on itself: $R \to RG$ (G runs through $SO(n)$). A one-parameter subgroup of this action is $R\exp(\Omega_c\alpha)$, $\alpha \in \mathbb{R}$, where Ω_c is an arbitrary fixed vector in the algebra. According to (1.1), on the group regarded as a configuration manifold, each subgroup generates the *left-invariant* vector field

$$\frac{d}{d\alpha} R\exp(\Omega_c\alpha)\bigg|_{\alpha=0} = R\Omega_c\,.$$

On the configuration manifold $\mathcal{R}$ specified above, the same subgroup generates the vector field $\{R\Omega_c\rho_1\,,\,\dots\,,\,R\Omega_c\rho_N\}$. Then, by the general formula (1.3), the kinetic (angular) momentum of the body relative to the *right* action of the group $SO(n)$ is defined as follows

$$\mathcal{I}_{SO(n)} = \sum_{i=1}^N \left(\frac{\partial T}{\partial \dot r_i}\,,\, R\Omega_c\rho_i\right) = \sum_{i=1}^N m_i(v_i\,,\,\Omega_c\rho_i) = \langle\Omega_c\,,\,M_c\rangle\,,$$

$$M_c = \sum_i^N m_i(v_i r_i^T - r_i v_i^T) \in so^*(n)\,.$$

Here the bilinear form $\langle\cdot\,,\,\cdot\rangle$ on $so(n)\times so^*(n)$ is also the Killing form of the algebra $so(n)$:

$$\langle\Omega_1\,,\,\Omega_2\rangle = -\frac{1}{2}\operatorname{tr}(\operatorname{ad}_{\Omega_1}\operatorname{ad}_{\Omega_2}) \equiv -\frac{1}{2}\operatorname{tr}(\Omega_1\Omega_2) \equiv \sum_{i<j}^n(\Omega_1)_{ij}(\Omega_2)_{ij}$$

for all $\Omega_1\,,\,\Omega_2 \in so(n)$. Since $v_i^T = \rho_i^T\Omega_c^T = -\rho_i^T\Omega_c$, we obtain

$$M_c = I_c\Omega_c + \Omega_c I_c\,, \qquad I_c = \sum_i^N m_i\rho_i\rho_i^T = \text{const}\,, \tag{2.5}$$

or, in vector form, $M_c = \mathcal{A}_I\Omega_c\,,\ \mathcal{A}_I\colon \mathbb{R}^m \to \mathbb{R}^m$.

The skew-symmetric matrix M_c and the symmetric matrix I_c are called the *angular momentum of the body* and its *mass tensor* in the *moving frame* respectively. Then it is natural to call the symmetric operator $\mathcal{A}_I$ the *inertia tensor* of the n-dimensional rigid body. Similarly to I_c, it is a constant tensor. In certain axes of the moving frame appropriately chosen tensors I_c and $\mathcal{A}_I$ transform simultaneously to diagonal form:

$$I_c = \operatorname{diag}(I_1\,,\,\dots\,,\,I_n)\,, \qquad (M_c)_{ij} = (I_i + I_j)(\Omega_c)_{ij}\,.$$

In a similar way, the one-parameter subgroups $\exp(\Omega_s\alpha)R$, $\alpha \in \mathbb{R}$, $\Omega_s \in so(n)$ of the *left* action $R \to GR$ of $SO(n)$ generates on the manifold $\mathcal{R}$ the vector field $\Omega_s r_i$. To this field there corresponds the kinetic momentum

$$\mathcal{I}_{SO(n)} = \sum_i^N \left(\frac{\partial T}{\partial \dot r_i}\,,\, R\Omega_s r_i\right) = \sum_i^N m_i(\dot r_i\,,\,\Omega_s r_i) = \langle\Omega_s\,,\,M_s\rangle_{so(n)}\,,$$

$$M_s = I_s\Omega_s + \Omega_s I_s \in so^*(n)\,, \qquad M_s = \mathcal{A}_I\Omega_s\,.$$

Here M_s, I_s, and $\mathcal{A}_I$ represent the angular momentum, the mass tensor, and the inertia tensor of the body in the fixed frame (in the space).

Since the Killing form on $so(n)$ is invariant with respect to Ad-transformations, M_s and M_c are transformed just as the angular velocities:

$$M_s = RM_cR^{-1} \quad \text{or} \quad M_s = \mathrm{Ad}_R\, M_c\,. \tag{2.6}$$

This enables us to regard Ω_c and M_c as vectors in the same space $so(n) = \mathbb{R}^m$. The operator $\mathcal{A}_I$ defines on $so(n)$ the nondegenerate scalar product $\langle\ ,\ \rangle_{\mathcal{A}}$:

$$\langle Z, Q\rangle_{\mathcal{A}} = \langle Z, \mathcal{A}_I Q\rangle = \langle \mathcal{A}_I Z, Q\rangle \qquad Z, Q \in so(n),$$

which, in turn, defines on the group the left-invariant metric $(\cdot\,,\cdot)_{\mathcal{A}}$

$$(v_1, v_2)_{\mathcal{A}} = \langle R^{-1}v_1, R^{-1}v_2\rangle_{\mathcal{A}} \qquad v_1, v_2 \in T_R SO(n)\,.$$

Then the kinetic energy (2.4) is represented in the form

$$T = \frac{1}{2}(\dot{R}, \dot{R})_{\mathcal{A}},$$

and, therefore, it is a function on $TSO(n)$ invariant with respect to the left action of the group.

The condition of free motion of the body implies the absence of torque relative to the action of the group $SO(n)$. Then, by Theorem 1, the angular momentum M_s relative to *left* action of the same group is constant.

Differentiating (2.6) and taking into account (2.2) along with the condition $M_s = \text{const}$, we obtain the generalized Euler equations describing the evolution of the angular velocity and of the angular momentum *in the body*

$$\dot{M}_c + [\Omega_c, M_c] = 0\,. \tag{2.7}$$

The substitution $M_c = I_c\Omega_c + \Omega_c I_c$ turns (2.7) into a closed system of m scalar equations

$$(I_i + I_j)\dot{\Omega}_{ij} = (I_i - I_j)\sum_{k=1}^{n} \Omega_{ik}\Omega_{kj}, \qquad i < j = 1, \ldots, n \tag{2.8}$$

(the subscript c is omitted).

These were first represented in the explicit form by Frahm (1874) [10], who considered them together with n^2 kinematic equations

$$\dot{R} = R\Omega_c\,. \tag{2.9}$$

The latter can be used to determine the position of the body in $\mathbb{R}^n$, i.e., the coordinates in the group $SO(n)$. Frahm also found a collection of first integrals of the joint system (2.8), (2.9) in the form

$$\sum_J |M_c|_J^J = \text{const}, \qquad k = 2, 4, \ldots, 2[n/2], \tag{2.10}$$

$$\sum_{\mu<s}^{n} (M_c)_{\mu s}\widehat{R}_{ij,\mu s} = \text{const}, \qquad i < j = 1, \ldots, n, \tag{2.11}$$

where $J = \{i_1, \dots, i_k\}$, $i_1 < \cdots < i_k$, is a multi-index of order k, $|M_c|_J^J$ is the corresponding k-order diagonal minor of M_c, and $\widehat{R}_{ij,\mu s}$ are the 2×2 minors of the rotation matrix R standing at the crossings of ith, jth columns and μth, sth rows. These integrals generalize the angular momentum integral and the area integral in the classical Euler problem.

Besides, it is easy to show that system (2.8) possesses the energy integral

$$\langle \Omega_c, M_c \rangle = l, \qquad l = \text{const}. \tag{2.12}$$

It is known that, as in the classical case $n = 3$, the Euler–Frahm equations (2.8), as well as the corresponding kinematic equations (2.9), are integrable, and the components Ω_{ij}, R_{ij} can be expressed in terms of theta-functions of complex time t. Moreover, as follows from [16], both systems remain integrable under the more general relation between the angular momentum and the angular velocity:

$$(M_c)_{ij} = \frac{a_i + a_j}{b_i + b_j} (\Omega_c)_{ij}, \tag{2.13}$$

where $a_1, \dots, a_n, b_1, \dots, b_n$ are arbitrary fixed parameters (when $b_i = I_i$, $a_i = I_i^2$, we come back to (2.5)). For the case $n = 4$, the integrability condition (2.13) was, in fact, obtained by Frahm, while the analytical study of the problem was performed by Schottky (1891) [21].

EXAMPLE 3. Using the construction introduced above, let us consider the generalized Chaplygin problem on an n-dimensional ball rolling without sliding on an $(n-1)$-dimensional hyperplane $\mathcal{H}$ in $\mathbb{R}^n$. As in the case $n = 3$, this is a nonholonomic system and its equation of motion cannot be obtained from any variational principle. The only way is to use the momentum theorem for the group $E(n)$, which is the configuration space for the ball. For the case of an unconstrained motion in $\mathbb{R}^n$, this theorem is the generalization of the classical theorem which asserts that the equations of motion of a rigid body split into the equations describing rotational motion of the body and the equations describing motion of its mass center as a mass point.

Let $\gamma \in \mathbb{R}^n$ be the unit vector normal to the hyperplane $\mathcal{H}$ and directed "upwards" (i.e., from $\mathcal{H}$ to the mass center C of the ball). Then, as before, ρ is the radius of the ball, Ω and $M = I\Omega + \Omega I \in so(n)$ are respectively its angular velocity and angular momentum matrices, $V_C \in \mathbb{R}^n$ is the velocity of the mass center which coincides with the geometric center of the ball, v_P is the velocity of the contact point P (here and below all the tensor objects are taken in the frame attached to the ball). Now replace the constraint $v_P = 0$ by the reaction $F \in \mathbb{R}^n$ acting at P. Then in the frame attached to the ball, we obtain the following equations

$$\dot{M} + [\Omega, M] = \mathcal{M}, \qquad m(\dot{v}_C + \Omega v_C) = F, \qquad \dot{\gamma} + \Omega\gamma = 0, \tag{2.14}$$

where $\mathcal{M} \in so^*(n)$, $\mathcal{M}_{ij} = \rho(F_i\gamma_j - F_j\gamma_i)$ is the torque of the reaction F relative to C.

The condition for rolling without sliding has the form

$$v_P = v_C - \Omega\gamma = 0. \tag{2.15}$$

Differentiating (2.15) and using the second equation in (2.14), we get

$$F = m\rho\dot{\Omega}\gamma\,.$$

Substituting this expression in (2.14), we obtain

$$\dot{M} + [\Omega\,,\, M] = D(\dot{\Omega}\Gamma + \Gamma\dot{\Omega})\,, \qquad D = m\rho^2\,,\ \Gamma_{ij} = \gamma_i\gamma_j\,,$$
$$\dot{\Gamma} + [\Omega\,,\,\Gamma] = 0\,.$$

It is easy to show that this system can be represented in the following compact commutative form

$$\text{(2.16)} \qquad \dot{K} + [\Omega\,,\, K] = 0\,, \qquad K = I\Omega + \Omega I + D(\Gamma\Omega + \Omega\Gamma) \in so^*(n)\,, \quad \dot{\Gamma} + [\Omega\,,\,\Gamma] = 0\,,$$

where K and $I + D\Gamma$ are respectively the angular momentum of the ball and its mass tensor relative to the contact point P. So, we may introduce the inertia operator $\hat{\mathcal{A}}\colon so(n) \to so(n)$ as follows:

$$\text{(2.17)} \quad K = \hat{\mathcal{A}}\Omega = (\mathcal{A}_I + \mathcal{A}_\Gamma)\Omega\,, \qquad \mathcal{A}_I\Omega = I\Omega + \Omega I\,,\ \mathcal{A}_\Gamma\Omega = \Gamma\Omega + \Omega\Gamma\,.$$

Equations (2.16) can be uniquely represented in terms of Ω, Γ. So they form a closed system for the determination of $\Omega(t)$, $\gamma(t)$.

It follows from (2.16) that $K_s = \mathrm{Ad}_R\, K$ is a constant vector in $so(n)$. Since γ is constant in the space, the system (2.16) has a set of geometric integrals

$$\text{(2.18)} \qquad \operatorname{tr} K^s = \text{const}\,, \quad \operatorname{tr}(K^s\Gamma^l) = \text{const}\,, \quad \operatorname{tr}\Gamma = 1\,, \quad s = 2, 4, 6, \dots,\ l = 1, 2, 3, \dots.$$

These are generalizations of the classical integrals (1.25).

PROPOSITION 3. *The system* (2.16) *has the integral invariant*

$$\int \mu\, d\Omega\, d\gamma\,, \qquad \mu = \sqrt{\det \hat{\mathcal{A}}}\,.$$

In the classical case $n = 3$, the density μ coincides with the density (1.27) found by Chaplygin.

PROOF OF PROPOSITION 3. First note that the system (2.16) can be rewritten in the form

$$\text{(2.19)} \qquad \hat{\mathcal{A}}\dot{\Omega} = -\operatorname{ad}_\Omega \mathcal{A}_I\Omega\,,$$

$$\text{(2.20)} \qquad \dot{\Gamma} = -\operatorname{ad}_\Omega \Gamma\,.$$

Besides, in view of (2.17), (2.20),

$$\text{(2.21)} \qquad \dot{\hat{\mathcal{A}}} = [\mathcal{A}_\Gamma\,,\, \operatorname{ad}_\Omega]\,.$$

In order to calculate the divergence of the system in the phase space (Ω, γ)

$$(2.22) \qquad \Delta = \sum_{i<j}^{n} \frac{\partial \dot{\Omega}_{ij}}{\partial \Omega_{ij}} + \sum_{i}^{n} \frac{\partial \dot{\gamma}_i}{\partial \gamma_i},$$

we define the $m \times m$ matrix

$$U_{QS} = \frac{\partial (\widehat{\mathcal{A}}\dot{\Omega})_Q}{\partial \Omega_S},$$

where the indices Q, S range over the pairs (i, j), $1 \leqslant i < j \leqslant n$. From (2.19) we obtain

$$(2.23) \qquad U = -\operatorname{ad}_\Omega \mathcal{A}_I + \operatorname{ad}_M, \qquad M = \mathcal{A}_I \Omega.$$

In view of the last equation in (2.14), the second sum in (2.22) vanishes. Therefore we may write

$$\Delta = \operatorname{tr}(\widehat{\mathcal{A}}^{-1}\{U\}), \qquad \{U\} = \frac{1}{2}(U + U^T).$$

(Since $\widehat{\mathcal{A}}$ is a symmetric matrix, the skew symmetric part of U does not contribute to Δ.) Then, in view of (2.21), (2.23), and the identity

$$\operatorname{tr}(\widehat{\mathcal{A}}^{-1}[\widehat{\mathcal{A}}, \operatorname{ad}_\Omega]) \equiv 0,$$

we obtain

$$\begin{aligned} \Delta &= \operatorname{tr}(\widehat{\mathcal{A}}^{-1}(U + U^T))/2 = \operatorname{tr}(\widehat{\mathcal{A}}^{-1}[\mathcal{A}_I, \operatorname{ad}_\Omega])/2 \\ &= -\operatorname{tr}(\widehat{\mathcal{A}}^{-1}[\mathcal{A}_\Gamma, \operatorname{ad}_\Omega])/2 = -\operatorname{tr}(\widehat{\mathcal{A}}^{-1}\dot{\widehat{\mathcal{A}}})/2. \end{aligned}$$

In conclusion, comparing this with the well-known identity

$$\frac{d}{dt} \det \widehat{\mathcal{A}} = \det \widehat{\mathcal{A}} \operatorname{tr}(\widehat{\mathcal{A}}^{-1}\dot{\widehat{\mathcal{A}}}),$$

we have

$$\Delta = -\frac{1}{2} \frac{d(\det \widehat{\mathcal{A}})/dt}{\det \widehat{\mathcal{A}}}.$$

Therefore, the function $\mu = \sqrt{\det \widehat{\mathcal{A}}}$ satisfies the equation for the density of the integral invariant:

$$\dot{\mu} + \Delta \mu = 0.$$

Finally, in addition to the geometric integrals (2.18), the equations of motion (2.16) possess the energy integral

$$(2.24) \qquad H(\Omega) = \frac{1}{2} \operatorname{tr}(\Omega \widehat{\mathcal{A}} \Omega)$$

(compare with (1.26)). Indeed, taking into account (2.19), (2.21), and the property $\widehat{\mathcal{A}}^T = \widehat{\mathcal{A}}$, we obtain

$$\begin{aligned} \dot{H} &= \operatorname{tr}(\Omega\widehat{\mathcal{A}}\dot{\Omega} + \Omega\dot{\widehat{\mathcal{A}}}\Omega/2) = \operatorname{tr}(\Omega[\mathcal{A}_I\Omega\,,\,\Omega]) - \langle\Omega\,,\,(\mathcal{A}_\Gamma \operatorname{ad}_\Omega - \operatorname{ad}_\Omega \mathcal{A}_\Gamma)\Omega\rangle \\ &= \operatorname{tr}(\Omega[\mathcal{A}_I\Omega\,,\,\Omega]) + \operatorname{tr}(\Omega[\mathcal{A}_\Gamma\Omega\,,\,\Omega])\,, \end{aligned}$$

which equals zero, since $\operatorname{ad}_\Omega \Omega = 0$ and $\operatorname{tr}(\Omega[X\,,\,\Omega]) = 0$ for any matrix X.

It is natural to suppose that, similarly to the Manakov system, the equations of motion of the n-dimensional Chaplygin ball are integrable, and, apart from (2.18), (2.24), there exist other nontrivial integrals (integrals (2.18), (2.24) form a complete set only for the classical case $n = 3$). However, the proof of this conjecture is still unknown.

§3. Generalized Poinsot model

How can one obtain a qualitative picture of the motion of n-dimensional rigid body in the integrable Euler–Frahm case? It is obvious that even if they are known, the functions $R_{ij}(t)$ can hardly be useful in this situation.

Recall that in the case $n = 3$ the remarkable Poinsot model of motion exists. Namely, consider the inertia ellipsoid with fixed center O

$$\mathcal{V}\colon \{(x\,,\,Jx) = l\}\,, \qquad x \in \mathbb{R}^3\,,$$

which is attached to the body with inertia tensor J. According to this model, the ellipsoid $\mathcal{V}$ rolls without sliding on a fixed plane π. The latter is perpendicular to the angular momentum vector $M = J\omega$, which is also fixed in the space and is located at constant distance $l/|M|$ from the center O (Figure 1(a)). The point P of contact of the ellipsoid with the plane traces the curve on $\mathcal{V}$ called the *polodia*.

Indeed, due to the energy integral $(\omega\,,\,J\omega) = l$, the end of the angular velocity vector ω always lies on $\mathcal{V}$. Let π be the tangent plane of $\mathcal{V}$ at the contact point P. Then the normal vector

$$n = \frac{1}{2}\operatorname{grad}(x\,,\,Jx)\Big|_{x=\omega}$$

coincides with the angular momentum vector M. Besides, the distance between π and the fixed center O is also constant: $(\omega\,,\,n) = l$. It follows from the condition $v_P = \omega \times \overrightarrow{OP} = 0$ that the plane π remains fixed in space.

This model does not admit an immediate generalization to the n-dimensional case, since for even n the skew symmetric operator Ω is generally nondegenerate and, as a consequence, an n-dimensional rigid body does not necessarily have a rotation axis as the set of instantly fixed points. Therefore, a rolling-without-sliding model cannot be realized in this situation. The way out is the following: instead of rotation in $\mathbb{R}^n$, we can consider rotation in the m-dimensional space of the Lie algebra $so(n)$. To carry out this approach, we note that, in view of (2.3), (2.5), (2.6), the mass tensor and the inertia tensor in the space and in the body are related as follows

$$\begin{aligned} I_s &= RI_cR^{-1}\,, & I\colon \mathbb{R}^n &\to \mathbb{R}^n\,, \\ \mathcal{A}_s &= \operatorname{Ad}_R \mathcal{A}_c \operatorname{Ad}_R^{-1}\,, & \mathcal{A}\colon \mathbb{R}^m &\to \mathbb{R}^m. \end{aligned}$$

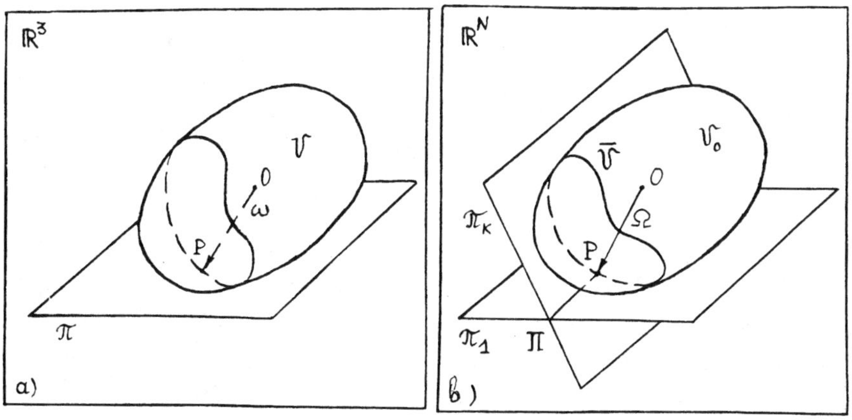

FIGURE 1

Comparing these two formulas, we see that it is natural to consider $\mathrm{Ad}_R \in SO(m)$ as the rotation matrix of an imaginary m-dimensional rigid body whose axes are attached to the eigen-axes of the tensor $\mathcal{A}_s$ in $\mathbb{R}^m$. Such an m-dimensional body is called the *kinematical body*.

Since the adjoint representation of the algebra $so(n)$ is exact, the position of the kinematical body in $\mathbb{R}^m$ uniquely defines the position of the "physical" n-dimensional body in $\mathbb{R}^n$.

Then from the equations

$$\dot{I}_s = [\Omega, I_s], \qquad \dot{\mathcal{A}}_s = [\mathrm{ad}_\Omega, \mathcal{A}_s]$$

we find that $\mathrm{ad}_\Omega \in so(m)$ plays the role of the angular velocity matrix of the kinematical body. In contrast to Ω, the operator ad_Ω is always degenerate and the "rotation axis" of the kinematical body is a whole r-dimensional linear space passing through the point O:

$$\mathrm{Ann}(\Omega) = \{d \in \mathbb{R}^m \mid \mathrm{ad}_\Omega d = 0\}, \qquad r = \mathrm{rank}(so(n)) = [n/2].$$

The space $\mathrm{Ann}(\Omega)$ is spanned by the matrices

$$\{\Omega, \Omega^3, \dots, \Omega^{2r-1}\}. \tag{3.1}$$

Using the Cayley–Hamilton theorem, one may show that the other odd powers Ω^i, $i > 2r-1$, can be represented as linear combinations of the vectors (3.1). Furthermore, the Euler–Frahm equations (2.7) imply that for any degree i

$$\dot{M}_c^i + [\Omega, M_c^i] = 0,$$

and, similarly to M_s, the skew symmetric matrices $M_s^3 = \mathrm{Ad}_R M_c^3$, $M_s^5 = \mathrm{Ad}_R M_c^5, \dots$ turn out to be fixed vectors in the algebra $so(n)$ (among them only r vectors are independent). Therefore, these equations have r independent trivial momentum integrals

$$\langle M^{2i-1}, M^{2i-1}\rangle = \mathrm{const}, \qquad i = 1, \dots, r, \tag{3.2}$$

which are functions of r Frahm integrals (2.10).

Now consider the generalized Poinsot model illustrating the motion of the kinematical body. Let us deal with the simplest model corresponding to the "nonphysical" case $\Omega = UM + MU$, $U = \operatorname{diag}(a_1, \dots, a_n)$ (the general case can be handled in an analogous but more complicated way). Then equations (2.7), apart from the integrals (3.2), also possess the following nontrivial independent integrals

$$\begin{aligned} &H_{mk}(M) = \operatorname{tr}\{M^m, U^k\} = h_{mk}, \qquad h_{mk} = \text{const}, \\ &m = 2, 4, \dots, 2[(n-1)/2], \ k = 1, 2, \dots, n-m, \end{aligned} \tag{3.3}$$

where $\{M^m, U^k\}$ are the homogeneous symmetric polynomials in the matrices M and U of degree m and k respectively (the integral H_{21}, up to the multiplication by a constant factor, coincides with the energy integral (2.12)). Now let X_{ij} be the projections of a vector $X \in \mathbb{R}^m = so(n)$ on the principal axes of $\mathcal{A}_s$. Let the inertia ellipsoid

$$\mathcal{V}_0 : \quad \left\{ \sum_{i<j}^{n} \frac{X_{ij}^2}{a_i + a_j} = h_{21} \right\}$$

be attached to the kinematical body as well as the surfaces $\mathcal{V}_{mk}$ $(k > 1)$, whose equations in the same principal axes arise from the integrals (3.3) as a result of the substitution

$$M_{ij} = \frac{X_{ij}}{a_i + a_j}.$$

The intersection of the surfaces $\mathcal{V}_0$, $\mathcal{V}_{mk}$ $(k > 1)$ define a surface $\widetilde{\mathcal{V}}$. Then we define the $(m-r)$-dimensional linear space $\Pi = \pi_1 \cap \dots \cap \pi_r$, where $\pi_i \subset \mathbb{R}^m$ is the hyperplane orthogonal to the fixed vector M_s^{2i-1} in the metric $\langle \cdot , \cdot \rangle$. The generalized Poinsot model is thus described by

THEOREM 4. *The set of all positions* $\{\mathrm{Ad}_R\}$ *corresponding to rotations* $\{R\}$ *of the n-dimensional rigid body coincides with the set of all positions in* $\mathbb{R}^m$ *of the surface* $\widetilde{\mathcal{V}}$ *with fixed center* $O = \{X = 0\}$, *which rolls without sliding on the fixed linear space* Π (*see the sketch in* Figure 1(b)). *The polodia, as a set on the surface* $\widetilde{\mathcal{V}}$ *traced by the point P, turns out to be not a single curve, but a whole d-dimensional torus, the joint level surface for the first integrals* (3.2), (3.3).

In the case $n = 3$ the surface $\widetilde{\mathcal{V}}$ coincides with the ellipsoid $\mathcal{V}_0$ and the linear space Π is merely the two-dimensional plane π_1, hence we come back to the classical Poinsot model.

COMMENTARY. The idea of the kinematical body was first discussed in [14]. It was proved that while this body rotates in the space $\mathbb{R}^m = so(n)$ according to the Euler–Frahm equations, the inertia ellipsoid $\mathcal{V}$ rolls without sliding on a fixed hyperplane $\pi \in so(n)$. However, in contrast to the three-dimensional case, for $n > 3$ the construction $\{\mathcal{V}, \pi\}$ fails to solve the "inverse problem", i.e., to recover the actual motion of the kinematical body. Namely, this construction admits superfluous positions of the inertia ellipsoid, and the contact point P on $\mathcal{V}_0$ covers the surface

$$\mathcal{V}_0 \cup \left\{ \sum_{i<j}^{n} \left(\frac{X_{ij}}{a_i + a_j} \right)^2 = \text{const} \right\},$$

which is more than the d-dimensional invariant torus. The construction $\{\widetilde{\mathcal{V}}, \Pi\}$ defined in Theorem 4 is free of this shortcoming. However, it also fails to reconstruct the motion of the kinematical body uniquely, since in the case $n > 3$ when we fix any instant rotation axis OP or even a k-dimensional linear space $\operatorname{Ann}(\Omega) \in \mathbb{R}^m$, we still leave some degrees of freedom for the surface $\widetilde{\mathcal{V}}$. Figuratively speaking, a certain "looseness" occurs. As a consequence, when dealing with the generalized Poinsot model based only on the first integrals, one cannot speak about the motion of the m-dimensional kinematical body but only about its positions.

PROOF OF THEOREM 4. In view of the definition of the surface $\widetilde{\mathcal{V}}$, the end of the vector $\Omega = \overrightarrow{OP}$ always belongs to this surface. Let the hyperplanes $\pi_1, \dots, \pi_r$ pass through P. The hyperplane π_1 is obviously tangent to $\mathcal{V}_0$ at P. Since $\Pi \subset \pi_1$, the linear space Π is also tangent to $\widetilde{\mathcal{V}} \subset \mathcal{V}_0$.

Let $n = M^{2i-1}/|M^{2i-1}|$, $|M^{2i-1}| = \sqrt{\langle M^{2i-1}, M^{2i-1}\rangle}$ be the unit normal vector of π. Then the latter is removed from the fixed center O at the distance $\langle \Omega, n_i\rangle = \langle UM + MU, M^{2i-1}\rangle/|M^{2i-1}|$, which, in view of the integrals (3.2), (3.3) for $k = 1$, is constant. Thus, by the condition $v_P = 0$, the linear space Π is fixed.

It follows that the end of the vector Ω in the principal axes of the ellipsoid $\mathcal{V}_0$ runs over the intersection of the surface $\widetilde{\mathcal{V}}$ with $\mathcal{V}_{m1}$, which is a d-dimensional invariant torus, the level surface for all first integrals (3.2), (3.3). The same holds for the trajectories of the Euler–Frahm equations in the phase space $\mathbb{R}^m = so(n)$.

§4. Euler–Poincaré equations with constraints

Let $v_1, \dots, v_n$ be linearly independent vector fields on M^n. Their commutators $[v_i, v_j]$ can be decomposed as follows

$$[v_i, v_j] = \sum c_{ij}^k(x) v_k, \qquad c_{ij}^k = -c_{ji}^k. \tag{4.1}$$

If $f(x)$ is a smooth function on M^n, then

$$\dot{f} = \left(\frac{\partial f}{\partial x}, \dot{x}\right) = \sum_{i=1}^{n} v_i(f)\omega_i,$$

where $v_i(f) = (\partial f/\partial x, v_i)$ is the derivative of f along v_i. The variables ω depend linearly on $\dot{x}$:

$$\dot{x}_k = \sum_{i=1}^{n} v_i(x_k)\omega_i, \qquad 1 \leqslant k \leqslant n. \tag{4.2}$$

The variables ω_i may not be derivatives of globally defined coordinate functions on M^n. For this reason the ω_i's are called *quasivelocities*.

The use of quasivelocities in mechanics is motivated by the fact that equations of motion written in terms of real velocities x often have a nonsymmetric and cumbersome form. As an example, we may recall the equations describing the rotation of a rigid body around a fixed point written in Euler's angles (θ, ψ, φ), which are globally defined coordinate functions on the group $SO(3)$.

Now rewrite the constraint equations (1.14) in the quasivelocities $\omega_1, \dots, \omega_n$:

$$f_1(\omega, x) = \dots = f_m(\omega, x) = 0 \qquad (m < n). \tag{4.3}$$

Then the Lagrange equations (1.17) for the case of the motion in a potential force field with potential function $V(x)$ take the form

$$\left(\frac{\partial \mathcal{L}}{\partial \omega_i}\right)^{\cdot} = \sum_{j,k=1}^{n} c_{ji}^{k} \omega_j \frac{\partial \mathcal{L}}{\partial \omega_k} + v_i(\mathcal{L}) + \sum_{1}^{m} \lambda_s \frac{\partial f_s}{\partial \omega_i}, \tag{4.4}$$

where $\mathcal{L}$ is the Lagrangian $L(x, \dot{x}) = T(x, \dot{x}) - V(x)$ expressed in terms of ω, x. These equations (in the absence of constraints) were first obtained by H. Poincaré [20]. If the gradients

$$\frac{\partial f_1}{\partial \omega}, \dots, \frac{\partial f_m}{\partial \omega}$$

are linearly independent, then the multipliers λ_s can be represented as functions of x and ω.

In general, equations (4.4) do not represent a closed system, and they must be considered together with the geometric equations (4.2).

Now let M^n be a Lie group $\mathfrak{G}$, and $v_1, \dots, v_n$ be the independent left-invariant vector fields on $\mathfrak{G}$ generated, according to (1.1), by basis vectors $Y_1, \dots, Y_n$ in the Lie algebra g of the group $\mathfrak{G}$. Then the coefficients c_{ji}^k, called the structure constants of the algebra g, are fixed:

$$[Y_i, Y_j] = \sum_k c_{ji}^k Y_k,$$

which is equivalent to relations (4.1). In this case the velocity vector $\dot{x} = (\dot{x}_1, \dots, \dot{x}_n) \in T_x\mathfrak{G}$ is generated by left translations of the vector $\Omega = \omega_1 Y_1 + \dots + \omega_n Y_n \in g$. Suppose that the Lagrangian $\mathcal{L}$ reduces to the kinetic energy ($L = T$) defined by a nondegenerate left-invariant metric $(\cdot, \cdot)_J$ on $\mathfrak{G}$. The latter is generated by the corresponding scalar product $\langle \cdot, \cdot \rangle_J$ on g. According to (4.2),

$$L = \frac{1}{2}(\dot{x}, \dot{x})_J = \frac{1}{2}\left(\sum_i v_i\omega_i, \sum_j v_j\omega_j\right)_J = \frac{1}{2}\sum_{i,j} J_{ij}\omega_i\omega_j.$$

Here $J_{ij} = (v_i, v_j)_J = \langle Y_i, Y_j\rangle_J = \text{const}$, because $v_1, \dots, v_n$ are left-invariant.

Consider the left action of the group $\mathfrak{G}$ on itself. Then one can consider the kinetic momentum relative to this action. According to the definition (1.2) and by (4.2), we obtain

$$\begin{aligned} K_s &= \mathcal{I}_{\mathfrak{G}}(x, \dot{x} \mid Y_s) = (\dot{x}, v_s)_J \\ &= \left\langle \sum_i \omega_i Y_i, Y_s \right\rangle_J = \sum_i \langle Y_i, Y_s\rangle_J \omega_i = \sum_i J_{si}\omega_i. \end{aligned}$$

Therefore, for an arbitrary vector $Y = y_1 Y_1 + \dots + y_n Y_n \in g$, we have

$$\mathcal{I}_{\mathfrak{G}}(x, \dot{x} \mid Y) = \langle Y, K\rangle, \qquad K = \partial\mathcal{L}/\partial\omega. \tag{4.5}$$

Thus, according to (1.3), the vector $K = (K_1, \dots, K_n)^T \in g^*$ is precisely the kinetic momentum which was to be found.

It is interesting to study the case when, in addition to the Lagrangian $\mathcal{L}$, the constraint functions $f_1, \dots, f_m$ are also left-invariant, i.e., they do not depend explicitly on x. Under this condition the equations (4.4), (4.3) form a closed system on the Lie algebra g:

$$\dot{K}_i = \sum_{j,k=1}^{n} c_{ji}^k K_k \omega_j + \sum_{s=1}^{m} \lambda_s \frac{\partial f_s}{\partial \omega_i}, \qquad f_1(\omega) = \cdots = f_m(\omega) = 0. \tag{4.6}$$

For the case $\mathfrak{G} = SO(3)$, systems with left-invariant constraints were first studied by G. Suslov (1900, [22]). He considered rotations of a rigid body about a fixed point under the action of the following nonholonomic constraint: the projection of the angular velocity vector $\omega \in \mathbb{R}^3$ to a certain unit vector e_l *fixed in the body* always equals zero:

$$(\omega, e_l) = 0. \tag{4.7}$$

The left action of the group $SO(3)$ leaves the kinetic energy of the body and the constraint (4.7) invariant. Thus, one may call the equations (4.6) the *Euler–Poincaré–Suslov* equations (EPS).

EXAMPLE 4. Consider the EPS equations on the Lie algebra $so(n)$. How can one write a multi-dimensional analog of the condition (4.7)? In order to answer this question recall that, instead of rotations *about an axis* in the three-dimensional case, in an n-dimensional case we have to speak about infinitesimal rotations in the two-dimensional planes $e_i \wedge e_j$ spanned by the vectors of an orthonormal frame $e_1, \dots, e_n$. Suppose, without loss of generality, that in the condition (4.7) $e_l = (1, 0, 0)$. Then this condition can be redefined as follows: only infinitesimal rotations in the planes $e_1 \wedge e_2$, $e_1 \wedge e_3$ are allowed. Hence, it is natural to define a multi-dimensional analog of Suslov's condition in the following way: for an n-dimensional body only infinitesimal rotations in the planes $e_1 \wedge e_j$, $j = 1, \dots, n$, i.e., in the planes containing the vector e_1, are allowed. Therefore, we have the following constraints imposed on the components of the angular velocity matrix *in the body*

$$\Omega_{ij} = 0, \qquad i, j \geqslant 2, \tag{4.8}$$

or

$$\Omega = \begin{pmatrix} 0 & \Omega_{12} & \dots & \Omega_{1n} \\ -\Omega_{12} & 0 & \dots & 0 \\ \vdots & \vdots & \ddots & \vdots \\ -\Omega_{1n} & 0 & \dots & 0 \end{pmatrix}. \tag{4.9}$$

According to (2.6), the angular momentum M is represented by the matrix $M = I\Omega + \Omega I$. By an appropriate orthogonal transformation which leaves the vector e_1

invariant, the symmetric mass tensor I can be reduced to the form

$$I = \begin{pmatrix} I_{11} & I_{12} & \dots & I_{1n} \\ I_{11} & I_{12} & \dots & 0 \\ \vdots & \vdots & \ddots & \vdots \\ I_{1n} & 0 & \dots & I_{nn} \end{pmatrix} \tag{4.10}$$

(the constraint equations (4.8) do not change under such a transformation). Therefore, equations (4.6) can be represented as follows

$$\dot{M} = [M\,,\,\Omega] + \Lambda\,, \qquad \Lambda = \begin{pmatrix} 0 & 0 & 0 & \dots & 0 \\ 0 & 0 & \lambda_{23} & \dots & \lambda_{2n} \\ 0 & -\lambda_{23} & 0 & \dots & \lambda_{3n} \\ \vdots & \vdots & \vdots & \ddots & \vdots \\ 0 & -\lambda_{2n} & -\lambda_{3n} & \dots & 0 \end{pmatrix}, \tag{4.11}$$

where λ_{ij} are Lagrange multipliers. Taking into account (4.8) and (4.10), we obtain from (4.11) the following closed system for the entries $\Omega_{12}\,, \dots\,, \Omega_{1n}$

$$\begin{aligned} (I_{11} + I_{22})\dot{\Omega}_{12} &= I_{12}(\Omega_{13}^2 + \Omega_{14}^2 + \dots + \Omega_{1n}^2) \\ &\quad - (I_{13}\Omega_{13} + I_{14}\Omega_{14} + \dots + I_{1n}\Omega_{1n})\Omega_{12}\,, \\ (I_{11} + I_{33})\dot{\Omega}_{13} &= I_{13}(\Omega_{12}^2 + \Omega_{14}^2 + \dots + \Omega_{1n}^2) \\ &\quad - (I_{13}\Omega_{13} + I_{14}\Omega_{14} + \dots + I_{1n}\Omega_{1n})\Omega_{13}\,, \\ &\dots\dots\dots\dots\dots\dots\dots\dots\dots\dots \\ (I_{11} + I_{nn})\dot{\Omega}_{1n} &= I_{1n}(\Omega_{12}^2 + \Omega_{13}^2 + \dots + \Omega_{1\,,n-1}^2) \\ &\quad - (I_{12}\Omega_{12} + I_{13}\Omega_{13} + \dots + I_{1\,,n-1}\Omega_{1\,,n-1})\Omega_{1n}\,. \end{aligned} \tag{4.12}$$

The equations for other Ω_{ij}'s can be used to determine the multipliers $\lambda_{ij}, i\,,\, j \geqslant 2$.

In the special case $I_{12} = \dots = I_{1n} = 0$, it follows from the equations (4.12) that $\Omega_{12}\,, \dots\,, \Omega_{1n} = \text{const}$ and the linear space in $so(n)$ spanned by the admissible velocities (4.9) is the eigenspace for the inertia operator $\mathcal{A}_I\colon so(n) \to so(n)^*$. In the general case, equations (4.12) have the energy integral

$$2H(\Omega) = (I_{11}+I_{22})\Omega_{12}^2+(I_{11}+I_{33})\Omega_{13}^2+\dots+(I_{11}+I_{nn})\Omega_{1n}^2 = h\,, \qquad h = \text{const}\,.$$

Thus, if $h > 0$, we obtain a dynamical system on an $(n-2)$-dimensional ellipsoid. Now put

$$F = (I_{11} + I_{22})I_{12}\Omega_{12} + (I_{11} + I_{33})I_{13}\Omega_{13} + \dots + (I_{11} + I_{nn})I_{1n}\Omega_{1n}\,.$$

Then the derivative of $F(\Omega)$ along the trajectories of the system (4.12) has the form

$$\dot{F} = \sum_{i<j}^{n}(I_{1i}\Omega_{1j} - I_{1j}\Omega_{1i})^2\,,$$

and it is positive everywhere in $so(n)$ except at the points of the line

$$\{\Omega_{12} = I_{12}\mu, \ldots, \Omega_{1n} = I_{1n}\mu, \ \mu \in \mathbb{R}\}. \tag{4.13}$$

These are the equilibrium points for the system (4.12). The line (4.13) meets the ellipsoid $H(\Omega) = h$ in the two diametrically opposed points S^-, S^+, in which the hyperplane $F(\Omega) = \text{const}$ is tangent to the ellipsoid. These points correspond to stable and unstable permanent rotations of the body in a certain 2-plane $v_1 \wedge v_2$ which is fixed in the space, as well as in the body (so called *plane rotations*).

Since $\dot{F} \geqslant 0$, all trajectories of the system (4.12) lying on one and the same ellipsoid are double-asymptotic: they tend to the points S^- and S^+ as $t \to \pm\infty$ (Figure 2(a)).

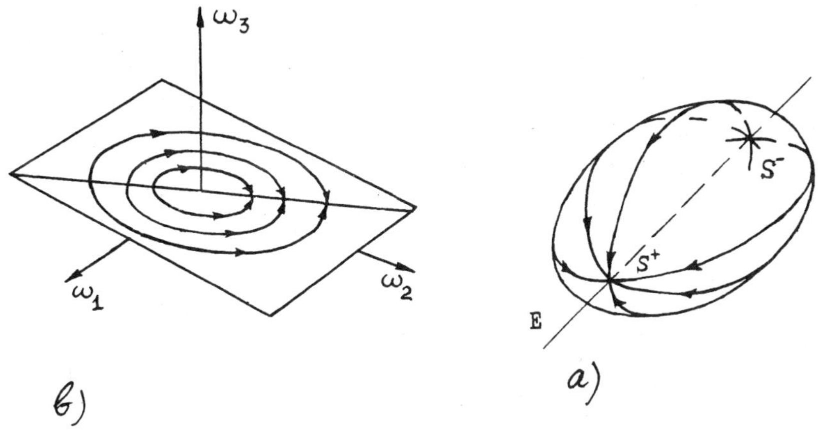

FIGURE 2

This picture represents an immediate generalization of the well-known phase portrait of the classical Suslov problem where the ellipsoid $\{H(\Omega) = h\}$ is reduced to an ellipse. The latter lies in the plane given by the equation (4.7) (Figure 2b). The motion of the n-dimensional rigid body is the asymptotic evolution from the permanent rotation in some 2-plane *fixed in the body* to the permanent rotation in the same 2-plane and with the same angular velocity, but in the opposite direction. We emphasize, however, that *in space* these rotations for $t \to -\infty$ and $t \to +\infty$ occur in different 2-planes.

As in the three-dimensional case, equations describing the position of the body in the space (i.e., the equations on the group $SO(n)$) seem to be nonintegrable.

What can one say about the analytic properties of the solutions of the system (4.12)? In the case $n = 3$, as it was noticed by Suslov, these solutions turn out to be meromorphic functions of the time t. More exactly, they are expressed in terms of fractional-rational functions of the form $\exp(bt)$ ($b = \text{const}$). (We leave the proof of this fact to the reader as an exercise.) In the case $n = 4$, the situation is different: the solutions of the system (4.12) generally branch in the complex plane. This follows from

PROPOSITION 5. *Let $n \geqslant 4$, $I_{12} \neq 0$, $I_{13} = \cdots = I_{1n} = 0$. Then the general solution of the system* (4.12) *is single-valued if and only if $I_{33} = \cdots = I_{nn} = 0$.*

PROOF OF THE NECESSITY. Let us use the asymptotic Kowalewski–Lyapunov method (see, for example, [24]). Note that if the conditions of Proposition 5 are fulfilled, the system (4.12) has the following particular solutions

$$\Omega_{12} = \alpha/t\,, \quad \Omega_{13} = \alpha/t\,, \quad \Omega_{14} = \cdots = \Omega_{1n} = 0\,, \qquad (4.14)$$
$$\alpha = (I_{11} + I_{33})/I_{12}\,, \quad \beta^2 = -(I_{11} + I_{22})(I_{11} + I_{33})/I_{12}^2\,.$$

The corresponding Kowalewski exponents are

$$-1\,,\ 2\,,\ 1 - \frac{I_{11} + I_{33}}{I_{11} + I_{44}}\,,\ \ldots\,,\ 1 - \frac{I_{11} + I_{33}}{I_{11} + I_{nn}}\,.$$

The existence of the exponent 2 follows from the existence of the quadratic energy integral. According to the Lyapunov theorem, if the solutions are single-valued, then all the ratios $(I_{11} + I_{33})/(I_{11} + I_{ss})$, $s \geqslant 4$, must be integers. Apart from (4.14), equations (4.12) have other meromorphic solutions in which all the Ω_{1s}'s, $s \geqslant 3$, except one, are zero. In this case, performing elementary computations and using the Lyapunov theorem again, we came to the conclusion that the numbers $(I_{11} + I_{ss})/(I_{11} + I_{kk})$ for $k\,,\ s \geqslant 3$ must be integers as well. Then, since $I_{ss} \geqslant 0$, the equalities $I_{33} = \cdots = I_{nn}$ must hold.

The proof of the sufficiency is left to the reader.

Thus, in the general case when $I_{33}\,, \ldots\,, I_{nn}$ are all different, the solutions of (4.12) cannot be single-valued functions of complex time.

EXAMPLE 5. Now consider the more general case when the left-invariant constraints are defined as follows

$$\Omega_{ij} = 0\,, \qquad i\,, j > 2\,.$$

Suppose that the mass tensor I is diagonal, $I = \operatorname{diag}(I_1\,, \ldots\,, I_n)$. Then, in view of (4.6), the EPS equations take on the following simple form

$$\begin{aligned}
(I_1 + I_2)\dot{\Omega}_{12} &= (I_2 - I_1)(\Omega_{13}\Omega_{23} + \cdots + \Omega_{1n}\Omega_{2n})\,, \\
(I_1 + I_3)\dot{\Omega}_{13} &= (I_1 - I_3)\Omega_{12}\Omega_{23}\,, \ldots\,, (I_1 + I_n)\dot{\Omega}_{1n} \\
&= (I_1 - I_n)\Omega_{12}\Omega_{2n}\,, \\
(I_2 + I_3)\dot{\Omega}_{23} &= (I_3 - I_2)\Omega_{12}\Omega_{23}\,, \ldots\,, (I_2 + I_n)\dot{\Omega}_{2n} \\
&= (I_n - I_2)\Omega_{12}\Omega_{1n}\,.
\end{aligned} \qquad (4.15)$$

These equations have $n - 1$ quadratic integrals

$$2H = \sum_{i \leqslant 2\,,\, j}^{n} (I_i + I_j)\Omega_{ij}^2\,,$$

$$F_1 = (I_1 + I_3)(I_2 - I_3)\Omega_{13}^2 + (I_2 + I_3)(I_1 - I_3)\Omega_{23}^2\,,$$

. . . .

$$F_{n-2} = (I_1 + I_n)(I_2 - I_n)\Omega_{1n}^2 + (I_2 + I_n)(I_1 - I_n)\Omega_{2n}^2\,,$$

which enable one to solve the system (4.15) by quadratures.

Suppose, without loss of generality, that $I_1 > I_2 > \cdots > I_n$. Then the functions $F_1(\Omega), \ldots, F_{n-2}(\Omega)$ are positive definite and we come to

PROPOSITION 6. *If* $c_1, \ldots, c_{n-2} > 0$ *and*

$$c_1/(I_2 + I_3) + \cdots + c_{n-2}/(I_2 + I_n) < h, \tag{4.16}$$

then the integral surface

$$\mathcal{T}: \quad \{\Omega_{ij} \mid 2H(\Omega) = h, \, F_1(\Omega) = c_1, \ldots, F_{n-2}(\Omega) = c_{n-2}\}, \\ c_1, \ldots, c_{n-2} = \text{const},$$

is the disjoint union of two $(n-2)$*-dimensional tori.*

PROOF. When $c_s > 0$, the curve $\{F_s = c_s\}$ on the plane $(\Omega_{1,s+2}, \Omega_{2,s+2})$ is an ellipse. Then, if condition (4.16) is fulfilled, Ω_{12} does not vanish on $\mathcal{T}$ and it is expressed in terms of the other Ω_{ij}'s up to sign flip. Therefore, a connected component of $\mathcal{T}$ is constructed as the topological product of circles.

Now define on $\mathcal{T}$ the angular variables $\varphi_1, \ldots, \varphi_{n-2}$ by putting

$$\begin{aligned} \Omega_{1,s+2} &= \sqrt{\frac{c_s}{(I_1 + I_{s+2})(I_2 - I_{s+2})}} \sin(\varphi_s), \\ \Omega_{2,s+2} &= \sqrt{\frac{c_s}{(I_2 + I_{s+2})(I_1 - I_{s+2})}} \cos(\varphi_s), \end{aligned} \qquad s = 1, \ldots, n-2, \tag{4.17}$$

and introduce the new time τ by the formula $d\tau = \Omega_{12}\, dt$. In view of the conditions of Proposition 1, $\Omega_{12} \neq 0$ on $\mathcal{T}$. Therefore, τ is a monotonic function of t, and, using (4.15), we obtain

$$\frac{d\varphi_1}{d\tau} = \nu_1, \ \ldots, \ \frac{d\varphi_{n-2}}{d\tau} = \nu_{n-2}, \tag{4.18}$$

$$\nu_1 = \sqrt{\frac{(I_1 - I_3)(I_2 - I_3)}{(I_1 + I_3)(I_2 + I_3)}}, \ \ldots, \ \nu_{n-2} = \sqrt{\frac{(I_1 - I_n)(I_2 - I_n)}{(I_1 + I_n)(I_2 + I_n)}}.$$

Equations (4.18) define a quasiperiodic motion on the $(n-2)$-dimensional tori with fixed frequencies $\nu_1, \ldots, \nu_{n-2}$.

Going back to the original time t, we obtain the following system

$$\frac{d\varphi_1}{dt} = \frac{\nu_1}{F}, \ \ldots, \ \frac{d\varphi_{n-2}}{dt} = \frac{\nu_{n-2}}{F}, \tag{4.19}$$

$$F^{-2} = \frac{1}{I_1 + I_2} \Omega_{12} \left(h - \sum_{s=1}^{n-2} \left(\frac{c_s}{I_2 - I_{s+2}} \cos^2(\varphi_s) - \frac{c_s}{I_1 - I_{s+2}} \sin^2(\varphi_s) \right) \right).$$

This system has the integral invariant

$$\operatorname{mes}(\mathcal{D}) = \int_{\mathcal{D}} F \, d\varphi_1 \ldots d\varphi_{n-2}.$$

As shown by Kolmogorov, for almost all frequencies $\nu_1, \dots, \nu_{n-2}$ equations (4.19) can be reduced to the form (4.18).

In conclusion, let us discuss the analytic properties of the solutions of system (4.15). Put $c_1, \dots, c_{n-2} = 0$. Then $\Omega_{14} = \Omega_{24} = \cdots = \Omega_{1n} = \Omega_{2n} = 0$, and equations (4.15) reduce to a closed system for $\Omega_{12}, \Omega_{13}, \Omega_{23}$, which coincides with the equations of the integrable Euler problem. The solutions of the latter equations are known to be elliptic, i.e., single-valued functions of time t. So one may ask whether or not this property is valid for *all* solutions of (4.15)? It turns out that in general the answer is negative.

PROPOSITION 7. *All the solutions of equations* (4.15) *are single-valued functions on complex plane iff* $\nu_1 = \cdots = \nu_{n-2}$.

PROOF. As above, we use the asymptotic Kowalewski–Lyapunov method. On the one hand, system (4.15) has the following particular solution

$$\Omega_{12} = \frac{\alpha_{12}}{t}, \quad \Omega_{13} = \frac{\alpha_{13}}{t}, \quad \Omega_{23} = \frac{\alpha_{23}}{t},$$
$$\Omega_{14} = \Omega_{24} = \cdots = \Omega_{1n} = \Omega_{2n} = 0,$$
$$\alpha_{12} = \sqrt{\frac{(I_1+I_3)(I_2+I_3)}{(I_1-I_3)(I_3-I_2)}}, \quad \alpha_{13} = \sqrt{\frac{(I_1+I_2)(I_2+I_3)}{(I_2-I_1)(I_3-I_1)}},$$
$$\alpha_{13} = \sqrt{\frac{(I_1+I_3)(I_2+I_1)}{(I_1-I_3)(I_2-I_1)}}.$$

The Kowalewski exponents for this solution are

$$\rho_1 = -1, \ \rho_2 = \rho_3 = 2, \ \rho_{4,5} = 1 \pm \nu_2/\nu_1, \ \dots, \ \rho_{2n-4,2n-3} = 1 \pm \nu_{n-2}/\nu_1.$$

If general solution of (4.15) is single-valued, then the ratios $\nu_2/\nu_1, \dots, \nu_{n-2}/\nu_1$ must be integers. On the other hand, for each $3 \leqslant i \leqslant n$ these equations have also the particular solution

$$\Omega_{12} = \frac{\beta_{12}}{t}, \quad \Omega_{1i} = \frac{\beta_{1i}}{t}, \quad \Omega_{2i} = \frac{\beta_{2i}}{t}, \qquad \Omega_{1k} = \Omega_{2k} = 0 \quad (k \neq i)$$

(the definition of the β's is analogous to that of the α's). Then the condition for the corresponding Kowalewski exponents to be integers implies that the ratios $\nu_1/\nu_i, \nu_2/\nu_i, \dots, \nu_{n-2}/\nu_i$ must be integers as well. Since $\nu_s > 0$ for all s, we come to the condition $\nu_1 = \cdots = \nu_{n-2}$.

Now let us prove the sufficiency of this condition. It turns out that if $\nu_1 = \cdots = \nu_{n-2} = \nu$, then the solutions of (4.15) are elliptic functions of t. To show this, we use the new time τ. Then, from (4.18) we have $\varphi_s = \nu\tau + \varphi_s^0$, $\varphi_s^0 = \text{const}$ $(1 \leqslant s \leqslant n-2)$. Since $d\tau = \Omega_{12}\, dt$, in view of (4.19) we obtain

$$t = \int \frac{c \, d\tau}{\sqrt{1 - \sum_i \mu_i \sin^2(\nu\tau + \varphi_i^0)}}, \tag{4.20}$$

where c , μ_i are constants, and $\sum_i |\mu_i| < 1$. This integral is known to be elliptic, while $\sin(\nu\tau)$, $\cos(\nu\tau)$, as well as radical in (4.20), are elliptic functions of t. Thus, by (4.17), $\Omega_{12}, \dots, \Omega_{1n}$ are single-valued functions of the complex time t.

It is interesting to note that in the most general case the square of Ω_{12}, as well as those of $\Omega_{1,s+2}$, $\Omega_{2,s+2}$ $(1 \leqslant s \leqslant n-2)$, is an entire function of the new time τ. This follows from (4.17) and the equation

$$(I_1 + I_2)\frac{d}{d\tau}\Omega_{12}^2 = 2(I_2 - I_1)\sum_{k=3}^{n}\Omega_{1k}\Omega_{2k}\,.$$

It may happen that the EPS equations for the group $SO(n)$ may also be integrable in the more general case when, considered in certain frame, the mass tensor is of diagonal form and the constraints are defined by any combination of equalities $\Omega_{ij} = 0$.

As it was mentioned above, in EPS equations the kinetic energy (metric) and the constraints are left-invariant on a Lie group. This gives the possibility to obtain a closed system on the corresponding Lie algebra.

Now consider another interesting case of the Euler–Poincaré equations on Lie groups when the kinetic energy is left-invariant, whereas the constraints are *right-invariant*. These are so called L–R systems, studied by Veselov and Veselova [23]. The right invariance condition means that in coordinates in the Lie algebra g (in quasivelocities) a constraint is determined by the relation

$$\langle \Omega, \mathcal{N} \rangle = \text{const}, \qquad \Omega \in g, \ \mathcal{N} \in g^*,$$

where, in contrast to the case of left-invariant constraints, $\mathrm{Ad}_x^* \mathcal{N} = \text{const}$.

EXAMPLE 6. The most descriptive illustration of an L–R system is the Veselova problem concerning the rotations of a rigid body with a fixed point under the action of the following nonholonomic constraint

$$(\omega, \gamma) = q, \qquad q = \text{const} \tag{4.21}$$

(the projection of the angular velocity vector $\omega \in \mathbb{R}^3$ to some unit vector γ *fixed in space* is constant) [9, 23]. Here $\mathfrak{G} = SO(3)$ and the Euler–Poincaré equations (4.4) along with the kinematic equations get the form

$$J\dot\omega = J\omega \times \omega + \lambda\gamma, \qquad \dot\gamma = \gamma \times \omega, \tag{4.22}$$

where, as before, J is the inertia tensor relative to the fixed point. Using this and differentiating (4.21), we find

$$\lambda = -\frac{(J\omega \times \omega, J^{-1}\gamma)}{(\gamma, J^{-1}\gamma)}.$$

Note that the system (4.22) admits the following representation

$$\begin{aligned} &\dot Q = Q \times \omega, && \dot\gamma = \gamma \times \omega, \\ &Q = J\omega - (J^0\omega, \gamma)\gamma, && J^0 = J - E, \end{aligned} \tag{4.23}$$

E being the unit matrix. Hence the vector Q, similarly to γ, is thus a fixed vector in space. Therefore, the system possesses three independent integrals

$$(Q, Q), \quad (Q, \gamma) = (\omega, \gamma), \quad (\gamma, \gamma) = 1. \tag{4.24}$$

The second integral coincides with the constraint equation. Since under the condition $q \neq 0$ the constraint (4.21) is nonstationary, the kinetic energy of the body cannot be constant. However, instead of an energy integral, there exists the following analog of the Jacobi–Painlevé integral

$$(J\omega, \omega) - 2(\omega, \gamma)(J^0\omega, \gamma). \tag{4.25}$$

Finally, the system considered in the phase space (ω, γ) has the integral invariant with density $\sqrt{(J^{-1}\gamma, \gamma)}$ and, therefore, it is integrable by the classical Jacòbi theorem.

In addition to the multi-dimensional generalization of the Suslov problem, it is worthwhile considering the generalization of the equations (4.22) which is an L–R system on the group $SO(n)$. In this case it is natural to take constraint equations in form (4.9) with the only difference that Ω_c must be replaced by the angular velocity in the space Ω_s. For the sake of generality, the zeros at the corresponding entries can be replaced by arbitrary constants:

$$\Omega_s = R\Omega_c R^{-1} = \begin{pmatrix} 0 & \Omega_{12} & \dots & \Omega_{1n} \\ -\Omega_{12} & & & \\ \vdots & & \widetilde{\Omega} & \\ -\Omega_{1n} & & & \end{pmatrix}, \tag{4.26}$$

$$\widetilde{\Omega}_{ij} = \text{const}, \qquad 2 \leqslant i < j \leqslant n.$$

Let $\{e_1, \dots, e_n\}$ be a fixed orthonormal basis in $\mathbb{R}^n$, $(e_{i1}, \dots, e_{in})$ be the projections of e_i to the axes of an orthonormal frame attached to the body, so that the rotation matrix is written in the form $R = (e_1 \dots e_n)^T$. Define 2-vectors $\mathcal{E}^{(rs)} = e_r \wedge e_s \in \bigwedge^2 \mathbb{R}^n$, $r, s = 1, \dots n$ represented by the $n \times n$ skew symmetric matrices $\mathcal{E}_{ij}^{(rs)} = e_{ri}e_{sj} - e_{rj}e_{si}$. These 2-vectors give an orthonormal basis in $so(n)$ with respect to the Killing form $\langle \cdot, \cdot \rangle$. Therefore, (4.26) is equivalent to the constraints

$$\langle \Omega, \mathcal{E}^{(rs)} \rangle = \widetilde{\Omega}_{ij}, \qquad 2 \leqslant r < s \leqslant n, \tag{4.27}$$

where the $\mathcal{E}^{(rs)}$'s can be regarded as generalizations of the vector γ in (4.21). The corresponding Euler–Poincaré equations with multipliers, as well as the kinematic equations, can be represented in the following matrix form

$$\dot{M} + [\Omega, M] = \sum_{\substack{(rs) \\ 2 \leqslant r < s \leqslant n}} \lambda_{(rs)} \mathcal{E}^{(rs)}, \qquad \dot{\mathcal{E}}^{(rs)} + [\Omega, \mathcal{E}^{(rs)}] = 0. \tag{4.28}$$

As above, $M = \mathcal{A}\Omega$ is the angular momentum of the body with the inertia tensor $\mathcal{A}$. Differentiating (4.27) and using (4.28), we obtain the following system of linear equations for determining the multipliers $\lambda_{(rs)}$

$$\sum_{\substack{(kl) \\ 2 \leqslant k < l \leqslant n}} \langle \mathcal{A}^{-1}\mathcal{E}^{(kl)}, \mathcal{E}^{(rs)} \rangle \lambda_{(k,l)} = \langle \mathcal{E}^{(rs)}, \mathcal{A}^{-1}[\Omega, M] \rangle, \; 2 \leqslant r < s \leqslant n. \tag{4.29}$$

Now let $H = \operatorname{span}(e_1 \wedge e_2, \dots, e_1 \wedge e_n)$ and H' be the orthogonal complement to H in $\bigwedge^2 \mathbb{R}^n$. It turns out that a part of the system (4.28) is separated and can be written in the following commutative form

$$
\text{(4.30)} \qquad \dot{Q} = [Q, \Omega], \qquad \dot{e}_1 = -\Omega e_1,
$$
$$
Q = M|_H + \Omega|_{H'} = \Omega + (M - \Omega)\Gamma - \Gamma(M - \Omega), \qquad \Gamma_{ij} = e_{1i}e_{1j},
$$

where $M|_H$, $\Omega|_{H'}$ are projections on the linear subspaces H, $H' \in \bigwedge^2 \mathbb{R}^n$. Equations (4.30) represent a closed system for determining Ω and e_1 as functions of time t. This system possesses multi-dimensional analogs of the integrals (4.24)

$$
\langle Q, Q \rangle = -\frac{1}{2}\operatorname{tr}(M\Gamma + \Gamma M)^2, \qquad (e_1, e_1),
$$
$$
Q_{ij}e_{1k} + Q_{jk}e_{1i} + Q_{ki}e_{1j} = \Omega_{ij}e_{1k} + \Omega_{jk}e_{1i} + \Omega_{ki}e_{1j}, \qquad 1 \leqslant i < j < k \leqslant n.
$$

Besides, it has an analog of the Jacobi–Painlevé integral (4.25), as well as the integral invariant. In the variables (Ω_{ij}, e_{1i}) the latter has the density $\mu = \sqrt{\det \mathcal{A}|_H}$, where $\mathcal{A}|_H$ denotes the restriction of the operator $\mathcal{A}$ to the linear space H.

Similarly to the multi-dimensional generalization of the Chaplygin problem discussed in §2, the system (4.28) or (4.30) also may be integrable.

§5. Historical comments

The reader, familiar with studies on the dynamics of multi-dimensional rigid bodies, may have already noticed that our references regarding the origins of the basic concepts of the theory seem quite unusual. In this connection, the authors would like to give some explanations.

In the current literature devoted to the formulation and the equations of the multi-dimensional analog of the Euler problem, the reader is usually referred to Arnold's well-known paper [1], in which it is also shown that these equations are Hamiltonian on the orbits of the coadjoint action of the group $SO(n)$. However, as it was mentioned above, more than a century ago, the idea of such a generalization was put forward by Cayley [6], and in 1873 Frahm [10] obtained dynamical and kinematic equations for the problem in explicit form. The latter also found a complete set of trivial integrals which are analogs of the classical angular momentum integral and the "area" integral. Moreover, for $n = 4$, he derived a condition on the coefficients of the inertia tensor for the dynamical equations to have an additional quadratic integral and, thereby, to be integrable "by quadratures". This condition, in fact, coincides with the general relation (2.13). (At present, condition (2.13) has become associated with the well-known Manakov paper [16]).

All this enables us to call the equations of free motion of the n-dimensional top the *Euler–Frahm equations*.

After Frahm, this dynamical system was discovered by H. Weyl [25], 1923; Blaschke [3], 1942 (of course, this list can be scarcely regarded as complete). Some authors tried in vain to perform the explicit integration of the Euler–Frahm equations in the four-dimensional case without knowing that the scheme of such an integration procedure had been given by Schottky as early as in 1891 [21]. The latter had noticed the remarkable connection (in fact an isomorphism) between the integrable cases found by Frahm and Clebsch's second integrable case of the Kirchhoff equations describing rigid body motion in an unbound volume of ideal liquid. (Both are six-dimensional dynamical systems). Explicit integration of the latter case had already been performed by that time by the brilliant analyst Kötter (see [15]). As Schottky asserted, Kötter's method of integration could be applied (almost without modifications) to Frahm's case. Later the Schottky paper was forgotten, and only in 1986 the explicit formulas describing the isomorphism mentioned above were obtained again by Bobenko [4], while the generalized Euler equations for the four-dimensional top, as well as the n-dimensional top, were integrated by using the recently discovered finite-gap integration method.

Generally, the problem of explicit integration of dynamical systems in theta-functions of time as a complex argument was very popular in Germany at the end of the last century.

According to some publications, the equations for geodesics of left-invariant metrics on Lie groups, regarded as natural generalizations of the Euler equations, came into use quite recently. However, the real history is different. As early as in 1901, Poincaré represented the Lagrange equations in "group" variables. He gave special attention to the case when the Lagrangian is a left-invariant function and mentioned that half of the equations form a closed system defined on the corresponding Lie algebra. As an example, he considered precisely the Euler equations for the n-dimensional top.

It is interesting to note that this paper of Poincaré's is well known to the physics community and is practically unknown to mathematicians dealing with the theory of *Euler–Poincaré equations on Lie algebras*!

It is also worth mentioning that physicists were definitely aware of the group structure involved in many dynamical systems. For instance, in some publications it is accepted that the interpretation of the Kirchhoff equations as the Euler–Poincaré equations on the Lie algebra $e(3)$ was first given by Novikov and Schmeltzer in [19]. But, as a matter of fact, the group structure of the Kirchhoff equations had been already realized and presented in explicit form by Birkhoff and Braquell in 1945 (see [2, 5]). Besides, even earlier, some researchers from the German applied mechanics community had used the so-called *Motor calculus* (*Motorrechnung*), which is, in fact, a matrix realization of the Lie algebra $e(3)$. The main object of the theory is represented by a second-order tensor (*Impulsmotor*), which turns out to be a skew-symmetric analog of the kinetic momentum relative to the group $E(3)$ defined in (1.7) (see [18]).

The generalization of the Euler–Poincaré equations to systems with left-invariant constraints (Euler–Poincaré–Suslov equations) seems to have been first considered in the paper [13] dealing with the existence of an invariant measure, while the same problem for systems with right-invariant constraints, as well as other relevant questions, were discussed in [9, 23].

References

1. V. I. Arnold, *Sur la géométrie différentielle des groupes de Lit de dimension infinie et ses applications à l'hydrodynamique des fluides parfaits*, Ann. Inst. Fourier (Grenoble) **XXVI** (1966), no. 1, 319–361.
2. G. Birkhoff, *Hydrodynamics. A study in logic, fact and similitude*, Princeton Univ. Press, Princeton, NJ, 1960.

3. W. Blaschke, *Nicht-Euclidische Geometrie und Mechanic*, Hamburger Mathematische Einrelshriften **34** (1942), 39–43.
4. A. I. Bobenko, *Euler equations on Lie algebras* $e(3)$ *and* $so(4)$. *An isomorphism of integrable cases*, Funktsional. Anal. i Prilozhen. **20** (1986), no. 1, 64–66; English transl. in Functional Anal. Appl. **20** (1986).
5. J. Braquell, Bull. Amer. Math. Soc. **52** (1946), 617.
6. A. Cayley, *Sur quelques propriétés des détérminant gauches*, J. Reine Angew. Math. **32** (1846), 119–123.
7. S. A. Chaplygin, *Selected works*, "Nauka", Moscow, 1981. (Russian)
8. Yu. Fedorov, *The motion of a rigid body in a spherical support*, Vestnik Moskov. Univ. Ser. I Math. Mekh. (1988), no. 5, 91-93; English transl. in Moscow Univ. Math. Bull. (1988).
9. ______, *Two integrable nonholonomic systems in classical dynamics*, Vestnik Moskov. Univ. Ser. I Math. Mekh. **105** (1989), no. 4, 38-41; English transl. in Moscow Univ. Math. Bull. (1989).
10. F. Frahm, *Ueber gewisse Differentialgleichungen*, Math. Ann. **8** (1874), 35–44.
11. M. Hölder, *Uber die explizite Form den dynamischen Gleihungen für die Bewegung eines starren Körpers relativ to einem geführten Bezugsystem*, Z. Angew. Math. Mech. **19** (1939), 166–198.
12. V. V. Kozlov, *Invariant measures of the Euler–Poincaré equations on Lie algebras*, Funktsional. Anal. i Prilozhen. **22** (1988), no. 1, 69–70; English transl. in Functional Anal. Appl. **22** (1988).
13. ______, *On integration theory of equations of nonholonomic mechanics*, Adv. in Mech. **8** (1985), no. 3, 85–107. (Russian)
14. V. Kozlov and D. Zenkov, *On geometric Poinsot interpretation for an n-dimensional rigid body*, Tr. Semin. Vectorn. Tenzorn. Anal. **23** (1988), 202–204. (Russian)
15. F. Kötter, *Ueber die Bewegung eines festen Körpers in einer Flüssigkeit* I, II, J. Reine Angew. Math. **109** (1892), 51–81; 89–111.
16. S. V. Manakov, *Remarks on the integrals of the Euler equations of the n-dimensional heavy top*, Funktsional. Anal. i Prilozhen. **10** (1976), no. 4, 93–94; English transl. in Functional Anal. Appl. **10** (1976).
17. A. Markeev, *The motion of the Chaplygin ball with cavities filled with ideal fluid*, Izv. Akad. Nauk SSSR Mekh. Tverd. Tela (1986), no. 1, 64–65; English transl. in Mechanics of Solids.
18. R. v. Mises, *Motorrechnung, ein neues Hilfsmittel der Mechanik*, Z. Angew. Math. Mech. **4** (1924), 196–232.
19. S. P. Novikov and I. Schmeltzer, *Periodic solutions of Kirchhoff equations for the free motion of a rigid body in an ideal fluid and extended Lyusternik–Schnirel'man theory*, Funktsional. Anal. i Prilozhen. **15** (1981), no. 3, 54–66; English transl. in Functional Anal. Appl. **15** (1981).
20. H. Poincaré, *Sur une forme nouvelle des équations de la Mécanique*, C. R. Acad. Sci. **132** (1901), 369–371.
21. F. Schottky, *Uber das analitische Problem der Rotation eines starren Körpers in Raume von vier Dimensionen*, Sitzungber. Königl. Preussischen Acad. Wiss. **XII** (1891), 227–232.
22. G. Suslov, *Theoretical mechanics*, "Gostekhizdat", Moscow–Leningrad, 1951. (Russian)
23. A. P. Veselov and L. E. Veselova, *The flows on Lie groups with nonholonomic constraints and integrable non-Hamiltonian systems*, Funktsional. Anal. i Prilozhen. **20** (1986), no. 4, 65–66; English transl. in Functional Anal. Appl. **20** (1986).
24. H. Yoshida, *Necessary condition for the existence of algebraic first integrals*, I. *Kowalewski exponents*, J. Celest. Mech. **31** (1983), 363–339.
25. H. Weyl, *Raum–Zeit–Materie*, Springer-Verlag, Berlin, 1923.

Translated by YU. FEDOROV

DEPARTMENT OF MATHEMATICS, MOSCOW STATE UNIVERSITY, MOSCOW 119899, RUSSIA

Amer. Math. Soc. Transl.
(2) Vol. **168**, 1995

Integrable Systems, Lax Representations, and Confocal Quadrics

YU. N. FEDOROV

§1. Introduction

The Jacobi problem on geodesics on an ellipsoid—one of the first integrable problem of classical dynamics solved by quadratures—still draws the attention of mathematicians. At present, a wide variety of integrable systems has been discovered, some of them being closely connected with this problem. Perhaps the best known example is the Neumann system describing the motion of a mass point on the unit sphere in a force field with a quadratic potential. The relation between the two systems was completely investigated by J. Moser [17]. Besides, at different times and in different ways, H. Minkowski [16] and V. V. Kozlov [14] indicated an isomorphism between the Jacobi problem and the second Clebsch integrable case of the Kirchhoff equations describing the motion of a rigid body in an ideal fluid.

The present paper displays a close connection between families of confocal quadrics—the geometrical background of the Jacobi problem—and some finite-dimensional integrable Hamiltonian systems discovered in the past twelve years using the well-known Lax pair representation.

First, let us recall the main ideas and results in this field. Consider the equations of motion of a mass point on the $(n-1)$-dimensional ellipsoid in $\mathbb{C}^n$

$$(X, qX) = 1, \tag{1.1}$$

where q is an $(n \times n)$-matrix of a nondegenerate quadratic form and $X = (X_1, \dots, X_n)$ are the Euclidean coordinates. Without loss of generality, we can put

$$q = \operatorname{diag}(1/a_1, \dots, 1/a_n), \qquad a_1 < a_2 < \cdots < a_n .$$

Using the method of Lagrange multipliers, we obtain the geodesic equation

$$\ddot{X} = -\mu q X, \qquad \mu = \frac{(\dot{X}, q\dot{X})}{(X, q^2 X)}, \tag{1.2}$$

which is equivalent to a $2n$-dimensional system of first-order differential equations.

1991 *Mathematics Subject Classification*. Primary 35Q58, 58F07.

The ellipsoid (1.1) can be included in the family of confocal quadrics

$$Q(s)\colon \left\{ \frac{X_1^2}{s-a_1} + \cdots + \frac{X_n^2}{s-a_n} = -1 \right\} \subset \mathbb{C}^n, \qquad s \in \mathbb{C}. \tag{1.3}$$

For $s = a_1, \dots, a_n$ this family contains n "plane" *focal quadrics* that are $(n-2)$-dimensional quadratic surfaces lying in the (focal) hyperplanes $\{X_i = 0\}$. On each of the quadrics $Q(s)$, $s \neq a_1, \dots, a_n$, as on an $(n-1)$-dimensional manifold, one can define curvilinear elliptic coordinates $\lambda_1, \dots, \lambda_{n-1}$ such that

$$X_i^2 = \frac{(a_i - \lambda_1)\cdots(a_i - \lambda_{n-1})(a_i - s)}{\prod_{j\neq i}(a_i - a_j)}. \tag{1.4}$$

Using the substitution (1.4) for $n = 3$ and $s = 0$, as well as first integrals of system (1.2), K. Jacobi [10] reduced the geodesic equation to an inversion problem for hyperelliptic integrals of first and second kind (the latter are so called *Abel transcendents*). For the motion on a quadric $Q(0)$ in $\mathbb{C}^n$, these integrals have the form

$$\begin{gathered} \int_{\lambda_0}^{\lambda_1} \frac{\lambda^i\, d\lambda}{\sqrt{R(\lambda)}} + \cdots + \int_{\lambda_0}^{\lambda_{n-1}} \frac{\lambda^i\, d\lambda}{\sqrt{R(\lambda)}} = \delta_i t + \varepsilon_i, \qquad i = 1, \dots, n-1, \\ R(\lambda) = (\lambda - a_1)\cdots(\lambda - a_n)(\lambda - c_1)\cdots(\lambda - c_{n-1}), \qquad c_1 = 0, \end{gathered} \tag{1.5}$$

where $\delta_1 = \cdots = \delta_{n-2} = 0$, $\delta_{n-1} = 1$, and ε_i are arbitrary constants. The constants $c_2, \dots, c_{n-1}$ depend only on initial conditions. In the real case they lie in intervals

$$[a_1; a_2], \quad \dots, \quad [a_{n-1}; a_n],$$

and at most two different c_i's belong to the same interval. The integrals (1.5) for $n = 3$ were inverted by K. Weierstrass [25] by means of hyperelliptic theta-functions.

The constants $c_2, \dots, c_{n-1}$ have a clear geometric interpretation. Consider the tangent line $l\colon \{X + \tau\dot X \mid \tau \in \mathbb{C}\}$ to a geodesic $X(t)$ on $Q(s)$. Then the qualitative behavior of the geodesic is very well described by the following generalization of the well-known Chasles theorem [5].

THEOREM 1. *All the tangent lines $l(t)$ of the geodesic on $Q(s)$ are also simultaneously tangent to the quadrics $Q(c_2), \dots, Q(c_{n-1})$.*

We add that if the geodesic lies in the hyperplane $\{X_i = 0\}$, then this theorem is valid for the quadrics $\widetilde{Q}(c_\alpha) = Q(c_\alpha) \cap \{X_i = 0\}$, $\alpha \in \{1, \dots, n\} \setminus i$ in the space $\mathbb{C}^{n-1} = \mathbb{C}^n \cap \{X_i = 0\}$.

In follows from this theorem that a general geodesic on a triaxial ellipsoid $Q(0)$ in $\mathbb{C}^3$ oscillates between the two lines of intersection of $Q(0)$ with the hyperboloid $Q(c_2)$, i.e., the curvature lines. When $c_2 = a_1$ or $c_2 = a_3$, the geodesic is the smallest or the biggest ellipse of the ellipsoid respectively, and for $c_2 = a_2$ it is either an asymptotic trajectory tending to the middle ellipse and passing through two of four diametrically opposite *ombilic points* (the intersections of $Q(0)$ with the focal hyperbola $Q(a_2)$) or is the middle ellipse itself.

Now we turn to the Neumann system describing the motion of a point mass on the unit sphere S^{n-1}: $\{a_1^2 + \cdots + a_n^2 = 1\}$, $\gamma \in \mathbb{C}^n$, in the force field with the quadratic potential $\mathcal{U} = (a_1\gamma_1^2 + \cdots + a_n\gamma_n^2)/2$ [19]. Again using the method of Lagrange multipliers, we obtain the equations

$$\dot{\gamma} = v, \qquad \dot{v} = -q\gamma + ((\gamma, q\gamma) - (v, v))\gamma. \tag{1.6}$$

This system is known to be Hamiltonian on the (co)tangent bundle TS^{n-1}, which is the joint level of the first integrals $(\gamma, \gamma) = 1$, $(\gamma, v) = 0$, with respect to the restriction of the standard symplectic structure in $\mathbb{R}^{2n} = (\gamma, v)$ to TS^{n-1}. Besides, the system possesses n integrals in involution

$$F_i = \gamma_i^2 + \sum_{j \neq i}^{n} \frac{(\gamma_i v_k - \gamma_k v_i)^2}{a_i - a_k}, \qquad i = 1, \dots, n-1, \tag{1.7}$$

which for $n = 3$ were found by H. Weber [24]. Among them $n - 1$ integrals are independent on TS^{n-1} (observe that $F_1 + \cdots + F_n = 1$). The Hamiltonian of the system (1.6) is the sum

$$H = \sum_{i=1}^{n} a_i F_i .$$

Thus, this system is integrable by the Liouville theorem. Under the substitution

$$\gamma_i^2 = \frac{(a_i - \nu_1)\cdots(a_i - \nu_{n-1})}{\prod_{j \neq i}(a_i - a_j)}, \qquad i = 1, \dots, n, \tag{1.8}$$

where $\nu_1, \dots, \nu_{n-1}$ are the spheroconic coordinates on S^{n-1}, the integrals (1.7) yield the following Abel equations defined by hyperelliptic integrals of the first kind

$$\int_{\nu_0}^{\nu_1} \frac{\nu^{i-1}\, d\nu}{\sqrt{R(\nu)}} + \cdots + \int_{\nu_0}^{\nu_{n-1}} \frac{\nu^{i-1}\, d\nu}{\sqrt{R(\nu)}} = \varepsilon_i t + d_i, \qquad i = 1, \dots, n-1, \tag{1.9}$$
$$d_i = \text{const}, \quad \varepsilon_1 = \cdots = \varepsilon_{n-2} = 0, \quad \varepsilon_{n-1} = 1,$$
$$R(\nu) = (\nu - a_1)\cdots(\nu - a_n)(\nu - c_1)\cdots(\nu - c_{n-1}).$$

Here, as above, $c_1, \dots, c_{n-1}$ are the constants depending on values of the integrals (1.7). As in the Jacobi problem on geodesics, in the real case the constants c_i in (1.9) may or may not separate $a_1, \dots, a_n$ (however, if all $F_i(v, \gamma) > 0$, then $a_1 < c_1 < a_2 < \cdots < c_{n-1} < a_n$). When $n = 3$, the trajectory $\gamma(t)$ covers in S^2 either a ring, or a curvilinear quadrangle.

There is a remarkable connection between the dynamics on TS^{n-1} and the Jacobi problem. It is described by

THEOREM 2 ([17, 18]). *Let $\gamma(t)$, $v(t)$ be a solution of the Hamiltonian system on TS^{n-1} with the Hamiltonian*

$$\mathcal{H} = \sum_{i=1}^{n} \frac{F_i}{a_i} = 0.$$

Then the line $l(t)\colon \{X = v + \tau\gamma \mid \tau \in \mathbb{C}\}$ *is always tangent to the ellipsoid*

$$Q(0)\colon \left\{\frac{X_1^2}{a_1} + \cdots + \frac{X_n^2}{a_n} = 1\right\},$$

while the contact point $p(t) = l(t) \cap Q(0)$ *with coordinates*

$$X_p = v - \frac{(v\,,\gamma)}{(\gamma\,, q\gamma)}\,\gamma$$

traces a geodesic on $Q(0)$.

Another remarkable relation between the two problems was discovered by H. Knörrer [12].

It follows from Theorems 1 and 2 that the common tangent lines of confocal quadrics play an important role in studying the integrable systems mentioned above. We are going to generalize this nice construction and to consider the varieties of common tangent linear spaces of confocal quadrics. It turns out that such varieties described in terms of the Plücker coordinates lead to all first integrals of the Clebsch–Perelomov system (the Neumann system is its special case) and the well-known Frahm–Manakov m-dimensional top system. Phase flows of these systems may induce geodesic or so called "quasi-geodesic" flows on a quadric (to avoid cumbersome explanations, we omitted proofs of some theorems). Unfortunately, many important questions concerning this connection remain unanswered. All this will be discussed in §4. Before, in §2, we consider the so-called hyperelliptic Lax pair for the Frahm–Manakov and Clebsch–Perelomov systems, which makes it possible to find an especially convenient form of their first integrals. In §3 we provide the necessary background for describing tangent linear spaces of a quadric in a projective space.

§2. Integrable systems and hyperelliptic Lax pairs

In modern approach to integrable Hamiltonian systems, their representation in Lax form (a Lax pair or L–A pair) plays a key role. Such a representation also makes it possible to construct and solve multi-dimensional integrable generalizations of various problems of dynamics. The best-known examples are the generalizations of the Euler and the Clebsch classical systems, whose Lax pairs were found by Manakov [15], Perelomov [20], and Bobenko [3]. These Lax pairs include an additional (spectral) parameter defined on the compactified complex plane or an elliptic curve (Riemannian surface of genus one). Until recently, there were no examples of L–A pairs representing physical systems with a spectral parameter ranging over an algebraic curve of genus > 1.

In this section for the two multi-dimensional generalizations mentioned above, we consider their Lax representations with a spectral parameter defined on an unramified covering of a hyperelliptic curve. A similarity of these Lax pairs enable us to establish a complete isomorphism between the invariant tori of the two multi-dimensional systems. This fact was also discovered in [2] by a different method.

First, recall that the equations of free motion of an m-dimensional rigid body first written by F. Frahm [8] have the Hamiltonian form

$$\dot{M} = [M, \Omega], \qquad \Omega = \frac{\partial H}{\partial M}, \tag{2.1}$$

where $\Omega \in so(m)$ is the angular velocity, $M \in so^*(m)$ is the angular momentum of the body (see also §2 of the article by Fedorov and Kozlov in this volume). These equations are known to be integrable provided the operators M and Ω are connected as follows

$$[M, V] = [\Omega, U], \tag{2.2}$$

where U, V are constant diagonal matrices:

$$U = \operatorname{diag}(a_1, \dots, a_m), \qquad V = \operatorname{diag}(b_1, \dots, b_m),$$

and all the eigenvalues of U and V are distinct. The integrability follows from the Lax pair of the system (2.1) with the rational spectral parameter λ

$$\dot{L}(\lambda) = [L(\lambda), A(\lambda)], \qquad L(\lambda) = M + \lambda U, \quad A(\lambda) = \Omega + \lambda V. \tag{2.3}$$

This Lax pair was found by Manakov [15].

Now suppose that $V = U^2$. Then, in view of (2.2), we have $\Omega = UM + MU$, and equations (2.1) take the simple form

$$\dot{M}_{ij} = (a_j - a_i) \sum_{k=1}^{m} M_{ik} M_{kj}, \qquad i, j = 1, \dots, m. \tag{2.4}$$

Apart from this "basic" system, there exists a whole hierarchy of "higher Manakov systems" commuting with (2.4). These systems are defined by the following hierarchy of A-operators in (2.3):

$$A_{l,r}(\lambda) = \{M^l, U^r\} + \lambda\{M^{l-1}, U^{r+1}\} + \cdots + \lambda^l U^{r+l},$$
$$r = 1, 2, \dots, \quad l = 1, 3, \dots,$$

where $\{M^l, U^r\}$ denotes a homogeneous symmetric matrix polynomial in M and U of degrees s and r respectively. For example:

$$\{M, U\} = MU + UM, \qquad \{M, U^2\} = MU^2 + UMU + U^2M,$$

etc. Expanding both sides of the Lax equations

$$\frac{d}{dt}(M + \lambda U) = [M + \lambda U, A_{l,r}(\lambda)]$$

in powers of λ near $\lambda = \lambda^0$, we obtain the higher Manakov systems

$$\dot{M} = [M, \{M^l, U^r\}], \qquad l = 1, 3, 5, \dots, \quad r \in \mathbb{N}, \tag{2.5}$$

and zero identities at the other powers.

One can show that the hierarchy (2.5) is complete, i.e., any system commuting with (2.4) corresponds to linear combination of the right-hand sides of (2.5).

Now turn to another Lax pair for the Frahm–Manakov equations (2.4). It was found in [7] and has the form

$$\dot{L}(s) = [L(s), A(s)], \tag{2.6}$$

$$L(s)_{ij} = \frac{\sqrt{\Phi(s)}}{\sqrt{(s-a_i)(s-a_j)}} M_{ij}, \qquad A(s)_{ij} = \sqrt{(s-a_i)(s-a_j)}\, M_{ij},$$
$$\Phi(s) = (s-a_1)(s-a_2)\cdots(s-a_m), \qquad i, j = 1, \dots, m,$$

where $s \in \mathbb{C}$ is a parameter. The functions $w_{ij} = \sqrt{(s-a_i)(s-a_j)}$ called *biradicals* are assumed to satisfy the relations

$$w_{ik} w_{kj} = (s-a_k) w_{ij}, \tag{2.7}$$

and then they are single-valued functions on a certain unramified covering of the hyperelliptic curve Γ: $\{w^2 = \Phi(s)\}$. This implies that the quotients w/w_{ij} in $L(s)$ are also single-valued on the same covering. In this connection we shall refer to the Lax pair (2.6) and the parameter s as *hyperelliptic*. Writing out equation (2.6) in scalar form, taking into account (2.7), and eliminating the radicals, we obtain just equations (2.4). The hierarchy of the systems (2.5) on $so(m)$ is generated by the operators

(2.8)
$$A = A_{l,r}(s) = SP_{l,r}(s)S, \qquad l = 1, 3, 5, \dots, \quad r \in \mathbb{N},$$
$$S = \operatorname{diag}(\sqrt{s-a_1}, \dots, \sqrt{s-a_m}),$$
$$P_{l,r}(s) = s^{r-1}M^l + s^{r-2}\{M^l, U\} + \cdots + s\{M^l, U^{r-2}\} + \{M^l, U^{r-1}\}.$$

In particular,

$$(A_{1,r}(s))_{ij} = (s-a_i)(s-a_j) \sum_{l=1}^{r} s^{r-l} \frac{a_j^l - a_i^l}{a_j - a_i} M_{ij}.$$

On $so(m)$ the latter matrices generate quadratic flows with the Hamiltonians

$$H_{1,r} = -\frac{1}{4}\operatorname{tr}(M\{M, U^r\}) = \frac{1}{2}\sum_{i<j}^{m} \frac{a_j^k - a_i^k}{a_j - a_i} M_{ij}^2, \qquad k = r+1. \tag{2.9}$$

PROPOSITION 3. *Biradicals w_{ij}, $i, j = 1, \dots, m$, satisfying conditions from(2.7), as well as the quotients w/w_{ij}, are single-valued functions on a 4^g-sheeted unramified covering $\widehat{\Gamma} \to \Gamma$ of the hyperelliptic curve Γ of genus g.*

By the Riemann–Hurwitz formula $\operatorname{gen}(\widehat{\Gamma}) = 4^g(g-1)+1$. Pick any canonical basis $\mathcal{B}$ in $H_1(\Gamma, \mathbb{Z})$. Then the passage from one Γ-sheet to another occurs after intersecting a cycle from $\mathcal{B}$. On second intersection of the same cycle, one returns to the original Γ-sheet. Thus the curve $\widehat{\Gamma}$ arises from Γ by doubling all cycles in $\mathcal{B}$.

In the case $m = 4$, due to the splitting $so(4) = so(3) + so(3)$, the operators L and A can be packed into (3×3)-matrices, and the Lax representation takes an especially symmetric form (we shall call it the $so(3)$-representation):

$$
\text{(2.10)} \qquad
\begin{gathered}
L_{\alpha\beta}(s) = \sqrt{(s-a_\alpha)(s-a_\beta)}\, M_{\gamma 4} + \sqrt{(s-a_\gamma)(s-a_4)}\, M_{\alpha\beta}\,, \\
A_{\alpha\beta}(s) = \sqrt{(s-a_\alpha)(s-a_\beta)}\, M_{\alpha\beta} + \sqrt{(s-a_\gamma)(s-a_4)}\, M_{\gamma 4}\,, \\
(\alpha\,, \beta\,, \gamma) = (1\,, 2\,, 3)\,.
\end{gathered}
$$

In this case the characteristic polynomial $|L(s) - zE|$ (E is the unit matrix) has the form $-z^3 - z\widetilde{Q}(s)$, where $\widetilde{Q}(s)$ is the following family of quadratic integrals of the Manakov systems on $so(4)$:

$$
\text{(2.11)} \qquad \widetilde{Q}(s) = \sum_{\alpha=1}^{3} \left(\sqrt{(s-a_\alpha)(s-a_4)}\, M_{\beta\gamma} + \sqrt{(s-a_\beta)(s-a_\gamma)}\, M_{\alpha 4}\right)^2.
$$

This family was first obtained by Schottky [23] and rediscovered again by Heine [9] as a curve of (affine) rank 3 quadrics in the projective complex linear span of the four independent quadratic integrals

$$
\begin{gathered}
(a_1 + a_4) M_1^2 + (a_2 + a_4) M_2^2 + (a_3 + a_4) M_3^2 \\
+ (a_2 + a_3) M_4^2 + (a_1 + a_3) M_5^2 + (a_1 + a_2) M_6^2 = d_0\,, \\
a_1 a_4 M_1^2 + a_2 a_4 M_2^2 + a_3 a_4 M_3^2 + a_2 a_3 M_4^2 + a_1 a_3 M_5^2 + a_1 a_2 M_6^2 = d_1\,, \\
M_1^2 + M_2^2 + M_3^2 + M_4^2 + M_5^2 + M_6^2 = d_2\,, \\
M_1 M_4 + M_2 M_5 + M_3 M_6 = d_3\,, \\
d_1\,, d_2\,, d_3\,, d_4 = \text{const}\,, \quad M_\alpha = M_{\alpha 4}\,, \quad M_{\alpha+3} = M_{\alpha\beta}\,, \quad (\alpha\,, \beta\,, \gamma) = (1\,, 2\,, 3)\,.
\end{gathered}
$$

In the case of arbitrary dimension m the characteristic polynomial for $L(s)$ in (2.6) is

$$
\text{(2.12)} \quad |L(s) - zE| = z^m + \sum_{k=2}^{m} z^{m-k} (\Phi(s))^{k/2-1} \widetilde{\mathcal{I}}_k(s\,, M) \qquad (k \text{ is even})\,,
$$

$$
\widetilde{\mathcal{I}}_k(s\,, M) = \sum_I \frac{\Phi(s)}{(s-a_{i_1})\cdots(s-a_{i_k})} |M|_I^I = \sum_{\mu=0}^{m-k} s^\mu H_{k\mu}(M)\,,
$$

where $|M|_I^I$ are order k diagonal minors corresponding to the multi-indices $I =$

$\{i_1, \dots, i_k\} \subset \{1, \dots, m\}$, $1 \leqslant i_1 < \dots < i_k \leqslant m$. In particular,

$$\begin{aligned}
\widetilde{\mathcal{I}}_2(s, M) &= \sum_{i<j}^{m} \frac{\Phi(s)}{(s-a_i)(s-a_j)} M_{ij}^2 \\
&= s^{m-2} \sum_{i<j}^{m} M_{ij}^2 + s^{m-3} \sum_{i<j}^{m} (a_i + a_j - \Delta_1) M_{ij}^2 + \dots \\
&\quad + \sum_{i<j}^{m} \frac{a_1 a_2 \cdots a_m}{a_i a_j} M_{ij}^2 , \\
\widetilde{\mathcal{I}}_4(s, M) &= \sum_{i<j<k<l}^{m} \Phi(s) \frac{(M_{ij}M_{kl} - M_{ik}M_{jl} + M_{il}M_{kj})^2}{(s-a_i)(s-a_j)(s-a_k)(s-a_l)} , \\
&\dots\dots\dots\dots\dots\dots\dots\dots , \\
\Delta_1 &= a_1 + a_2 + \dots + a_m , \dots .
\end{aligned}$$

The coefficients of the polynomial $\widetilde{\mathcal{I}}_k(s, M)$ are homogeneous integrals of degree k in M. The leading coefficients $H_{k, m-k}(M)$ are the invariants (Casimir functions) $\mathcal{I}_k(M) = \sum_I |M|_I^I$ of the algebra $so(m)$. The polynomials $H_{k\mu}(M)$, $k = 2, 4, \dots \leqslant m$, $\mu = 0, 1, \dots, m-k$ form a complete set of independent first integrals in involution. In particular, $H_{2\mu}(M)$ are linear combinations of the Hamiltonians (2.9) generating quadratic flows on $so(m)$.

For example, the Manakov system on the 15-dimensional algebra $so(6)$ possesses 9 independent integrals $H_{k\mu}(M)$ such that

$$\begin{aligned}
\widetilde{\mathcal{I}}_2(s, M) &= H_{20} + sH_{21} + s^2 H_{22} + s^3 H_{23} + s^4 H_{24} , \\
\widetilde{\mathcal{I}}_4(s, M) &= H_{40} + sH_{41} + s^2 H_{42} , \\
\widetilde{\mathcal{I}}_6(s, M) &= H_{60} .
\end{aligned}$$

Therefore, the dimension of general invariant tori of this system equals $15 - 9 = 6$. This is consistent with the general formula

$$\text{dim of general tori} = \frac{1}{2}[\dim so(m) - \operatorname{corank} so(m)] = \frac{1}{2}\left(\frac{m(m-1)}{2} - \left[\frac{m}{2}\right]\right).$$

The angular momentum M can be naturally associated with the 2-vector

$$\omega = \sum_{i<j}^{m} M_{ij}\, dx_i \wedge dx_j , \tag{2.13}$$

which, in general, has maximal rank $2[m/2]$. In the degenerate case when rank $\omega = k < 2[m/2]$, the invariant $H_{k, m-k}(M)$, as well as *all* integrals of this and higher degrees, vanishes, and the dimension of the corresponding general invariant tori drops to the number of nonzero nontrivial integrals, i.e., to $k(m - (k+2)/2)/2$. For the maximal degeneration $k = 2$ all the integrals can be represented as quadratic ones, and the dimension of the corresponding invariant tori equals $m - 2$.

REMARK 1. Fix some values of all the first integrals of the Frahm–Manakov system by putting in (2.12)

$$\widetilde{\mathcal{I}}_k(s\,,\,M) = c_0^{(k)}(s - c_1^{(k)})\cdots(s - c_{m-k}^{(k)})\,, \qquad k = 2\,,\,4\,,\,\ldots \leqslant m\,,$$

$$c_0^{(k)}\,,\,c_1^{(k)}\,,\,\ldots\,,\,c_{m-k}^{(k)} = \text{const}\,.$$

Consider the spectral curve $\mathcal{K} \subset \mathbb{C}^2 = (z\,,\,\lambda)$ defined by the characteristic equation $|L(\lambda) - zE| = 0$ for Manakov's Lax pair (2.3). According to the method of finite-gap integration, the phase flow of the Frahm–Manakov system can be realized as a straight line motion on the Jacobi variety $\operatorname{Jac}(\mathcal{K})$ of the curve $\mathcal{K}$ (see [6]). Besides, in the case $m = 4$, Adler and van Moerbecke [1] linearized the flow on a two-dimensional Prym variety $\operatorname{Prym}(\mathcal{K}/\mathcal{K}_0) \subset \operatorname{Jac}(\mathcal{K})$, where $\mathcal{K}_0 = \mathcal{K}/\tau$ is the quotient of $\mathcal{K}$ by the involution $\tau\colon (z\,,\,\lambda) \to (-z\,,\,-\lambda)$ related to skew-symmetry of M. The curve $\mathcal{K}$ has genus 3 and can be viewed as a double covering of the elliptic curve $\mathcal{K}_0$ ramified at four points on $\mathcal{K}_0$. However, Heine [9] showed that only squares of M_{ij}, but not M_{ij} themselves, are Abelian functions on $\operatorname{Prym}(\mathcal{K}/\mathcal{K}_0)$, i.e., the affine part of $\operatorname{Prym}(\mathcal{K}/\mathcal{K}_0)$ does not characterize the joint level $\mathcal{A}$ of the four integrals of the system. On the other hand, he considered another genus 3 curve

$$\mathcal{C}\colon \{z^2 + \widetilde{\mathcal{I}}_2(s) + 2d_3\sqrt{\Phi(s)} = 0 \mid \Phi(s) = (s - a_1)\cdots(s - a_4)\}\,,$$

which is the double ramified covering of the elliptic curve $\Gamma\colon \{w^2 = \Phi(s)\}$ with four branch points on Γ, and concluded that $\mathcal{A}$ coincides with the affine part of $\operatorname{Prym}(\mathcal{C}/\Gamma)$. Thus, the Abelian varieties $\operatorname{Prym}(\mathcal{K}/\mathcal{K}_0)$ and $\operatorname{Prym}(\mathcal{C}/\Gamma)$ are not isomorphic, but only *isogeneous*. In the special case $d_3 = 0$ the torus $\operatorname{Prym}(\mathcal{C}/\Gamma)$ is isogeneous to the Jacobi variety of the genus 2 hyperelliptic curve $\Gamma'\colon \{y^2 = \Phi(s)\widetilde{\mathcal{I}}_2(s)\}$ (three of four periods of $\operatorname{Jac}(\Gamma')$ are multiplied by 2).

It is remarkable that the "right" curve $\mathcal{C}$ appears as a part of the spectral curve $\overline{\mathcal{C}}$ for the elliptic Lax pair (2.10) provided its spectral parameter ranges over the curve Γ. More precisely, $\overline{\mathcal{C}}$ is the union of $\mathcal{C}$ and the line $\{z = 0\}$.

Trying to find an analog of this phenomenon for higher-dimensional case, consider the spectral curve defined by the characteristic polynomial (2.12) for the hyperelliptic Lax pair on $so(m)$ and suppose that $\operatorname{rank} M = 4$. Then this curve decomposes into the line $\{z^{m-4} = 0\}$ and the curve given by the equations

$$z^4 + \widetilde{\mathcal{I}}_2(s)z^2 + \widetilde{\mathcal{I}}_4(s)w^2 = 0\,, \qquad w^2 = \Phi(s)\,.$$

CONJECTURE. *For the degenerate case* $\operatorname{rank} M = 4$, *the joint level of all first integrals of the system on* $so(m)$ $(m \geqslant 4)$ *is the affine part of the Abelian variety* $\operatorname{Prym}(\mathcal{E}/\mathcal{E}_0)$, *where the genus* $3m - 9$ *curve*

$$\mathcal{E}\colon \left\{u^2 + \widetilde{\mathcal{I}}_2(s) + 2\sqrt{\Phi(s)\widetilde{\mathcal{I}}_4(s)} = 0\right\}$$

is the double ramified covering of the genus $m - 3$ *hyperelliptic curve*

$$\mathcal{E}_0\colon \{t^2 = \Phi(s)\widetilde{\mathcal{I}}_4(s)\}\,,$$

the factor of $\mathcal{E}$ *by the involution* $u \to -u$.

Another hierarchy of integrable systems arises as the result of multi-dimensional generalizations of the Clebsch case in the Kirchhoff equations describing the motion of a rigid body in an ideal fluid. These equations turn out to be equivalent to the equations of the motion of a rigid body with a fixed point in a potential field with a quadratic potential function (Brun's problem [4]). For the n-dimensional case, we write this potential function in the form

$$Y(\gamma) = (\gamma, B\gamma)/2, \qquad B = \operatorname{diag}(b_1, \dots, b_n), \; b_1, \dots, b_n = \text{const}.$$

Here γ denotes a constant vector in $\mathbb{R}^n$. Its evolution in the moving frame attached to the body is described by the Poisson equations

$$\dot{\gamma} = -\Omega\gamma, \tag{2.14}$$

where, as before, Ω is the angular velocity of the body.

Let $K \in so^*(n)$ be the angular momentum and

$$\Omega_{ij} = c_{ij} K_{ij}, \qquad c_{ij} = \text{const}, \; i, j = 1, \dots, n.$$

Then the equations of motion can be written in the form

$$\begin{aligned} \dot{K} &= [K, \Omega] + [\Gamma, B], \qquad \Gamma_{ij} = \gamma_i\gamma_j, \\ \dot{\gamma} &= -\Omega\gamma. \end{aligned} \tag{2.15}$$

These equations are Hamiltonian with respect to the Lie–Poisson–Kirillov bracket of the Lie algebra $e(n)$ with the Hamilton function

$$H = -\frac{1}{2}\operatorname{tr}(K\Omega) + Y(\gamma) = \frac{1}{2}\sum_{i<j}^{n} c_{ij}K_{ij}^2 + \frac{1}{2}\sum_{i=1}^{n} b_i\gamma_i^2.$$

The system (2.15) always has $[(n+1)/2]$ independent trivial integrals, the Casimir functions of the algebra $e(n)$

$$\mathcal{J}_k(K, \gamma) = \sum_{J'} \begin{vmatrix} K & \gamma \\ -\gamma^T & 0 \end{vmatrix}_{J'}^{J'}, \qquad k = 2, \dots, 2[n+1/2],$$

i.e., the sums of all diagonal minors of order k corresponding to the multi-indices $J' = \{j_1, \dots, j_k\}$ containing the last index $n+1$. For example, $\mathcal{J}_2 = \gamma_1^2 + \dots + \gamma_n^2$, and for $n = 3$

$$\mathcal{J}_4(K, \gamma) = (K_{12}\gamma_3 + K_{23}\gamma_1 + K_{31}\gamma_2)^2,$$

which is square of the well-known area integral in the classical rigid body problem.

As was shown by Perelomov [20], system (2.15) is completely integrable provided K, Ω are related as follows

$$[K, B] + [D, \Omega] = 0,$$

where $D = \operatorname{diag}(D_1, \dots, D_n)$, $D = \text{const}$, and in this case it admits the Lax pair with a rational spectral parameter λ

$$\frac{d}{dt}(\Gamma/\lambda + K + \lambda D) = [\Gamma/\lambda + K + \lambda D, \Omega + \lambda B].$$

For λ^{-1} we obtain the equation $\dot{\Gamma} = [\Gamma, \Omega]$, which is equivalent to the Poisson equation (2.14).

In the sequel we restrict ourselves to the case $B = D^2$. Thus $\Omega = DK + KD$. The hierarchy of "higher" Clebsch–Perelomov systems is generated by the Lax pairs

$$\dot{L}(\lambda) = [L(\lambda), A_{l,r}(\lambda)], \qquad l, r \in \mathbb{N}, \; l - r \text{ is odd}, \tag{2.16}$$

where $A_{l,r}(\lambda)$ is the result of the elimination of the principal part of $\lambda^{-r}L^l(\lambda)$.

The L–A pair (2.16) with the matrix $A_{1,0}(\lambda)$ gives rise to the simplest system of the hierarchy

$$\dot{K} = [\Gamma, D], \quad \dot{\Gamma} = [\Gamma, K], \qquad \Gamma_{ij} = \gamma_i\gamma_j. \tag{2.17}$$

As noticed by Ratui [21], the restriction of this system to the joint level of the invariant manifolds $\mathcal{J}_4(K, \gamma) = 0$, $\mathcal{J}_6(K, \gamma) = 0, \ldots$ under the substitution

$$K_{ij} = \dot{\gamma}_i\gamma_j - \gamma_i\dot{\gamma}_j \tag{2.18}$$

reduces to the n-dimensional Neumann system (1.6) where a_i are replaced by D_i.

Now let us consider the hyperelliptic Lax pair for the Clebsch–Perelomov systems for the same case $B = D^2$:

$$\begin{gathered} \dot{L}(r) = [L(r), A(r)], \qquad r \in \mathbb{C}, \\ L(r) = \begin{pmatrix} W(r) & P(r) \\ -P^T(r) & 0 \end{pmatrix}, \qquad W(r)_{ij} = \sqrt{\Psi(r)}/\sqrt{(r - D_i)(r - D_j)}\, K_{ij}, \\ P_i(r) = \sqrt{\Psi(r)}/\sqrt{r - D_i}\,\gamma_i, \\ \Psi(r) = (r - D_1)\cdots(r - D_n), \qquad i, j = 1, \ldots, n. \end{gathered} \tag{2.19}$$

Here for the system (2.17) we have

$$A(r) = \begin{pmatrix} & & & -\sqrt{r - D_1}\,\gamma_1 \\ & * & & \vdots \\ & & & -\sqrt{r - D_n}\,\gamma_n \\ \sqrt{r - D_1}\,\gamma_1 & \cdots & \sqrt{r - D_n}\gamma_n & 0 \end{pmatrix}$$

whereas for the Brun system (2.15) with $B = D^2$

$$\begin{gathered} A(r) = \begin{pmatrix} \widehat{W}_{ij}(r) & \widehat{P}(r) \\ -\widehat{P}^T(r) & 0 \end{pmatrix}, \\ \widehat{P}_i(r) = \sqrt{r - D_1}\,(r + D_i)\gamma_i, \qquad \widehat{W}_{ij}(r) = \sqrt{(r - D_i)(r - D_j)}\, K_{ij}, \\ i, j = 1, \ldots, n. \end{gathered}$$

The functions $\sqrt{\Psi(r)}$, $\sqrt{(r-D_i)(r-D_j)}$ satisfy a condition analogous to (2.7). Writing out the characteristic polynomial for $L(r)$, we obtain

(2.20)

$$|L(r)-zE| = z^{n+1} + \sum_{k=2}^{n} z^{n+1-k}(\Psi(s))^{k/2-1}\widetilde{\mathcal{J}}_k(r\,,K) \qquad (k \text{ is even}),$$

$$\begin{aligned}\widetilde{\mathcal{I}}_k(r\,,K\,,\gamma) &= \sum_J \frac{\Psi(r)}{(r-D_{j_1})\cdots(r-D_{j_k})}|K|_J^J \\ &\quad + \sum_{J'} \frac{\Psi(r)}{(r-D_{j_1})\cdots(r-D_{j_{k-1}})}\begin{vmatrix} K & \gamma \\ -\gamma^T & 0\end{vmatrix}_{J'}^{J'} \\ &= \sum_{\mu=0}^{n+1-k} r^\mu H_{k\mu}(K\,,\gamma), \qquad J=\{j_1\,,\dots\,,j_k\}\subset\{1\,,\dots\,,n\},\end{aligned}$$

where, as in (2.12), $|\cdot|_J^J$, $|\cdot|_{J'}^{J'}$ are diagonal minors of order k, and $J'=\{j_1\,,\dots\,,j_{k-1}\,,n+1\}$, $1\leqslant j_1<\cdots<j_{k-1}\leqslant n$. The leading coefficients $H_{k\,,n-k}$ are the invariants $\mathcal{I}_k(K\,,\gamma)$ of the algebra $e(n)$. As for the hyperelliptic Lax pair (2.6) for the Manakov system, all integrals $H_{k\mu}(K\,,\gamma)$ in (2.20) are independent and in involution with respect to the Lie–Poisson bracket of the algebra $e(n)$. Their linear combinations are the Hamiltonians of the Clebsch–Perelomov systems (2.15). For example[1],

$$\widetilde{\mathcal{J}}_2(r) = \sum_{j=1}^{n}\frac{\Psi(r)}{r-D_j}\gamma_j^2 + \sum_{j<i}^{n}\frac{\Psi(r)}{(r-D_i)(r-D_j)}K_{ij}^2\,. \tag{2.21}$$

Putting successively $r=D_j$, $j=1\,,\dots\,,n$, into (2.21) and dividing by

$$\prod_{l\neq j}(D_j-D_l)\,,$$

we get n independent integrals of the special form

$$F_j(K\,,\gamma) = \gamma_i^2 + \sum_{i\neq j}^{n}\frac{K_{ij}^2}{D_j-D_i}\,. \tag{2.22}$$

Thus, we can write

$$\widetilde{\mathcal{J}}_2(r) = \sum_{j=1}^{n}\frac{\Psi(r)}{r-D_j}F_j(K\,,\gamma)\,.$$

It is obvious that after the substitutions (2.18) and $D_i=a_i$, the functions F_j coincide with the integrals (1.7) of the Neumann problem. The sums

$$H_1 = \frac{1}{2}\sum_{j=1}^{n} D_jF_j\,, \qquad H_2 = \frac{1}{2}\sum_{j=1}^{n} D_j^2F_j$$

[1] For simplicity here and below we suppose that $|\gamma|=1$.

are the Hamiltonians of the system (2.17) and the system (2.15) with $B = D^2$, $\Omega = DK + KD$ respectively. We note that for the case $n = 3$ the integrals (2.22) were first found by Weber [24].

As in (2.13), the variables K, γ can be associated with the 2-vector

$$\omega = \sum_{i<j}^{n} K_{ij} dx_i \wedge dx_j + \sum_{i}^{n} \gamma_i dx_i \wedge dx_0 , \tag{2.23}$$

which, in general, has the maximal rank $2[(n+1)/2]$.

Note that for the case $n = 3$ the operators $L(r)$ and $A(r)$ can be written in the $so(3)$-representation, which is similar to the representation (2.10) for the Frahm–Manakov system, namely

$$\begin{aligned} L_{ij}(r) &= \sqrt{(r - D_i)(r - D_j)}\,\gamma_k + \sqrt{r - D_\gamma}\,K_{ij} ,\\ A_{ij}(r) &= \sqrt{(r - D_i)(r - D_j)}\,K_{ij} + \sqrt{r - D_k}\,\gamma_k , \end{aligned} \qquad (i, j, k) = (1, 2, 3).$$

The characteristic polynomial for this special Lax pair has the form $-r^3 - \widetilde{Q}(r) r$, where

$$\widetilde{Q}(r) = \sum_{i=1}^{3} \left(\sqrt{r - D_i}\,K_{jk} + \sqrt{(r - D_j)(r - D_k)}\,\gamma_i \right)^2 . \tag{2.24}$$

The family (2.24) of quadratic integrals was first found by Kötter [13].

The similarity of the operators $L(s)$ and $L(r)$ in the hyperelliptic Lax pairs (2.6), (2.19) indicates that there must exist a connection between the Manakov system on $so(m)$ and the Clebsch–Perelomov system on $e(n)$, $n = m - 1$. Indeed, let the parameters in these Lax pairs be related as follows

$$r = 1/(s - a_m), \quad a_i - a_m = 1/D_i , \qquad i = 1, \ldots, n. \tag{2.25}$$

Therefore

$$\begin{aligned} \sqrt{\Phi(s)} &= \frac{\sqrt{\Psi(r)}}{r\sqrt{\Psi(0)}}, \qquad \sqrt{(s - a_i)(s - a_j)} = \frac{\sqrt{(r - D_i)(r - D_j)}}{r^{m/2}\sqrt{D_i D_j}} ,\\ &\sqrt{(s - a_i)(s - a_m)} = \frac{\sqrt{r - D_i}}{r\sqrt{-D_i}} . \end{aligned} \tag{2.26}$$

Each of these relations is determined up to a flip of sign. However, they turn to be single-valued on the corresponding Riemann surface, where all the radicals are single-valued as well. Then, in view of (2.26), under the substitution

$$M_{ij} = \frac{\sqrt{D_i D_j}}{\sqrt{\Psi(0)}} K_{ij} , \quad M_{im} = \frac{\sqrt{D_i}}{\sqrt{\Psi(0)}} \gamma_i , \qquad i, j = 1, \ldots, n, \tag{2.27}$$

the operator $L(s)$ transforms to $L(r)$ for $m = n - 1$ up to multiplication by r^ν, $\nu = m/2 - 1$. Since the characteristic equations for two L-operators that differ by a scalar factor lead to the same set of first integrals, we arrive at the following

THEOREM 4. *Let the parameters of the Frahm–Manakov system on* $so(m)$ *and the Clebsch–Perelomov system on* $e(m-1)$ *be related as described in* (2.25). *Then there is a global isomorphism between the invariant tori of these systems described by* (2.27).

For the case $m = n+1 = 4$ this isomorphism was first noticed by Schottky [23], who compared the families (2.11), (2.24), and was described in an explicit form by Bobenko [3]. For arbitrary dimension m formulas (2.27) were also found by Bolsinov [2] by a different method.

§3. Linear spaces tangent to quadrics

As we have seen (cf. Theorem 1), tangent lines to quadrics play an important role in the qualitative description of the geodesics on an ellipsoid in the Jacobi problem as well as of the trajectories on an unit sphere in the Neumann problem. In order to set the background for further constructions, let us turn to more general objects, $(k-1)$-dimensional linear spaces ($(k-1)$-planes) Λ in the n-dimensional projective space $\mathbb{P}^n = (x_0 : x_1 : \ldots : x_n)$, where $x_0, x_1, \ldots, x_n$ are homogeneous coordinates. The set of all such $(k-1)$-planes is known as the *Grassmann variety* or *Grassmannian* $G(k, n+1)$, which is also equivalent to the set of all k-planes $\overline{\Lambda}$ in $\mathbb{C}^{n+1}$ passing through the origin. (It is obvious that the map

$$\mathbb{C}^{n+1} = (x_0, x_1, \ldots, x_n) \to \mathbb{P}^n = (x_0 : x_1 : \ldots : x_n)$$

gives a one-to-one correspondence between the planes Λ and $\overline{\Lambda}$, and the planes $\overline{\Lambda}$ lying in the hyperplane $\{x_0 = 0\} \subset \mathbb{C}^{n+1}$ correspond to planes Λ at infinity of $\mathbb{P}^n$.) The Grassmannian $G(k, n+1)$ is a smooth compact manifold of dimension $k(n+1-k)$.

Any k-plane $\overline{\Lambda}$ can be represented as a linear span of independent vectors $v_1, \ldots, v_k \in \mathbb{C}^{n+1}$ forming the factorable polyvector (k-vector)

$$v_1 \wedge \ldots \wedge v_k = \sum G_I e_I,$$

where $I = \{i_1, \ldots, i_k\} \subset \{0, 1, \ldots, n\}$, the vectors $e_0, e_1, \ldots, e_n$ are basis vectors in $\mathbb{C}^{n+1}$, and $e_I = e_{i_1} \wedge \cdots \wedge e_{i_k}$ are basis k-vectors in the exterior power $\bigwedge^k \mathbb{C}^{n+1}$, i.e., the linear space of all k-vectors in $\mathbb{C}^{n+1}$. Let $v_l = (v_{l0}, v_{l1}, \ldots, v_{ln})^T$. The coefficients G_I are determinants of the corresponding $k \times k$ minors of the $k \times (n+1)$ matrix

$$\begin{pmatrix} v_{10} & \cdots & v_{k0} \\ \vdots & \ddots & \vdots \\ v_{1n} & \cdots & v_{kn} \end{pmatrix}.$$

Thus we have defined the *Plücker embedding*

$$G(k, n+1) \to \mathbb{P}(\textstyle\bigwedge^k \mathbb{C}^{n+1}) = \mathbb{P}^{N-1}, \qquad N = \binom{n+1}{k},$$

which maps a $(k-1)$-plane $\Lambda \subset \mathbb{P}^n$ to the point in $\mathbb{P}^{N-1}$ with the homogeneous Plücker coordinates $\{G_I\}$. It is obvious that homogeneous coordinates in $\mathbb{P}^{N-1}$ define a $(k-1)$-plane Λ in $\mathbb{P}^n$ only if they are coefficients of a factorable polyvector.

In particular, the Grassmannian $G(2, n+1)$ is the set of lines in $\mathbb{P}^n$. To each line in $\mathbb{C}^n \subset \mathbb{P}^n$ determined by the parametric equation

$$X = d + \gamma\tau, \qquad \tau \in \mathbb{C},\ d, \gamma \in \mathbb{C}^n,\ \gamma \neq 0,$$

there corresponds the point in $\mathbb{P}^{(n+1)n/2-1}$ with the Plücker coordinates

$$\begin{aligned}&(G_{12} : \dots : G_{n-1\,n} : \gamma_1 : \dots : \gamma_n),\\ &G_{ij} = d_i\gamma_j - d_j\gamma_i, \quad i < j = 1, \dots, n.\end{aligned} \tag{3.1}$$

For lines at infinity $\mathbb{P}^n \setminus \mathbb{C}^n$, we must put $\gamma = 0$ in (3.1).

In the space $\mathbb{C}^{n+1}$ a line $l \subset \mathbb{P}^n$ is represented by the factorable 2-vector

$$\bar{l} = \sum_{0\leqslant i<j\leqslant n} \overline{G}_{ij} e_i \wedge e_j = \bar{d} \wedge \bar{\gamma} = \sum_{1\leqslant i<j\leqslant n} G_{ij} e_i \wedge e_j + \sum_{i=1}^{n} \gamma_i e_i \wedge e_0,$$

$$\bar{d} = (1, d_1, \dots, d_n)^T, \ \bar{\gamma} = (0, \gamma_1, \dots, \gamma_n)^T.$$

In this case the factorability condition has the form $\bar{l} \wedge \bar{l} = 0$. It follows that points in $\mathbb{P}^{(n+1)n/2-1}$ corresponding to lines in $\mathbb{P}^n$ are exactly those points that belong to the intersection of the $\binom{n+1}{4}$ quadrics

$$\sum \varepsilon_{i_1 i_2 i_3 i_4} \overline{G}_{i_1 i_2} \overline{G}_{i_3 i_4} = 0, \tag{3.2}$$

where $\varepsilon_{i_1 i_2 i_3 i_4}$ is the Levi–Civita tensor. For the classical case $n = 3$, we have only one condition

$$G_{23}\gamma_1 + G_{31}\gamma_2 + G_{12}\gamma_3 = 0,$$

which defines the well-known *Klein quadric*.

The Grassmannian $G(1, n+1)$ is the projective space $\mathbb{P}^n$ itself, while $G(n, n+1)$ is the *dual projective space* $*\mathbb{P}^n = (y_0 : y_1 : \dots : y_n)$, i.e., the space of all hyperplanes $\{x_0 y_0 + x_1 y_1 + \dots + x_n y_n = 0\} \subset \mathbb{P}^n$. Here $y_i = G_I$, $I = \{0, 1, \dots, n\} \setminus i$ are the Plücker coordinates of a hyperplane, also known as its *tangential coordinates*.

Recall that any quadratic surface (a quadric) Q in $\mathbb{P}^n$ can be thought of as the set of zeros of the quadratic form

$$\sum_{i,j=0}^{n} \bar{q}_{ij} x_i x_j$$

with a symmetric $(n+1) \times (n+1)$ matrix $\bar{q}$. In $\mathbb{C}^{n+1} = (x_0, x_1, \dots, x_n)$ this set is represented by a cone $\overline{Q}$ with the vertex at the origin. In the Euclidean space $\mathbb{C}^n \subset \mathbb{P}^n$ the affine part of Q is given by the equation

$$Q(X) = \sum_{i,j=1}^{n} \bar{q}_{ij} X_i X_j + \dots + \sum_{i=1}^{n} \bar{q}_{i0} X_i + \bar{q}_{00} = 0. \tag{3.3}$$

The quadric Q is called the *projective closure* of the quadric (3.3). It is a nondegenerate quadric if the corresponding matrix $\bar{q}$ is nondegenerate.

Let us consider the set of all $(k-1)$-planes tangent to a nondegenerate quadric Q in $\mathbb{P}^n$. It turns out that this set is also described as a certain quadric in $\bigwedge^k \mathbb{C}^{n+1}$.

THEOREM 5. *The set T^kQ of all k-planes (factorable polyvectors) $\overline{\Lambda} = \sum G_I e_I$ that are tangent to the nondegenerate quadric $\{(x, \overline{q}x) = 0\}$ is the intersection of the Grassmannian $G(k, n+1) \subset \bigwedge^k \mathbb{C}^{n+1}$ with the quadric*

$$\sum_{I,J} |\overline{q}|^I_J G_I G_J = 0, \tag{3.4}$$

where $|q|^I_J$ is the minor of $\overline{q}$ corresponding to rows with indices I and columns with indices J.

COROLLARY 6. *In the case $k = n$, when Λ is the hyperplane $\{x_0y_0 + x_1y_1 + \cdots + x_ny_n = 0\}$ in $\mathbb{P}^n$, and its k Plücker coordinates G_I coincide with its tangential coordinates y_i, $i = \{0, 1, \dots, n\} \setminus I$, the quadric (3.4) is reduced to the well-known dual quadric*

$$(y, \overline{q}^{-1}y) = 0 \tag{3.5}$$

in the dual projective space $*\mathbb{P}^n$.

PROOF OF THEOREM 5. Note that on the product $\mathbb{C}^{n+1} \otimes \mathbb{C}^{n+1}$ the nondegenerate matrix $\overline{q}$ defines the symmetric pairing $\langle v, w\rangle = (v, \overline{q}w)$, which can be extended to the direct product $\bigwedge^k \mathbb{C}^{n+1} \otimes \bigwedge^k \mathbb{C}^{n+1}$. Namely, for any two basis k-vectors $e_I = e_{i_1} \wedge \cdots \wedge e_{i_k}$, $e_J = e_{j_1} \wedge \cdots \wedge e_{j_k}$, we put

$$\langle e_I, e_J\rangle = \sum (-1)^r \langle e_{i_1}, e_{p_1}\rangle \cdots \langle e_{i_k}, e_{p_k}\rangle = |\overline{q}|^I_J, \tag{3.6}$$

which is the sum over all permutations $(p_1, \dots, p_k)$ of $(j_1, \dots, j_k)$, and r is the parity of the permutation, One sees easily that this operation is in fact symmetric in e_I, e_J and under a permutation of any two indices in I or J the both sides of (3.6) changes signs simultaneously. By linearity, this operation extends to $\bigwedge^k \mathbb{C}^{n+1} \otimes \bigwedge^k \mathbb{C}^{n+1}$. For factorable k-vectors

$$\overline{\Lambda} = v_1 \wedge \cdots \wedge v_k, \qquad \Xi = w_1 \wedge \cdots \wedge w_k \tag{3.7}$$

we have

$$\langle \overline{\Lambda}, \Xi\rangle = \sum (-1)^r \langle v_{i_1}, w_{p_1}\rangle \cdots \langle v_{i_k}, w_{p_k}\rangle, \tag{3.8}$$

$(p_1, \dots, p_k)$ being a permutation of $(1, \dots, k)$.

Now let Λ be a $(k-1)$-plane tangent to Q. Then the polyvector $\overline{\Lambda} = \sum G_I e_I$ can be represented in the form (3.7), such that, for example $\langle v_1, v_1\rangle = 0$ (the vector v_1 lies on a generator of the cone $\overline{Q}$) and $\langle v_1, v_i\rangle = 0$ for other i's. Therefore, taking (3.8) into account, we obtain

$$\langle \overline{\Lambda}, \overline{\Lambda}\rangle = \sum_{I,J} |\overline{q}|^I_J G_I G_J = 0.$$

Thus, the Plücker coordinates of the tangent $(k-1)$-plane satisfy the condition (3.4).

It remains to prove that if a factorable k-vector $\overline{\Lambda}$ satisfies (3.4), then the corresponding $(k-1)$-plane Λ is tangent to Q. Note that the expression $\langle\overline{\Lambda}, \overline{\Lambda}\rangle$ up to multiplication by a nonzero factor is the determinant of the quadratic form $q|_{\overline{\Lambda}}$, i.e., the restriction of $\overline{q}$ to the k-dimensional linear space $\overline{\Lambda} \subset \mathbb{C}^{n+1}$. The form $q|_{\overline{\Lambda}}$ is degenerate if and only if $\overline{\Lambda}$ is tangent to the cone $\overline{Q}$, i.e., Λ is tangent to Q. □

For $k = n$ we have

$$|\overline{q}|^I_J = (\overline{q}^{-1})_{ij} \det \overline{q}, \qquad i = \{0, 1, \dots, n\} \setminus I, \; j = \{0, 1, \dots, n\} \setminus J,$$

which establishes Corollary 6.

Using the form $q|_{\overline{\Lambda}}$, one can find the homogeneous coordinates

$$x^* = (x_0^*, x_1^*, \dots, x_n^*)$$

of the contact point $p = \Lambda \cap Q$, or, which is the same, the coordinates of the vector lying on the tangent line $\overline{\Lambda} \cap \overline{Q}$ in $\mathbb{C}^{n+1}$. Namely, for $\overline{\Lambda} = v_1 \wedge \cdots \wedge v_k$ we look for this vector in the form $x^* = t_1^* v_1 + \cdots + t_k^* v_k$, where $t^* = (t_1^*, \dots, t_k^*)$ is the one-dimensional kernel of $q|_{\overline{\Lambda}}$ (in the special cases when Λ and Q do not intersect at a single point, but along an l-plane, the kernel is $(l+1)$-dimensional).

Suppose without loss of generality that $\overline{q} = \operatorname{diag}(1/a_0, 1/a_1, \dots, 1/a_n)$. Then, after some calculations, we obtain

$$x_i^* = \sum_{J'} \frac{1}{a_{J'}} G_{iJ'} (v_1 \wedge \cdots \wedge v_{k-1})_{J'}, \qquad a_{J'} = a_{i_1} \cdots a_{i_{k-1}}, \tag{3.9}$$

where the multi-indices $J' = \{i_1, \dots, i_{k-1}\}$ range over all $k-1$ combinations from $\{0, 1, \dots, n\} \setminus i$, and $(v_1 \wedge \cdots \wedge v_{k-1})_{J'}$ are the Plücker coordinates in $G(k-1, n+1)$ defining an arbitrary $k-1$-dimensional linear subspace $\overline{\Lambda'} = \operatorname{span}(v_1, \dots, v_{k-1}) \subset \overline{\Lambda}$. Suppose that $\overline{\Lambda'}$ is the section of $\overline{\Lambda}$ by the hyperplane $\{x_r = 0\}$. Then in (3.9) we can put

$$v_l = (v_{l0}, \dots, v_{l,r-1}, 0, v_{l,r+1}, \dots, v_{ln})^T, \qquad l = 1, \dots, k-1.$$

The result is described by

THEOREM 7. *Let the $(k-1)$-plane Λ be tangent to the nondegenerate quadric $Q\colon \{x_0^2/a_0 + x_1^2/a_1 + x_2^2/a_2 + \cdots + x_n^2/a_n = 0\}$ at a single point p. Then the homogeneous coordinates of p are determined in terms of the Plücker coordinates G_I of the k-plane $\overline{\Lambda}$ as follows*

$$x_i^* = \sum_{J'} \frac{1}{a_{J'}} G_{iJ'} G_{rJ'}, \qquad a_{J'} = a_{i_1} \cdots a_{i_k}, \; i = 0, 1, \dots, n, \tag{3.10}$$

where r is any index, provided the hyperplane $\{x_r = 0\} \subset \mathbb{C}^{n+1}$ does not contain the plane $\overline{\Lambda}$.

We add that if Λ is tangent to Q along a whole linear space, then all x_i^* vanish.

Let us illustrate this formula for the important examples of lines and hyperplanes tangent to Q. In the first case, using the Plücker coordinates G_{ij}, γ_i introduced in (3.1) and putting in (3.10) $r = 0$, we find the Euclidean coordinates $X_i^* = x_i^*/x_0^*$ of the contact point in the form

$$X_i^* = \frac{\sum_l G_{il}\gamma_l/a_l}{(\gamma, q\gamma)}, \qquad q = \operatorname{diag}(1/a_1, \dots, 1/a_n). \tag{3.11}$$

The same expressions are also obtainable by elementary calculation of the minimum of the function $Q(X(\tau))$, $X(t) = d + \tau\gamma$. In the second case, using the tangency condition

$$\sum_{i=0}^{n} a_i G_I^2 = 0, \qquad I = \{0, 1, \dots, n\} \setminus i,$$

(see Corollary 6) up to multiplication by a common factor we have $x_i^* = a_i G_I = a_i y_i$, which follows from the duality principle for points and hyperplanes in $\mathbb{P}^n$.

§4. Integrable systems and common tangent linear spaces of confocal quadrics

In studying the family of confocal quadrics in projective space, it is useful to consider the family of dual quadrics, because it is a *pencil* of quadrics. For example, the projective closure of quadrics of the family (1.3)

$$Q(s)\colon \left\{\frac{x_1^2}{s-a_1} + \dots + \frac{x_n^2}{s-a_n} + x_0^2 = 0\right\} \subset \mathbb{P}^n, \qquad s \in \mathbb{C} \cup \infty, \tag{4.1}$$

is dual to the quadrics of the pencil

$$Q^*(s)\colon \{Q_0^*(y) - sQ_\infty^*(y) = 0\} \subset *\mathbb{P}^N,$$
$$Q_0^*(y) = a_1 y_1^2 + \dots + a_n y_n^2 + y_0^2 = 0, \qquad Q_\infty^*(y) = y_1^2 + \dots + y_n^2 = 0.$$

A general family $Q(r)$ of confocal quadrics (a confocal family) arises by dualization of the general pencil $Q^*(r)\colon \{Q_0^*(y) - rQ_1^*(y) = 0\} \subset *\mathbb{P}^n$, where Q_0^*, Q_1^* are any distinct quadrics. Each focal quadric in $Q(r)$ is dual to a degenerate quadric in $Q^*(r)$. A pencil $Q^*(r)$ in $\mathbb{P}^n$ is called *nonsingular* if it contains $n+1$ distinct degenerate quadrics.

The family (4.1) enjoys some well-known nice properties. Namely, in $\mathbb{C}^n \subset \mathbb{P}^n$ the quadrics $Q(s)$ intersect at right angles (with respect to the Euclidean metric) at every point. Another property is described by

PROPOSITION 8. *Let l be a common tangent line of confocal quadrics $Q(c_1)$, $\dots$, $Q(c_{n-1})$ of the family* (4.1), *γ a vector along l, and $p_1, \dots, p_{n-1}$ the points of contact of l with these quadrics. Then the normal vectors to $Q(c_1), \dots, Q(c_{n-1})$ at $p_1, \dots, p_{n-1}$ together with γ form an orthogonal basis in $\mathbb{C}^n \subset \mathbb{P}^n$.*

This property enables us to give a geometrically invariant definition of a geodesic on a quadric which is not based on the geodesic equation (1.2).

PROPOSITION 9 [11]. *Let $X(t)$ be a differentiable curve on the quadric $Q(c_\alpha)$ of the family* (4.1) *such that $\dot{X}(t) \neq 0$ and $l(t) = \operatorname{span}(X(t), \dot{X}(t))$ is a common tangent line of the quadrics $Q(c_1), \ldots, Q(c_{n-1})$ for all t. By $\pi_i(t)$ we denote the tangent hyperplane of $Q(c_i)$ at the contact point $p_i(t) = l \cap Q(c_i)$. Then the curve $X(t) = p_\alpha(t)$ is a geodesic on $Q(c_\alpha)$ if and only if*

$$\ddot{X}(t) \in \bigcap_{i \neq \alpha} \pi_i(t) \quad \textit{for all } t. \tag{4.2}$$

PROOF. By Proposition 8, the intersection $\bigcap_{i\neq\alpha} \pi_i(t)$ is the 2-plane through $p_\alpha(t)$ spanned by the tangent vector $\dot{X}(t)$ and the normal vector n_α to $Q(c_\alpha)$ at $X(t)$. Therefore $\ddot{X}(t) \in \operatorname{span}(X, \dot{X}, n_\alpha)$ which is the definition of a geodesic.

In particular, when $X(t)$ describes the motion of a mass point on $Q(c_\alpha)$, the acceleration vector $\ddot{X}(t)$ is obviously parallel to n_α. In the sequel we will refer to the family (4.1) as the *canonical* confocal family.

It also makes sense to consider general confocal families in $\mathbb{P}^n$. It is easy to show that by an appropriate nondegenerate projective transformation $\mathbb{P}^n \to \mathbb{P}^n$ any nonsingular confocal family can be reduced to the *homogeneous* family

$$Q(s)\colon \left\{ \frac{x_1^2}{s-a_1} + \cdots + \frac{x_n^2}{s-a_n} + \frac{x_0^2}{s-a_0} = 0 \right\}, \qquad s \in \mathbb{C} \cup \infty, \tag{4.3}$$

which is dual to the pencil

$$Q(s)^*\colon \{Q_0^*(y) - s(y, y) = 0\} \subset *\mathbb{P}^n, \quad Q_0^*(y) = a_1 y_1^2 + \cdots + a_n y_n^2 + a_0 y_0^2 = 0.$$

Note that Proposition 8 is not valid in this case.

Each fixed point x in $\mathbb{P}^n$ belongs to exactly n distinct quadrics of the homogeneous confocal family for $s = \lambda_1, \ldots, \lambda_n$. The Euclidean coordinates of this point are represented as follows

$$X_i^2 = \frac{(a_i - \lambda_1)\cdots(a_i - \lambda_n)\prod_l (a_0 - a_l)}{(a_0 - \lambda_1)\cdots(a_0 - \lambda_n)\prod_{l \neq i}(a_i - a_l)}, \qquad i = 1, \ldots n. \tag{4.4}$$

This gives us a projective generalization of elliptic coordinates.

Now we are ready to describe the set of all common tangent linear $(k-1)$-planes of several confocal quadrics in terms of their Plücker coordinates. First, consider the quadrics of the homogeneous confocal family. Then, using Theorem 5, we obtain the following condition for a $(k-1)$-plane to be tangent to the quadric $Q(s) \subset \mathbb{P}^n$, or equivalently, for a k-plane in $\mathbb{C}^m$ to be tangent to the cone $\overline{Q}(s) \subset \mathbb{C}^m$ $(m = n+1)$

$$\text{(4.5)}\quad \mathcal{D}_k(s) = \sum_I \frac{\Phi(s)}{(s - a_{i_1})\cdots(s - a_{i_k})} G_I^2 = c_0(s - c_1)\cdots(s - c_{m-k}) = 0,$$

$$\Phi(s) = (s - a_0)(s - a_1)\cdots(s - a_n), \qquad I = \{i_1, \ldots, i_k\} \subset \{0, 1, \ldots, n\}.$$

Here $c_0, c_1, \dots, c_{m-k}$ are functions of Plücker coordinates G_I of the $(k-1)$-plane. Since $\mathcal{D}_k(s)$ is a polynomial of degree $m-k$ in s, a $(k-1)$-plane Λ is simultaneously tangent to exactly $m-k$ confocal quadrics in $\mathbb{P}^n$. (Note that this conclusion is also valid for quadrics from a general confocal family that is dual to a nonsingular pencil of quadrics in $*\mathbb{P}^n$.) Of course, the same result can also be obtained by purely geometric arguments.

Given a fixed $(k-1)$-plane Λ with the Plücker coordinates G_I, condition (4.5) can be considered as an algebraic equation, whose roots are the parameters $s = c_1, \dots, s = c_{m-k}$ of $m-k$ confocal quadrics tangent to Λ. If $\Lambda \subset \mathbb{R}^n$, then these parameters lie in the intervals

$$]-\infty; a_1], \ [a_1; a_2], \ \dots, \ [a_{n-1}; a_n].$$

Conversely, let $m-k$ confocal quadrics for $s = c_1, \dots, s = c_{m-k}$ be fixed. Then (4.5) gives rise to $m-k$ equations on the coordinates G_I

$$\mathcal{D}_k(s = c_l, G_I) = 0, \qquad l = 1, \dots, m-k.$$

They define cones in $\mathbb{C}^N = \bigwedge^k \mathbb{C}^{n+1}$, whose joint intersection with the Grassmannian $G(k, n+1)$ represents the set C_k of all common tangent $(k-1)$-planes of the quadrics $Q(c_1), \dots, Q(c_{m-k}) \subset \mathbb{P}^n$.

An analogous approach to the description of the set C_k of all common tangent linear $(k-1)$-planes of confocal quadrics of the canonical family (4.1) in $\mathbb{P}^n$ leads to the condition

$$\begin{aligned} \mathcal{D}'_k(s) &= \sum_J \frac{\Psi(s)}{(s-a_{j_1})\cdots(s-a_{j_k})} G_J^2 + \sum_{J'} \frac{\Psi(s)}{(s-a_{j_1})\cdots(s-a_{j_{k-1}})} G_{J'}^2 \\ &= c_0(s-c_1)\cdots(s-c_{n+1-k}) = 0, \qquad (4.6) \\ \Psi(s) &= (s-a_1)\cdots(s-a_n), \\ J &= \{j_1, \dots, j_k\} \subset \{1, \dots, n\}, \ J' = \{0, j_1, \dots, j_{k-1}\}. \end{aligned}$$

In this case the following generalization of Proposition 8 holds.

PROPOSITION 10. *Let $p_1, \dots, p_{m-k}$ $(m = n+1)$ be the points of contact of a $(k-1)$-plane $\Lambda \subset \mathbb{C}^n$ with the quadrics $Q(c_1), \dots, Q(c_{m-k})$ of the canonical confocal family. Then the normal vectors at $p_1, \dots, p_{m-k}$ to $Q(c_1), \dots, Q(c_{m-k})$ together with some orthogonal basis in Λ form an orthogonal basis in $\mathbb{C}^n$.*

As shown by Knörrer [11] (see also Moser [17]), the set C_2 can be identified with an unramified covering of the Jacobi variety of the hyperelliptic curve

$$\{w^2 = (s-a_1)\cdots(s-a_n)(s-c_1)\cdots(s-c_{n-1})\}.$$

It would be interesting to study the algebraic structure of the general sets C_k.

In order to describe the connection between C_k and the integrable systems considered in §2, we need the following

PROPOSITION 11. *Let the 2-vector ω associated with the angular momentum $M \in so(n+1)$ or the pair $(K, \gamma) \in e(n)$ as in (2.13), (2.23) have rank $k < n+1$. Then $\overline{\Lambda} = \omega^{k/2}$ is a factorable k-vector corresponding to the $(k-1)$-plane $\Lambda \subset \mathbb{P}^n$ with the Plücker coordinates*

$$(4.7) \qquad G_I = (|M|_I^I)^{1/2} \quad \text{or, respectively, } G_I = \left(\begin{vmatrix} K & \gamma \\ -\gamma^t & 0 \end{vmatrix}_I^I \right)^{1/2},$$

where the last index $m = n+1$ is replaced by zero.

In particular, for rank $\omega = 2$ the coordinates G_{ij} are the components of M or of the pair (K, γ), and for rank $\omega = 4$ we have

$$G_{ijlr} = M_{ij}M_{lr} - M_{il}M_{jr} + M_{ir}M_{lj}.$$

Now we have the following important result.

THEOREM 12. *Suppose that the 2-vector ω defined above has rank $k < m = n+1$. Fix integral constants of the Frahm–Manakov or Clebsch–Perelomov systems by putting*

$$\begin{aligned} \widetilde{\mathcal{I}}_k(s, M) &= c_0(s-c_1)\cdots(s-c_{m-k}), \\ \widetilde{\mathcal{J}}_k(r, K, \gamma) &= c_0(r-c_1)\cdots(r-c_{m-k}), \end{aligned} \qquad c_0, c_1, \dots, c_{m-k} = \text{const}$$

in the polynomials (2.12), (2.20) (the superscript (k) is omitted). Then to each trajectory $M(t)$ or $(K(t), \gamma(t))$ there corresponds a $(k-1)$-plane $\Lambda(t)$ determined by (4.7) such that it is simultaneously tangent to $m-k$ quadrics $Q(c_1), \dots, Q(c_{m-k})$ of the homogeneous confocal family (a_0 replaced by a_m) or, respectively, of the canonical confocal family (a_i replaced by D_i).

The proof is obvious: under the substitution (4.7) the polynomials $\mathcal{D}_k(s)$, $\mathcal{D}'_k(s)$ in the conditions (4.5), (4.6) coincide with the polynomials $\widetilde{\mathcal{I}}_k(s, M)$, respectively $\widetilde{\mathcal{J}}_k(r, K, \gamma)$ (r replaced by s and D_i by a_i) defined in (2.12), (2.20). By abuse of terminology, we may say that *the integrals of the Frahm–Manakov and Clebsch–Perelomov systems for the integral constants fixed are the tangency conditions themselves.*

REMARK 2. Apart from the confocal families (4.1), (4.3) with a diagonal quadratic form $\overline{q}(s)$, one may consider a general confocal family of the form

$$Q(s)\colon \{(x, (sE-A)^{-1}x) = 0\},$$

where A is a nondiagonal symmetric $(n+1)\times(n+1)$ matrix. (For $n = 3$ such families were studied by Schläfli [22].) In this case, the polynomial $\mathcal{D}_k$ (the left hand side of the condition (3.4)) takes the form

$$\mathcal{D}_k(s) = \sum_{I,J}(-1)^r|sE-A|_{J'}^{I'}G_IG_J = \sum_{\mu=0}^{n+1-k} s^\mu H'_{k\mu}(M), \qquad k = 2, 4, \dots,$$

$$r = i_1' + \cdots + i_m' + j_1' + \cdots + j_m',$$

$$I' = \{0, 1, \dots, n\} \setminus I, \quad J' = \{0, 1, \dots, n\} \setminus J, \quad \#I = \#J = k.$$

After the identification $G_I = (|M|_I^I)^{1/2}$ the coefficients $H'_{k\mu}$ can be regarded as a complete set of first integrals of the integrable Hamiltonian system describing the geodesic flow on the algebra $so(n+1)$ for the nondiagonal metric

$$\widetilde{\mathcal{I}}_2(s, M) = \sum_{i<j, p<q} (-1)^r |sE - A|_{J'}^{I'} M_{ij} M_{pq}, \tag{4.8}$$

$$I' = \{0, 1, \dots, n\} \setminus \{ij\}, \quad J' = \{0, 1, \dots, n\} \setminus \{pq\}.$$

Again, the leading coefficients $H'_{k, n+1-k}$ are invariants of the algebra. However, this does not give us a new integrable system on $so(n+1)$. Indeed, note that any orthogonal transformation $R\colon \mathbb{C}^{n+1} \to \mathbb{C}^{n+1}$, $R \in SO(n+1)$ extends naturally to the transformation $\bigwedge^k R\colon \bigwedge^k \mathbb{C}^{n+1} \to \bigwedge^k \mathbb{C}^{n+1}$. For $k = 2$, when $\bigwedge^2 \mathbb{C}^{n+1}$ and the Lie algebra $so(n+1)$ coincide as linear spaces, we have $\bigwedge^2 R = \mathrm{Ad}_R$. Now let R take the matrix A to a diagonal form. Then Ad_R does the same for the metric (4.8) on $so(n+1)$. As a result, we come again to the Manakov metric. An analogous construction for the Clebsch–Perelomov systems can be developed.

Now consider the hierarchy of the Clebsch–Perelomov systems for the case of maximal degeneration $\operatorname{rank}\omega = 2$. Then the dimension of the general invariant tori equals $n-1$ and the quadratic integrals $H_{2\mu}(K, \gamma)$, $\mu = 0, 1, \dots, n-2$ generate, as Hamiltonians, independent commuting vector fields on the tori.

THEOREM 13. *For the case* $\operatorname{rank}\omega = 2$, *the general solution* $K(t), \gamma(t)$ *of the Clebsch–Perelomov systems with the Hamiltonian*

$$\sum_{\mu=0}^{n-1} f_\mu H_{2\mu}(K, \gamma), \qquad f_\mu = \text{const},$$

can be found in terms of the spheroconic coordinates (1.8) *by inverting the hyperelliptic integrals* (1.9) *for* $\varepsilon_\mu = f_\mu$, $\mu = 0, 1, \dots, n-2$ *(a_i replaced by D_i).*

For the same case $\operatorname{rank}\omega = 2$ we also have

THEOREM 14. *Let* $K(t)$, $\gamma(t)$ *be a solution of the Clebsch–Perelomov system with the Hamiltonian*

$$\widetilde{\mathcal{J}}_2(c_\alpha) = \sum_{j=1}^{n} \frac{\Psi(c_\alpha)}{c_\alpha - D_j} F_j(K, \gamma) = 0,$$

and $l(t)$ *be a line (1-plane) associated with* $K(t), \gamma(t)$ *as in Proposition* 11 *and tangent to the quadrics* $Q(c_1), \dots, Q(c_{n-1})$ *of the canonical confocal family at the points* $p_1(t), \dots, p_{n-1}(t)$ *respectively (a_i replaced by D_i). Then*

(1) *the point* $p_\alpha(t)$ *traces a geodesic* $X(t)$ *on* $Q(c_\alpha)$ *such that* $l(t)$ *is the tangent line of the curve* $X(t)$;
(2) *this solution can be obtained in terms of the spheroconic coordinates in* (1.8) *by inversion of the hyperelliptic integrals* (1.9) *for* $\varepsilon_\mu = c_\alpha^\mu$, $\mu = 0, 1, \dots, n-2$.

In particular, considering the system with the Hamiltonian

$$\widetilde{\mathcal{J}}_2(r=0) = D_1 \cdots D_n \sum_{j=1}^{n} \frac{F_j(K, \gamma)}{D_j} = 0, \tag{4.9}$$

taking into account (2.17), and putting $a_i = D_i$, we arrive at Theorem 2 that relates the dynamics on TS^{n-1} to the geodesic motion on the ellipsoid $Q(0)$.

REMARK 3. Note that for $n = 3$ the Hamiltonian (4.9) is precisely the Hamiltonian of the classical Brun problem considered above. Thus, according to Theorem 14, there is a connection between the geodesics on the triaxial ellipsoid $Q(0)$ and the motion of a rigid body with inertia tensor D^{-1} in the force field with quadratic potential $Y(\gamma) = (\gamma, D^2\gamma)/2$. For example, the geodesics that are the smallest or the biggest ellipses on $Q(0)$ correspond to stable permanent rotations of the body about its principal axes perpendicular to the symmetry axis of the force field, while the asymptotic geodesics passing through the ombilic points on $Q(0)$ correspond to the double-asymptotic motion starting as unstable rotation about the middle principal axis and tending to the rotation about the same axis, but in the opposite direction.

As we mentioned in the Introduction, an analogous connection between the Jacobi problem and the problem on the motion of a rigid body in an ideal fluid was studied by Minkowski in [16].

PROOF OF THEOREM 14. 1) On the one hand, the Clebsch–Perelomov system with the Hamiltonian $H = \widetilde{\mathcal{J}}_2(r)$ has the form

$$
\begin{aligned}
\dot{K}_{ij} &= \frac{\Psi(r)(D_j - D_i)}{(r - D_i)(r - D_j)} \left[\sum_{l=1}^{n} \frac{1}{r - D_l} K_{il} K_{lj} + \gamma_i \gamma_j \right], \\
\dot{\gamma}_i &= \frac{\Psi(r)}{(r - D_i)} \sum_{l=1}^{n} \frac{1}{r - D_l} K_{il} \gamma_l .
\end{aligned}
\tag{4.10}
$$

On the other hand, let the curve $X(t) \subset \mathbb{C}^n$ be a geodesic on a fixed quadric $Q(s)$ of the canonical confocal family. Then, according to (3.11),

$$
X_i(t) = \frac{\sum_{l=1}^{n} G_{il} q(s) \gamma_l}{(\gamma, q(s)\gamma)}, \qquad q(s) = \operatorname{diag}\left(\frac{1}{s - a_1}, \dots, \frac{1}{s - a_n} \right),
$$

where G_{ij}, γ_i are the Plücker coordinates of the line $l(t)$ tangent to $X(t)$: $\gamma_i = \dot{X}_i(t)$, $G_{ij} = X_i \dot{X}_j - \dot{X}_i X_j$. Without loss of generality, we can think of $X(t)$ as a solution of the geodesic equation (1.2). Then we obtain the differential equations for G, γ

$$
\begin{aligned}
\dot{G}_{ij} &= -\lambda' \frac{q_j - q_i}{(\gamma, q\gamma)} \sum_{l=1}^{n} G_{il} q_l \gamma_l \sum_{l=1}^{n} q_l G_{jl} \gamma_l , \\
\dot{\gamma}_i &= -\lambda' q_i \sum_{l=1}^{n} G_{il} q_l \gamma_l , \\
q_i &= q_{ii}(s), \quad \lambda' = 1/(X, q^2(s) X) .
\end{aligned}
\tag{4.11}
$$

Now put $G_{ij} = K_{ij}$, $s = r$, $a_i = D_i$. Then, taking into account the conditions $K_{ij}\gamma_k + K_{jk}\gamma_i + K_{ki}\gamma_j = 0$ which follows from the factorability conditions (3.2), as well as the tangency condition $\widetilde{\mathcal{J}}_2(r) = 0$, using also the time change

$$
dt \to \frac{\Psi(r)}{\lambda'} dt ,
$$

one can transform the equations (4.11) precisely to the system (4.10). Therefore, in the case of maximal degeneration ($\operatorname{rank}\omega = 2$) any solution of (4.10) under the condition $\widetilde{\mathcal{J}}_2(K, \gamma, r) = 0$ defines a line $l(t)$ tangent to a geodesic on $Q(r)$. Finally, putting $s = r = c_\alpha$ establishes item 1.

2) Note that $\widetilde{\mathcal{J}}_2(c_\alpha) = \sum_{\mu=0}^{n-1} c_\alpha^\mu H_{2\mu}$ which, in view of Theorem 13, establishes item 2.

Now the following question naturally arises: can the Frahm–Manakov system (2.4), as well as the higher Manakov and Clebsch–Perelomov systems (2.5), (2.16) be related in any way to the geodesic motion on a quadric? The answer seems to be negative. However, one may consider a certain modification of the concept of a geodesic by considering Proposition 9 as a definition of such a modification. More precisely:

DEFINITION. Let $\mathcal{Q} = \{Q(c_1), \dots, Q(c_{n-1})\}$ be quadrics of a general confocal family. Suppose that conditions of Proposition 9 are fulfilled. We say that the curve $X(t)$ on $Q(c_\alpha) \in \mathcal{Q}$ is a *quasigeodesic associated with the set* $\mathcal{Q}$ if $X(t)$ satisfies condition (4.2).

It follows that the qualitative behavior of a quasigeodesic is just the same as that of an actual geodesic. It is also described by Theorem 1, the multi-dimensional version of the Chasles theorem. It is clear that quasigeodesics associated with quadrics of the canonical confocal family are true geodesics.

Now we can formulate an analog of Theorem 14 for the Frahm–Manakov system on $so(m)$. Namely, let the 2-vector ω defined in (2.13) have the minimal rank 2. Fix all integral constants of this system. By Theorem 12 this determines the root constants $c_1, \dots, c_{m-2}$. Then we have

THEOREM 15. *Let $M(t)$ be a solution of the Frahm–Manakov system with the Hamiltonian $H = \widetilde{\mathcal{I}}_2(c_\alpha) = 0$, and $l(t)$ be the line associated with $M(t)$ and tangent to the quadrics $Q(c_1), \dots, Q(c_{m-2})$ of the homogeneous confocal family at points $p_1, \dots, p_{m-2}$ respectively. Then*

(1) *the point $p_\alpha(t)$ describes a trajectory $X(t)$ on $Q(c_\alpha)$ such that $l(t)$ is a tangent line of $X(t)$ and, after a change of the parameter t, this trajectory is a solution of the differential equations*

$$\ddot{X}_i = -\mu'\left(\frac{1}{c_\alpha - a_i} - \frac{1}{c_\alpha - a_0}\right)X_i, \qquad i = 1, \dots, n, \tag{4.12}$$

$$\mu' = \frac{\sum_i \dot{X}_i^2/(c_\alpha - a_i)}{\sum_i X_i^2/(c_\alpha - a_i)^2 + 1/(c_\alpha - a_0)^2};$$

(2) *$X(t)$ is a quasigeodesic associated with the set $Q(c_1), \dots, Q(c_{m-2})$,*

(3) *the general solution of* (4.12) *can be found by inversion of the Abelian integrals* (1.5), *where $\lambda_1, \dots, \lambda_{m-2}$ are the curvilinear elliptic coordinates on $Q(c_\alpha)$ defined in* (4.4) ($\lambda_n = \lambda_{m-1}$ *being replaced by* c_α).

REMARK 4. Equations (4.12) can be interpreted as those describing the motion of a mass point on the quadric $Q(c_\alpha)$. However, unlike the Jacobi problem, it is clear that the kinetic energy is not conserved in this case.

PROOF OF THEOREM 15. Part (1) can be established just as part (1) of Theorem 14. Let, as above, π_l be the hyperplane $\{v_0^{(l)}x_0 + v_1^{(l)}x_1 + \cdots + v_n^{(l)}x_n = 0\}$ tangent to $Q(c_l)$ at $p_l = l \cap Q(c_l)$. To prove part (2), we must show that the vector

$$n_\alpha = (n_{\alpha 1}, \ldots, n_{\alpha n})^T, \qquad n_{\alpha i} = -\left(\frac{1}{c_\alpha - a_i} - \frac{1}{c_\alpha - a_0}\right)X_i$$

with origin at p_α lies in each hyperplane π_l $(l \neq \alpha)$. This occurs if and only if for all $l = 1, \ldots, n$ $(l \neq \alpha)$ the following relation holds

$$h_l = v_1^{(l)}n_{\alpha 1} + \cdots + v_n^{(l)}n_{\alpha n} = 0,$$

that is

$$h_l = -\frac{1}{c_\alpha - a_0}\left(v_1^{(l)}\frac{a_1 - a_0}{c_\alpha - a_1}X_1 + \cdots + v_n^{(l)}\frac{a_n - a_0}{c_\alpha - a_n}X_n\right) = 0.$$

Since $p_\alpha \in l \subset \pi_l$, we have $v_1^{(l)}X_1 + \cdots + v_n^{(l)}X_n + v_0^{(l)} = 0$. It follows that

$$h_l = -\frac{1}{c_\alpha - a_0}\left(v_1^{(l)}\frac{c_\alpha - a_0}{c_\alpha - a_1}X_1 + \cdots + v_n^{(l)}\frac{c_\alpha - a_0}{c_\alpha - a_n}X_n + v_0^{(l)}\right) = 0,$$

which in the homogeneous coordinates takes the form

$$\begin{aligned} h_l = &-\frac{1}{x_0}\left(v_0^{(l)}\frac{x_0}{c_\alpha - a_0} + v_1^{(l)}\frac{x_1}{c_\alpha - a_1} + \cdots + v_n^{(l)}\frac{x_n}{c_\alpha - a_n}\right) \\ &- \frac{1}{x_0}[v_0^{(l)}v_0^{(\alpha)} + v_1^{(l)}v_1^{(\alpha)} + \cdots + v_n^{(l)}v_n^{(\alpha)}]. \end{aligned}$$

It can easily be seen that the sum in the square brackets equals zero. Indeed, since $l \subset \pi_l \cap \pi_\alpha$ and l is tangent to the quadrics $Q(c_\alpha)$, $Q(c_l)$, in the dual space $*\mathbb{P}^n$ the line $L = \operatorname{span}(\pi_l^*, \pi_{(\alpha)}^*)$ touches the dual quadrics $Q^*(c_\alpha)$, $Q^*(c_l)$ at the points π_l^*, π_α^*. This implies that

$$(c_\alpha - a_0)v_0^{(l)}v_0^{(\alpha)} + (c_\alpha - a_1)v_1^{(l)}v_1^{(\alpha)} + \cdots + (c_\alpha - a_n)v_n^{(l)}v_n^{(\alpha)} = 0,$$

and

$$(c_l - a_0)v_0^{(l)}v_0^{(\alpha)} + (c_l - a_1)v_1^{(l)}v_1^{(\alpha)} + \cdots + (c_l - a_n)v_n^{(l)}v_n^{(\alpha)} = 0.$$

Subtracting these two equations, we get

$$v_0^{(l)}v_0^{(\alpha)} + v_1^{(l)}v_1^{(\alpha)} + \cdots + v_n^{(l)}v_n^{(\alpha)} = 0.$$

Therefore, $h_l = 0$, which establishes part (2).

Finally, as far as the higher systems (2.5) and (2.16) are concerned, we come to the following analog of Theorems 14 and 15:

THEOREM 16. *Let $M(t)$, $(K(t), \gamma(t))$ be solutions of the higher Frahm–Manakov and Clebsch–Perelomov systems for the case* $\operatorname{rank}\omega = k < m = n+1$ *with the Hamiltonian* $\widetilde{\mathcal{I}}_k(s = c_\alpha, M) = 0$, *respectively*, $\widetilde{\mathcal{J}}_k(r = c_\alpha, K, \gamma) = 0$, *and* $\Lambda(t)$ *be the* $(k-1)$*-plane associated with these solutions as in Proposition* 11 *and tangent to* $m-k$ *quadrics* $Q(c_1), \dots, Q(c_{m-k})$ *at the points* $p_1, \dots, p_{m-k}$. *Then the contact point* p_α *defined by* (3.10) *traces a curve* $X(t)$ *on* $Q(c_\alpha)$ *such that* $l(t) = \operatorname{span}(X(t), \dot{X}(t)) \subset \Lambda(t)$ *and*

$$\ddot{X}(t) \subset \bigcap_{i \neq \alpha} \pi_i(t), \qquad i = 1, \dots, m-k \quad \text{for all } t, \tag{4.13}$$

where $\pi_i(t)$ *is the tangent hyperplane of* $Q(c_i)$ *at* p_i. *Besides, for the higher Clebsch–Perelomov systems we have*

$$\ddot{X}(t) \subset \operatorname{span}(X(t), \Lambda(t), n_\alpha), \tag{4.14}$$

where n_α *is the normal vector of* $Q(c_\alpha)$ *at* $p_\alpha(t)$.

We see that (4.13) and (4.14) are weaker versions of definition (4.2) of a quasi-geodesic and the definition of a true geodesic respectively. As we have seen above, the condition for the Frahm–Manakov and Clebsch–Perelomov systems to be degenerate is a necessary condition for these systems to be associated with linear spaces tangent to confocal quadrics and, thereby, to be connected with the motion on a quadric. Reducing the problem to geodesic or quasigeodesic motion on a quadric allows us to use the elliptic coordinates defined by (1.4), (4.4) and to obtain the solution immediately as an inversion of the hyperelliptic integrals. So, the following question arises: can the elliptic coordinates or their analogs be applied to solve the systems considered above for the case $\operatorname{rank}\omega > 2$ as well as for the nondegenerate case?

References

1. M. Adler and P. van Moerbecke, *Linearization of Hamiltonian systems, Jacobi varieties and representation theory*, Adv. Math. **38** (1980), 318–379.
2. A. V. Bolsinov, *Commutative families of functions related to consistent Poisson brackets*, Acta Appl. Math. **24** (1991), 253–274.
3. A. I. Bobenko, *Euler equations on Lie algebras* $e(3)$ *and* $so(4)$. *An isomorphism of integrable cases*, Funktsional. Anal. i Prilozhen. **20** (1986), no. 1, 64–66; English transl. in Functional Anal. Appl. **20** (1986).
4. F. Brun, *Rotation kring fix punkt*, Ofers. Kongl. Svenska Vetensk. Acad. Forhandl. Stockholm (1893), no. 7, 455–468.
5. M. Chasles, *Les lignes géodesiques et les lignes de courbure des surfaces du second degré*, J. Math. **11** (1846), 5–20.
6. B. A. Dubrovin, *Completely integrable Hamiltonian systems associated with matrix operators and Abelian varieties*, Funktsional. Anal. i Prilozhen. **11** (1977), no. 4, 21–48; English transl. in Functional Anal. Appl. **11** (1977), 28–41.
7. Yu. Fedorov, *Lax representations with a spectral parameter defined on coverings of hyperelliptic curves*, Mat. Zametki **54** (1993), no. 1, 94–109; English transl. in Math. Notes **54** (1993).
8. F. Frahm, *Ueber gewisse Differentialgleichungen*, Math. Ann. **8** (1874), 35–44.
9. L. Heine, *Geodesic flow on* $so(4)$ *and Abelian surfaces*, Math. Ann. **263** (1983), 435–472.
10. C. Jacobi, *Vorlesungen über Dynamik. Gesammelte Werke, Supplementband*, Berlin, 1884.
11. H. Knörrer, *Geodesics on the ellipsoid*, Invent. Math. **59** (1980), 119–143.

12. ______, *Geodesics on quadrics and a mechanical problem of C. Neumann*, J. Reine Angew. Math. **334** (1982), 69–78.
13. F. Kötter, *Uber die Bewegung eines festen Körpers in einer Flüssigkeit.* I, II, J. Reine Angew. Math. **109** (1892), 51–81; 89–111.
14. V. Kozlov, *Two integrable problems of classical dynamics*, Vestnik Moskov. Univ. Ser. I Mat. Mekh. **1981**, no. 4, 80–83; English transl. in Moscow Univ. Math. Bull. **35** (1981).
15. S. V. Manakov, *Remarks on the integrals of the Euler equations of the n-dimensional heavy top*, Funktsional. Anal. i Prilozhen. **10** (1976), no. 4, 93–94; English transl. in Functional Anal. Appl. **10** (1976).
16. H. Minkowski, *Uber die Bewegung eines festen Körpers in einer Flüssigkeit*, Sitzungberichte der Königlich Preussischen Academie der Wissenshaften zu Berlin **XXX** (1888), 1095–1110.
17. J. Moser, *Various aspects of Hamiltonian systems*, Bikrhäuser, Boston, 1978.
18. D. Mumford, *Tata lectures on Theta.* I, II, Bikrhäuser, Boston, 1983, 1984.
19. C. Neumann, *De problemate quodam mechanica, quod ad primam integralium ultraellipticorum classem revocatur*, J. Reine Angew. Math. **56** (1859).
20. A. Perelomov, *Some remarks on integrability of equations describing a rigid body movement in an ideal liquid*, Funktsional. Anal. i Prilozhen. **15** (1981), no. 2, 83–85; English transl. in Functional Anal. Appl. **15** (1981).
21. T. Ratiu, *The C. Neumann problem as a completely integrable system on an adjoint orbit*, Trans. Amer. Math. Soc. **264** (1981), no. 2, 321–329.
22. L. Schläfli, J. Reine Angew. Math. **43** (1852).
23. F. Schottky, *Uber das analitische Problem der Rotation eines starren Körpers in Raume von vier Dimensionen*, Sitzungberichte der Königlich Preussischen Academie der Wissenshaften zu Berlin **XII** (1891), 227–232.
24. H. Weber, *Anwendung der Theta functionen zweir Veränderlicher auf die Theorie der Bewegung eines festen Körpers in einer Flüssigkeit*, Math. Ann. **14** (1878), 173–206.
25. K. Weierstrass, *Über die geodätischen Linien auf dem dreiachsigen Ellipsoid*, Math. Werke I, 257–266.

Translated by THE AUTHOR

DEPARTMENT OF MATHEMATICS, MOSCOW STATE UNIVERSITY, MOSCOW 119899, RUSSIA

Amer. Math. Soc. Transl.
(2) Vol. **168**, 1995

Recent Results on Asymptotic Behavior of Integrals of Quasiperiodic Functions

N. G. MOSHCHEVITIN

§1. Returns of an integral

1.1. Preliminary results. Let $f\colon \mathbb{T} \to \mathbb{R}$ be a smooth function defined by a Fourier series:

$$f(x_1, \dots, x_s) = \sum_{|m_1|+\dots+|m_s|\neq 0} f_{m_1, \dots, m_s} e^{2\pi i(m_1x_1+\dots+m_sx_s)}. \tag{1}$$

Let $\omega_1, \dots, \omega_s \in \mathbb{R}$ (the frequencies) be linearly independent over $\mathbb{Z}$. We shall consider the integral $I(t) = \int_0^t f(\omega_1 t, \dots, \omega_s t)\, dt$. According to the well-known Weyl theorem [3], $I(t) = o(t)$, $t \to \infty$.

DEFINITION 1. The integral $I(t)$ *returns* if

$$\forall \varepsilon > 0, \ \forall T > 0, \ \exists t > T: \ |I(t)| < \varepsilon.$$

CONJECTURE 1 (V. V. Kozlov). Let $f(x_1, \dots, x_s)$ be an analytic function. Then the integral $I(t)$ returns for any frequencies $\omega_1, \dots, \omega_s$ linearly independent over $\mathbb{Z}$.

The conjecture is obviously true when f is a trigonometric polynomial.

THEOREM 1. *Let f be analytic. Then $I(t)$ returns under the following condition on the frequencies*:

$$\begin{gathered} \exists C, \eta > 0, \ \forall (m_1, \dots, m_s) \in \mathbb{Z}^s \setminus \{0\}: \\ |m_1\omega_1 + \dots + m_s\omega_s| > C(|m_1| + \dots + |m_s|)^{-\eta}. \end{gathered} \tag{2}$$

(*In applications such a relation is called "strict incommensurability" of the set* $\omega_1, \dots, \omega_s$.)

Theorem 1 together with a simple statement of the metric theory of Diophantine approximations (see [4]) implies

Research is supported by the Pro-mathematicaFoundation, SMF, and by the International Scientific Foundation (grant No. MAK-000).

THEOREM 2. *Let f be analytic. Then $I(t)$ returns for almost all sets of frequencies $(\omega_1, \dots, \omega_s) \in \mathbb{R}^s$.*

REMARK. In Theorems 1 and 2 we can assume f to be a function of a certain finite smoothness.

The results of Theorems 1 and 2 are determined by the fact that under the conditions of Theorem 1 there are no "small denominators" $|m_1\omega_1 + \dots + m_s\omega_s|^{-1}$ and hence the sum

$$\sum_{m_1, \dots, m_s} (|m_1| + \dots + |m_s|)^{-\eta-1-\varepsilon} |m_1\omega_1 + \dots + m_s\omega_s|^{-1}$$

is bounded. More precise conditions on f are given in [5–6]. The main idea of the last paper will be discussed later in §2.

In the general case $s \geqslant 3$, Conjecture 1 has not yet been proved.

1.2. The case $s = 2$. This case was considered in [7, Chapter 8], [8, 9]. Let us formulate the most exact result from [9]:

THEOREM 3. *Let $s = 2$ and $f(x_1, x_2)$ be an absolutely continuous function on the torus $\mathbb{T}^2(x_1, x_2)$. Then $I(t)$ returns for any frequencies (ω_1, ω_2) if ω_1/ω_2 is irrational.*

The statement of Theorem 3 is not true for a continuous function f. We shall present an appropriate counterexample in §4.

Some results on the boundedness of the integral $I(t)$ can be found in [30].

1.3. The case $s = 3$. The following statement from the paper [10] shows that in the case $s = 3$ we can replace condition (2) by a weaker condition.

THEOREM 4. *Let the function $f(x_1, x_2, x_3)$ defined by (1) be analytic. Let the linearly independent (over $\mathbb{Z}$) frequencies ω_1, ω_2, ω_3 satisfy the condition*

$$(3) \qquad \exists C, \eta_1, \ \forall p \in \mathbb{N}, \ a_1, a_2 \in \mathbb{Z}: \quad \max_{j=1,2} |\omega_j/\omega_3 - a_j/p| > Cp^{-\eta_1}.$$

Then $I(t)$ returns.

The idea of the proof is to show that under the condition (3) small linear forms $m_1\omega_1 + m_2\omega_2 + m_3\omega_3$ that fail to satisfy the inequality from (2) appear very seldom. Theorem 4 implies that for $s = 3$ Conjecture 1 is not yet proved only when the real numbers $\alpha_1 = \omega_1/\omega_3$ and $\alpha_2 = \omega_2/\omega_3$ admit very good simultaneous rational approximations.

1.4. The multi-dimensional case. We construct an example of sets of incommensurable frequencies $(\omega_1, \dots, \omega_s)$ for which the integral $I(t)$ returns in spite of the absence of any lower bound for the linear form $m_1\omega_1 + \dots + m_s\omega_s$ (see [11–12]).

THEOREM 5. *The integral $I(t)$ returns under the following conditions:*

A) $|f_{m_1, \dots, m_s}| < C(\overline{m}_1 \cdots \overline{m}_s)^{-s-\varepsilon}$, $C, \varepsilon > 0$, $\overline{m}_j = \max\{|m_j|, 1\}$;

B) *the frequencies ω_j admit specific rational approximations: the system of inequalities*

$$|\omega_j - a_j(p_j\tau_j)^{-1}| < M(p\tau_1 \cdots \tau_s)^{-1}\varphi(\tau), \qquad j = 1, \dots, s,$$

where $p = [p_1, \dots, p_s]$, $\tau = \min_j \tau_j$, M *is a constant and* $\varphi(\tau) \to 0$ *as* $\tau \to \infty$ *has an infinite number of solutions in the numbers* a_j, $p_j \in \mathbb{Z}$, $\tau_j \in \mathbb{N}$, *such that* τ *takes arbitrary large values and*

$$(4) \qquad \begin{gathered} (a_i, \tau_i) = 1, \quad \forall i, \quad (p_1, \tau_j) = (\tau_i, \tau_j) = 1, \quad \forall i \neq j, \\ \tau_j \leqslant \gamma_1 \tau \ln^{\gamma_2} \tau, \quad \forall j, \quad p \leqslant \gamma_1 \ln^{\gamma_2} \tau. \end{gathered}$$

One can easily construct many other simple examples of sets of frequencies ω for which $I(t)$ returns in spite of the fact that the linear form $\|m_1\alpha_1 + \dots + m_s\alpha_s\|$ does not admit a lower bound similar to (2).

§2. Mean values

Let $I(\tau, \varphi) = \int_0^t f(\omega_1 t + \varphi_1, \dots, \omega_s t + \varphi_s)\, dt$. Consider the mean values

$$J_\infty^s(t) = \sup_{\varphi \in \mathbb{R}^s} |I(t, \varphi)| \quad \text{and} \quad J_2^s(t) = \left(\int_{\mathbb{T}^s} |I(t, \varphi)|^2\, dt \right)^{1/2}.$$

Obviously, $J_\infty^s(t) > J_2^s(t)$.

COROLLARY. *Under the conditions of Theorems* 1, 3, *and* 4, *the mean values* $J_\infty^s(t)$ *and* $J_2^s(t)$ *return, in particular,* $J_\infty^2(t)$ *and* $J_2^2(t)$ *do return for any* ω_1, ω_2 *linearly independent over* $\mathbb{Z}$ *and any smooth* f.

The case $s \geqslant 3$ differs from the case $s = 2$. Suppose that $\Phi(y) > 0$ and the series $\sum_{m_1, \dots, m_s \in \mathbb{Z}} \Phi((1 + |m_1|) \cdots (|1 + |m_s|))$ converges. We define f to be of type Φ if

$$|f_{m_1, \dots, m_s}| < \Phi((1 + |m_1|) \cdots (|1 + |m_s|)).$$

THEOREM 6. *For any given type* Φ *one can find a function* $f: \mathbb{T}^3 \to \mathbb{R}$ *of type* Φ *and frequencies* ω_1, ω_2, ω_3 *linearly independent over* $\mathbb{Z}$ *satisfying the following condition: the mean value* $J_2^3(t)$ (*and hence* $J_\infty^3(t)$) *does not return.*

§3. Discretization

We shall reformulate the results from §1 in terms of the distribution of fractional parts of the system of functions

$$\{\alpha_1 x\}, \dots, \{\alpha_{s-1} x\}, \qquad x = 1, \dots, q,$$

where $\alpha_1, \dots, \alpha_{s-1}$, together with 1, are linearly independent over $\mathbb{Z}$ (see [13–16]).

3.1. Notation. Denote by $N_q(\gamma_1, \dots \gamma_{s-1})$ the number of integers x satisfying

$$\{\alpha_1 x\} < \gamma_1, \dots, \{\alpha_{s-1} x\} < \gamma_{s-1}, \qquad 1 \leqslant x \leqslant q$$

(here $\{\cdot\}$ denotes the fractional part). Let D_q be the discrepancy of the set

$$\{(\{\alpha_1 x\}, \dots, \{\alpha_{s-1} x\}) : x = 1, \dots, q\} \subset [0, 1)^{s-1},$$

i.e., let

$$D_q = \sup_{\gamma_1, \dots, \gamma_{s-1} \in [0, 1)} |N_q(\gamma_1, \dots, \gamma_{s-1}) - \gamma_1, \dots, \gamma_{s-1} q|.$$

According to the well-known Weyl theorem [3], $D_q = o(q)$ as $q \to \infty$.

3.2. Results.

THEOREM 1′. *Let* $\alpha_1, \dots, \alpha_{s-1}$ *be such that*

$$(5)\qquad \begin{gathered}\exists\, C_1, \gamma_1 > 0, \quad \forall\, (m_1, \dots, m_{s-1}) \in \mathbb{Z}^{s-1} \setminus \{0\}: \\ \|m_1\alpha_1 + \dots + m_{s-1}\alpha_{s-1}\| > C_1(\overline{m}_1 \cdots \overline{m}_{s-1})^{-\gamma_1 - 1}\end{gathered}$$

(*here* $\|\zeta\| = \min_{a\in\mathbb{Z}} |\zeta - a|$ *is the distance to the nearest integer*). *Then*

$$D_q = O(q^{1-(\gamma_1(s-1)+1)^{-1}} \ln q), \quad \forall q.$$

Since for almost all $\alpha_1, \dots, \alpha_{s-1}$, we have

$$\begin{gathered}\|m_1\alpha_1 + \dots + m_{s-1}\alpha_{s-1}\| \geqslant C_1(\overline{m}_1 \cdots \overline{m}_{s-1} \ln^\beta(\overline{m}_1 \cdots \overline{m}_{s-1} + 1))^{-1}, \\ \forall\, (m_1, \dots, m_{s-1}) \in \mathbb{Z}^{s-1} \setminus \{0\}\end{gathered}$$

(see [4]), it follows that Theorem 1 implies

THEOREM 2′. *For almost all sets* $(\alpha_1, \dots, \alpha_{s-1})$ *the discrepancy* D_q *admits the following upper bound*:

$$D_q \ll q^\varepsilon, \quad \forall q$$

for any positive ε.

Theorem 1′ is proved, for example, in [13–14] (see also [15]). It has an interesting elementary improvement in the case $s = 2$, $\gamma_1 = 0$ (see [13]). The ideas of the proof of Theorem 1′ are similar to those from the paper [6]: to estimate the sum

$$\sideset{}{'}\sum_{m_1, \dots, m_{s-1} = -p}^{p} (\overline{m}_1 \cdots \overline{m}_{s-1} \|m_1\alpha_1 + \dots + m_{s-1}\alpha_{s-1}\|)^{-1}$$

by the sums

$$\sum_{\substack{n_1 \dots n_{s-1} \\ |n_i| < m_i}} \|n_1\alpha_1 + \dots + n_{s-1}\alpha_{s-1}\|^{-1}$$

with the help of the Abel transformation, and then use (5).

THEOREM 3′. *Let* $s = 2$. *Then for any* $\alpha_1 \notin \mathbb{Q}$ *one can find a sequence* q_ν *satisfying* $D_{q_\nu} < 3$. (*It is sufficient to take as* q_ν *the denominators of fractions convergent to* α_1.)

The one-dimensional situation was considered by many authors ([31–33] and others). There are interesting statistical results in [29].

THEOREM 4′. *Suppose* $s = 3$ *and* α_1, α_2 *satisfy the following conditions*:

(i) 1, α_1, α_2 *are linearly independent over* $\mathbb{Z}$,
(ii) $\exists\, C, \eta$: $\forall\, p \in \mathbb{N}$, $\|p\alpha_1\| + \|p\alpha_2\| > Cp^{-\eta}$.

Let $\rho = (\eta + 1)/(8\eta^2 + 4\eta + 1)$. *Then for any* $\varepsilon > 0$ *the inequality* $D_q < q^{1-\rho+\varepsilon}$ *has infinitely many solutions in natural numbers* q.

THEOREM 5'. *Let* $\alpha_1, \dots, \alpha_{s-1}$ *have the following property: the system of inequalities*

$$|\alpha_j - a_j(p_j\tau_j)^{-1}| < M(p\tau_1 \cdots \tau_{s-1})^{-1-2/(s-1)}, \qquad j = 1, \dots, s-1,$$

where p, τ, *and* M *are the same as in Theorem 5, has infinitely many solutions in numbers* a_j, p_j, τ_j *satisfying* (4). *Then for some constant* K_1 *the inequality*

$$D_q \leqslant K_1 q^{1-1/(s-1)} (\ln q)^{\gamma_2+s-1}$$

has infinitely many solutions in natural numbers q.

Theorems 4' and 5' are proved in [16].

The case $s \geqslant 3$ differs from the case $s = 2$ radically. Recently, we obtained the following result:

THEOREM 6'. *For any positive function* $\varphi(y)$ *satisfying* $\lim_{y\to\infty} \varphi(y) = 0$ *one can find real numbers* α_1, α_2 *linearly independent over* $\mathbb{Z}$ *together with* 1 *such that for the discrepancy* D_q *the inequality*

$$D_q \gg q\varphi(q), \quad \forall q \in \mathbb{N},$$

is valid.

The proof of Theorems 6' and 6 uses some ideas connected with multidimensional generalizations of the continuous fraction algorithm.

§4. Nonreturn and distribution of algebraic numbers

4.1. Nonreturn. Now we shall show that the condition that the function $f(x_1, \dots, x_s)$ defined by (1) is smooth enough is necessary for returns. First of all we shall discuss the examples from [1, Chapter 9], [2] (see also [7, Chapter 8], [8]).

THEOREM 7. *Let* $\omega_1 = 1$, $\omega_2 = \sqrt{2}$. *Then there exists a continuous function* $f(x_1, x_2)$ *such that* $I(t) \to \infty$ *as* $t \to \infty$.

Theorem 6 shows that under the condition that f is continuous there may be no returns.

THEOREM 8. *Let* $\omega_1, \dots, \omega_s$ *satisfy the following inequality*:

$$(6) \qquad \begin{gathered} \exists C > 0, \quad |m_1\omega_1 + \cdots + m_s\omega_s| > C\Big(\max_{j=1,\dots,s} |m_j|\Big)^{1-s}, \\ \forall (m_1, \dots, m_s) \in \mathbb{Z}^s \setminus \{0\}. \end{gathered}$$

Then there exists a function $f(x_1, \dots, x_s) \in C^{s-2}(\mathbb{T}^s) \setminus C^{s-1}(\mathbb{T}^s)$ *such that* $I(t) \geqslant C'$ *for some positive constant* C' *and for all* $t \geqslant t_0$.

The last theorem was proved in [17]. The proof is based on the fact that under condition (6) the linear forms for which the same upper bound holds appear often and regularly (see the Dirichlet theorem [18, Chapter 2]). Theorem 7 enables one to conjecture that Theorem 7 may be strengthened:

CONJECTURE 2. There exist frequencies such that for some function $f \in C^{s-2}(\mathbb{T}^s) \setminus C^{s-1}(\mathbb{T}^s)$ one has $I(t) \to \infty$ as $t \to \infty$.

We shall refer to paper [9] once more, since some additional ideas on the lower bound of our integral $I(t)$ are discussed there.

4.2. Algebraic numbers. Condition (6) holds for numbers $\omega_1 = \alpha_1, \dots, \omega_{s-1} = \alpha_{s-1}, \omega_s = 1$, which form a basis of some real algebraic field $\mathbb{K}$, $\deg \mathbb{K} = s$ (see [18, Chapter 2]). Hence one may assume that Conjecture 2 holds for such a set of frequencies. We note that the integral $I(t)$ with such frequencies was considered in [19], where the question of boundedness of the integral $I(t)$ is discussed. It seems to us that the conjecture from paper [19] follows from Schmidt's inequality

$$\begin{gathered} \|m_1\alpha_1 + \dots + m_{s-1}\alpha_{s-1}\| > C(\alpha, \varepsilon)(\overline{m}_1 \cdots \overline{m}_{s-1})^{-1-\varepsilon}, \\ \forall (m_1, \dots, m_{s-1}) \in \mathbb{Z}^{s-1} \setminus \{0\} \end{gathered} \tag{7}$$

(see [16, Chapter 8]).

Inequality (7) together with Theorem 1′ shows that in our case the discrepancy of the sequence $\{\alpha_1 x\}, \dots, \{\alpha_{s-1} x\}$, $x = 1, \dots, q$ satisfies $D_q < C_1(\alpha, \varepsilon) q^{\varepsilon}$, for any $\varepsilon > 0$. In the one-dimensional case, when α_1 is a quadric irrationality, we have (see [13, Chapter 2]) $D_q \leqslant C \ln q$.

4.3. Lower bounds for D_q. First of all we shall give some general estimates that hold for any sequence of points from $[0, 1)^{s-1}$. According to Roth's theorem [13, Chapter 2] for any finite sequence of q points from $[0, 1)^{s-1}$, we have the inequality $D_q \geqslant C_s \ln^{(s-2)/2} q$ and for any infinite sequence of points from $[0, 1)^{s-1}$ there exist natural numbers q_ν ($q_\nu \to \infty$) such that the discrepancy D_{q_ν} of the first q_ν points of our sequence satisfies the estimate $D_{q_\nu} \geqslant C_s' \ln^{(s-1)/2} q$. These results where sharpened by W. Schmidt:

1) case $s = 2$: $D_{q_\nu} \gg \log q_\nu$ for some infinite sequence $\{q_\nu\} \subset \mathbb{N}$;
2) case $s = 3$: $D_q \gg \log q$ for any $q \in \mathbb{N}$.

PROBLEM. To construct effectively natural numbers q_ν such that the discrepancy D_{q_ν} of the sequence $\{\alpha_1 x\}, \dots, \{\alpha_{s-1} x\}$, $x = 1, \dots, q_\nu$ (where $1, \alpha_1, \dots, \alpha_{s-1}$ form the basis of an algebraic field) increases faster than $\ln^{(s-2)/2} q_\nu$.

The case $s = 2$ seems to be very simple: the inequality $D_{q_\nu} \gg \log q_\nu$ for the sequence $\{\alpha_1 x\}$ has been known for a long time. We consider a weaker result for a certain q.

THEOREM 9. *Let*

$$\|q\alpha_1 - 1/2\| < Cq^{-1}. \tag{8}$$

(Here α_1 is a quadratic irrationality or more generally a continued fraction for $\alpha_1 = [a_0; a_1, \dots, a_l, \dots]$ has bounded elements: $a_r < M$, where M is a constant.) Then the discrepancy of the sequence $\{\alpha_1 x\}$, $x = 1, \dots, q$, satisfies $D_q \geqslant C_1(\alpha) \ln^{1/2} q$.

According to the Chebyshev theorem (see [20]), natural numbers q satisfying condition (8) appear regularly. The idea of the proof is to consider the diaphony F_2

of the sequence, which is defined as

$$F_2(q) = \left(\sum_{m=-\infty}^{+\infty}{}' \frac{1}{m^2} \left| \sum_{k=1}^{q} e^{2\pi i k\alpha} \right|^2 \right)^{1/2},$$

to obtain the lower bound for $F_2(q)$, and to use inequality $F_2(q) \ll D_q$ (see [21–22]). One can find the results on the exact order of the diaphony $F_2(q)$ in [22–23]. No multi-dimensional generalization of Theorem 8 has been obtained yet. Apparently, Theorem 9 admits a multi-dimensional generalization and the ideas from [24–26] can be useful.

§5. One generalization

We shall consider one generalization constructed by S. A. Dovbysh.

Suppose G is a matrix Lie group, g is its algebra, and $A(t) \subset g$ is a smooth function of real variable t. Let $X(t)$ be the trajectory of the equation $\dot{X} = AX$ under the initial condition $X(0) = E$, $E = \operatorname{diag}(1, \dots, 1)$. We may assume $G = SO(n)$.

DEFINITION 2. The trajectory $X(t)$ returns if

$$\forall \varepsilon > 0, \quad \forall T > 0, \quad \exists t > T, \quad X(t) - E = O(\varepsilon).$$

5.1. Periodic case. The following theorem seems obvious.

THEOREM 10. *Let $A(t+1) \equiv A(t)$. Then*

$$\forall q \in \mathbb{N}, \quad \exists t_q \in [1, q]: \ X(t_q) = E + O(q^{-1/n}).$$

This theorem has various proofs. It follows from the reducibility of equation $\dot{X} = AX$ when A is periodic, or from Poincaré's return theorem (see [27]), or from Kronecker's theorem (see [18, Chapter 2]).

5.2. Quasiperiodic case. Let us formulate the main result from [28].

THEOREM 11. *Let $s \geqslant 1$,*

$$A(t) = \sum_{m_1, \dots, m_s = -\infty}^{+\infty} A_{m_1, \dots, m_s} e^{2\pi i (m_1\omega_1 + \dots + m_s\omega_s)t},$$

where $A_{m_1, \dots, m_s} \in so(n)$ are constants,

$$\sum |A_{m_1, \dots, m_s}| \max_{j=1, \dots, s} |m_j| < \infty$$

($|\cdot|$ *denotes the norm*). *Let $q \in \mathbb{N}$ satisfy*

$$\|q\omega_j/\omega_s\| \leqslant q^{-1}\varphi(q), \qquad j = 1, \dots, s-1.$$

Then there exists $t_q \in (0, \omega_s^{-1} q\varphi(q)^{-n/(2n+1)})$ such that

$$X(t_q) = E + O(\varphi(q)^{-n/(2n+1)}).$$

Let $s = 2$. Then by the Khinchin metric theorem [20], for almost all $\alpha = \omega_1/\omega_2$ the inequality $\|\alpha q\| \ll q^{-1}(\ln q)^{-1}$ has infinitely many solutions in natural numbers q. Hence we have

THEOREM 12. *In the two-dimensional case there are returns for almost all frequencies* (ω_1, ω_2).

CONJECTURE 3. Let $A(t) \subset so(n)$ be a quasiperiodic smooth function. Then the occurence of returns does not depend on the arithmetic properties of the frequencies, i.e., for all real $\omega_1 \dots, \omega_s$ the trajectory $X(t)$ returns.

Simple examples constructed by S. A. Dovbysh show that in the case of noncompact noncommutative group G there may be no returns even under the condition $\int_{\mathbb{T}} A(y)\,dy = 0$.

The author thanks V. V. Kozlov, N. M. Korobov, N. I. Feldman, S. A. Dovbysh, and D. V. Treshchev for their attention and help.

The author is grateful to the Foundation Pro-Mathematica and to the International Scientific Foundation for partial support.

References

1. H. Poincaré, *On curves defined by differential equations*, H. Poincaré, Oeuvres, t. I, Gauthier-Villars, Paris, 1916–1956, pp. 3–223.
2. ———, *Sur les series trigonometriques*, C. R. Acad. Sci. Paris **101** (1885), no. 2, 1131–1134.
3. H. Weyl, *Über die Gleichverteilung von Zahlen mod. Eins*, Math. Ann. **77** (1916), 313–352.
4. V. G. Sprindzhuk, *Metric theory of Diophantine approximation*, Moscow, 1977. (Russian)
5. I. L. Verbitskiĭ, *Estimates for the remainder term in Kronecker theorem*, Teor. Funktsiĭ Funktsional. Anal. **5** (1967), 84–98. (Russian)
6. V. G. Sprindzhuk, *Asymptotic behavior of integrals of quasiperiodic functions*, Differentsial'nye Uravneniya **3** (1967), no. 5, 862–868; English transl. in Differential Equations **3** (1967).
7. V. V. Kozlov, *Methods of qualitative analysis in solid body dynamics*, Moscow State University, Moscow, 1980. (Russian)
8. ———, *On integrals of quasiperiodic functions*, Vestnik Moskov. Univ. Ser. I Mat. Mekh. (1978), no. 1, 106–115; English transl. in Moscow Univ. Math. Bull. **33** (1978).
9. E. A. Sidorov, *Conditions of uniform Poisson stability of cylindrical systems*, Uspekhi Mat. Nauk **34** (1979), no. 6, 184–188; English transl. in Russian Math. Surveys **34** (1979).
10. N. G. Moshchevitin, *To the problem on behavior of integral of conditionally periodic function*, Mat. Zametki **50** (1991), no. 3, 97–106; English transl., Math. Notes **50** (1991), no. 3, 945–952.
11. ———, *On the behavior of integral of conditionally periodic function*, Mathematical Methods in Mechanics (V. V. Kozlov, ed.), Moscow State University, 1990, pp. 63–66. (Russian)
12. ———, *Final properties of integrals of quasiperiodic functions related to the problems of small divisors*, Vestnik Moskov. Univ. Ser. I Mat. Mekh. (1988), no. 5, 96–98; English transl. in Moscow Univ. Math. Bull. **43** (1988).
13. L. Kuipers and H. Niederreiter, *Uniform distribution of sequences*, Wiley, New York and London, 1974.
14. H. Niederreiter, *Application of Diophantine approximation to numerical integration*, Diophantine Approximations and Applications (Osgood, ed.), Academic Press, New York, 1973, pp. 129–199.
15. Hua Loo Keng and Wang Yuan, *Application of number theory to numerical analysis*, Springer-Verlag, Berlin, Heidelberg, and New York, 1981.
16. N. G. Moshchevitin, *On distribution of fractional parts for a system of linear functions*, Vestnik Moskov. Univ. Ser. I Mat. Mekh. (1990), no. 4, 26–30; English transl. in Moscow Univ. Math. Bull. **45** (1990).
17. ———, *On non-return of integrals of conditionally-periodic functions*, Mat. Zametki **49** (1991), 138–140; English transl., Math. Notes **49**, no. 6, 650–652.
18. W. M. Schmidt, *Diophantine approximations*, Lecture Notes in Math., vol. 785, Springer-Verlag, Heidelberg and Berlin, 1980.
19. L. G. Peck, *On uniform distribution of algebraic numbers*, Proc. Amer. Math. Soc. **4** (1953), no. 1, 440-443.
20. A. Ya. Khinchine, *Chain fractions*, "Nauka", Moscow, 1978. (Russian)

21. W. Fleischer and H. Stegbuchner, *Über eine Undleichung in der Theorie der Gleichverteilung mod.* 1, Österreich. Akad. Wiss. Math.-Natur. Kl. Sitzungsber. II **191** (1982), 133–139.
22. V. F. Lcv, *Crosstalk and quadric deviations of multi-dimensional cells,* Mat. Zametki **47** (1990), no. 6, 45–55; English transl., Math. Notes (1990), no. 5–6, 556–564.
23. V. A. Bykovskiĭ, *On the exact order of error of optimal cubature formulae in spaces with dominant derivative and L_2-discrepancies of nets* (1985), Dalnevost. Vysh. Nauchnyĭ Tsentr Akad Nauk SSSR (preprint), Vladivostok. (Russian)
24. L. J. Peck, *Simultaneous rational approximations to algebraic numbers,* Bull. Amer. Math. Soc. **67** (1961), 197–201.
25. T. W. Cusick, *Diophantine approximations of linear forms over an algebraic number field,* Mathematica **20** (1973), 16–23.
26. B. F. Skubenko, *On the generalized Roth-Schmidt theorem,* Zap. Nauchn. Sem. Leningrad. Otdel. Mat. Inst. Steklov. (LOMI) **134** (1984), 226–231; English transl. in J. Soviet Math. **36** (1987), no. 1.
27. V. I. Arnold, *Mathematical methods of classical mechanics,* Springer-Verlag, New York and Berlin, 1989.
28. S. E. Lapin, *On return of trajectories of systems of linear differential equations with quasiperiodic coefficients,* Mat. Zametki **54** (1993), no. 1, 52–56; English transl. in Math. Notes **54** (1993).
29. J. Beck, *Randomness of* $n\sqrt{2}$ mod 1 *and a Ramsey property of the hyperbola,* Sets, Graphs and Numbers (Budapest 1991), Colloq. Math. Soc. János Bolyai, vol. 60, North-Holland, Amsterdam, 1992, pp. 23–66.
30. P. Hellekalek, *Weyl sums over irrational rotations,* Österreich. Ung. Slow. Koll. über Zahlentheorie,, Grazer Math. Ber., vol. 318, Karl-Franzens-Univ., Graz, 1992, pp. 77–98.
32. G. Hardy and J. Littlewood, *The lattice-points of a right angled triangle.* I, Proc. London Math. Soc. **3** (1920), 15–36.
33. V. T. Sos, *On the discrepancy of the sequence* $\{n\alpha\}$, Topics in Number Theory (P. Turan, ed.), Colloq. Math. Soc. János Bolyai, vol. 13, Akademiai Kiado, Budapest, 1974, pp. 359–367.

Translated by THE AUTHOR

DEPARTMENT OF MATHEMATICS, MOSCOW STATE UNIVERSITY, MOSCOW 119899, RUSSIA

Amer. Math. Soc. Transl.
(2) Vol. **168**, 1995

The Method of Pointwise Mappings in the Stability Problem of Two-Segment Trajectories of the Birkhoff Billiards

A. A. MARKEEV

§1. Stability of critical points under area preserving mappings

1.0. Statement of the problem. We consider a pointwise mapping $x_0, y_0 \to x_1, y_1$ that preserves the area. Let the expansion of the mapping in a series in the neighborhood of the critical point $x_0 = 0$, $y_0 = 0$ be of the form

$$\left\| \begin{matrix} x_1 \\ y_1 \end{matrix} \right\| = A \left\| \begin{matrix} x_0 \\ y_0 \end{matrix} \right\| + \left\| \begin{matrix} X^{(2)} \\ Y^{(2)} \end{matrix} \right\| + \left\| \begin{matrix} X^{(3)} \\ Y^{(3)} \end{matrix} \right\| + \dots,$$

where

$$A = \left\| \begin{matrix} a & b \\ c & d \end{matrix} \right\|,$$

and $X^{(2)}$, $Y^{(2)}$ and $X^{(3)}$, $Y^{(3)}$ are second- and third-order terms with respect to x_0, y_0. The area preservation condition is:

$$\begin{vmatrix} \dfrac{\partial x_1}{\partial x_0} & \dfrac{\partial x_1}{\partial y_0} \\ \dfrac{\partial y_1}{\partial x_0} & \dfrac{\partial y_1}{\partial y_0} \end{vmatrix} = 1.$$

This implies that $\det A = 1$.

We are concerned with the stability problem of the critical point.

Consider the characteristic equation for the matrix A

$$\rho^2 - \rho \operatorname{tr} A + 1 = 0,$$

where $\operatorname{tr} A = a + d$. Setting the absolute values of both multiplicators ρ_1 and ρ_2 equal to 1 yields necessary stability condition in the linear approximation. It can be written in the form

$$|\operatorname{tr} A| \leqslant 2.$$

1991 *Mathematics Subject Classification*. Primary 58F10.

Within the region of linear stability defined by the inequality $|\operatorname{tr} A| < 2$, the multiplicators are pure imaginary conjugates, whose absolute value is equal to 1. The boundary of the region of linear stability is specified by the relation $|\operatorname{tr} A| = 2$. Here the multiplicators are real numbers and are both equal to either -1 or $+1$.

If $|\operatorname{tr} A| > 2$, the critical point of the mapping is unstable not only in the linear approximation but also when arbitrary nonlinear terms are taken into account. As for the region of linear stability, i.e., for $|\operatorname{tr} A| \leqslant 2$, the addition of higher order terms may lead to either stability or instability.

The stability of a critical point under the mapping that preserves phase volume was investigated earlier: by C. L. Siegel [1] in the case of mappings of the plane at third-order resonance and by A. P. Markeev [2] for multi-dimensional mappings at third- and fourth-order resonance. Our aim is to obtain stability and instability conditions under an area preserving mapping for arbitrary resonance of up to the fourth order (inclusive) and to express these conditions in terms of the coefficients of the power series expansion of the mapping.

1.1. The case $|\operatorname{tr} A| < 2$.

1.1.1. *Linear normalization.* Within the region of linear stability there exists a canonical univalent change of variables

$$\left\| \begin{matrix} x \\ y \end{matrix} \right\| = N \left\| \begin{matrix} q \\ p \end{matrix} \right\|$$

transforming the linear term of the mapping to a rotation by the angle α, where $0 < |\alpha| < \pi$. The computations show that the change-of-variables matrix N is of the form

$$N = \left\| \begin{matrix} 0 & -\dfrac{b}{\sqrt{b \sin \alpha}} \\ \dfrac{\sin \alpha}{\sqrt{b \sin \alpha}} & \dfrac{a - \cos \alpha}{\sqrt{b \sin \alpha}} \end{matrix} \right\|,$$

where $\operatorname{sgn} \alpha = \operatorname{sgn} b$. We shall assume that this transformation has already been carried out, and the mapping under study is written as

$$\begin{aligned} q_1 &= (\cos \alpha) q_0 + (\sin \alpha) p_0 + Q^{(2)} + Q^{(3)} + \dots, \\ p_1 &= (-\sin \alpha) q_0 + (\cos \alpha) p_0 + P^{(2)} + P^{(3)} + \dots, \end{aligned} \tag{1.1.1}$$

where

$$\begin{aligned} Q^{(2)} &= a_{20} q_0^2 + a_{11} q_0 p_0 + a_{02} p_0^2, \\ P^{(2)} &= b_{20} q_0^2 + b_{11} q_0 p_0 + b_{02} p_0^2, \\ Q^{(3)} &= a_{30} q_0^3 + a_{21} q_0^2 p_0 + a_{12} q_0 p_0^2 + a_{03} p_0^3, \\ P^{(3)} &= b_{30} q_0^3 + b_{21} q_0^2 p_0 + b_{12} q_0 p_0^2 + b_{03} p_0^3. \end{aligned}$$

The area preservation condition

$$\frac{\partial q_1}{\partial q_0} \frac{\partial p_1}{\partial p_0} - \frac{\partial q_1}{\partial p_0} \frac{\partial p_1}{\partial q_0} = 1 \tag{1.1.2}$$

implies the following relation for the coefficients of the expansion (1.1.1):

$$
\begin{aligned}
&(b_{11}+2a_{20})\cos\alpha-(2b_{20}-a_{11})\sin\alpha=0\,,\\
&(2b_{02}+a_{11})\cos\alpha-(b_{11}-2a_{02})\sin\alpha=0\,,\\
&(b_{21}+3a_{30})\cos\alpha-(3b_{30}-a_{21})\sin\alpha+2a_{20}b_{11}-2a_{11}b_{20}=0\,,\\
&(b_{12}+a_{21})\cos\alpha-(b_{21}-a_{12})\sin\alpha+2a_{20}b_{02}-2a_{02}b_{20}=0\,,\\
&(3b_{03}+a_{12})\cos\alpha-(b_{12}-3a_{13})\sin\alpha+2a_{11}b_{02}-2a_{02}b_{11}=0\,.
\end{aligned}
\tag{1.1.3}
$$

To simplify the computation, we pass to canonically adjoint complex variables u and v according to the relations

$$q=\frac{u+v}{2}\,,\qquad p=i\,\frac{u-v}{2}\,.$$

In the new variables, the mapping takes the form

$$u_1=e^{i\alpha}u_0+u^{(2)}+u^{(3)}+\dots\,,\qquad v_1=e^{-i\alpha}v_0+v^{(2)}+v^{(3)}+\dots\,, \tag{1.1.4}$$

where

$$u^{(n)}=\sum_{i+j=n}c_{ij}u_0^iv_0^j\,,\quad v^{(n)}=\sum_{i+j=n}d_{ij}u_0^iv_0^j\,,\qquad n=2\,,3\,.$$

The coefficients c_{ij} and d_{ij} can be expressed linearly in terms of the expansion coefficients (1.1.1). The results of our subsequent investigation do not depend on the sign of α; hence, to be definite, we assume $0<\alpha<\pi$.

1.1.2. *The case* $\alpha=2\pi/3$ (*third-order resonance*). We simplify the nonlinear part of the mapping (1.1.4) as much as possible by an appropriate canonical change of variables $u\,,\,v\to\xi\,,\,\eta$ given by the generating function

$$S=u\eta+S_{30}u^3+S_{12}u\eta^2+S_{21}u^2\eta+S_{03}\eta^3+\dots\,.$$

Carrying out the necessary computations, we see that the entire quadratic part of the mapping can be eliminated by a suitable choice of the coefficients S_{ij} except for the case when $\alpha=2\pi/3$. For this value, third-order resonance occurs, and the normal mapping form in the variables ξ and η up to second-order terms (inclusive) is the following:

$$\xi_1=e^{i\alpha}\xi_0+c_{02}\eta_0^2+\dots\,,\qquad \eta_1=e^{-i\alpha}\eta_0+d_{20}\xi_0^2+\dots\,. \tag{1.1.5}$$

Here the coefficients c_{02} and d_{20} are expressed in terms of the coefficients of the mapping (1.1.1) by the following relations:

$$
\begin{aligned}
c_{02}&=A-iB\,, & d_{20}&=A+iB\,,\\
A&=\tfrac14(a_{20}-a_{02}-b_{11})\,, & B&=\tfrac14(a_{11}-b_{02}+b_{20})\,.
\end{aligned}
$$

Now we shall try to find a Hamiltonian system with 2π-periodic Hamiltonian such that the mapping over a period induced by the system coincides with the normal form (1.1.5). We shall seek the Hamiltonian function in the form

$$H = i\,\frac{\alpha}{2\pi}\,\xi\eta + h_{30}(t)\xi^3 + h_{21}(t)\xi^2\eta + h_{12}(t)\xi\eta^2 + h_{03}(t)\eta^3 + \dots, \tag{1.1.6}$$

where $h_{ij}(t)$ are the desired 2π-periodic functions in time t.

Let us write down the appropriate differential equations with Hamiltonian (1.1.6) and find their solution as a series in the initial values ξ_0 and η_0. For $t = 2\pi$, the coefficients of this series must be identical with their counterparts in the normal mapping form (1.1.5). The resulting Hamiltonian will be as follows:

$$H = i\,\frac{\alpha}{2\pi}\,\xi\eta + \frac{c_{02}}{6\pi}\,e^{-i\alpha}e^{it}\eta^3 - \frac{d_{20}}{6\pi}\,e^{i\alpha}e^{-it}\xi^3 + \dots, \qquad \alpha = \frac{2\pi}{3}. \tag{1.1.7}$$

Now we introduce to the real variables φ, r by the formula

$$\xi = -i\sqrt{2r}\,e^{i\varphi}, \qquad \eta = i\sqrt{2r}\,e^{-i\varphi}. \tag{1.1.8}$$

This change of variables is canonical of rank $1/2i$. In the new variables, the Hamiltonian (1.1.7) can be written as

$$H = \frac{\alpha}{2\pi}\,r + \frac{\varkappa}{3\pi}\,\sqrt{2r}\cos(3\varphi + \alpha + \delta - t) + O(r^2), \tag{1.1.9}$$

where

$$\varkappa = \sqrt{A^2 + B^2}, \qquad \cos\delta = \frac{A}{\sqrt{A^2 + B^2}}, \qquad \sin\delta = \frac{B}{\sqrt{A^2 + B^2}}.$$

The solution $r = 0$ to the equations with Hamiltonian (1.1.9) is unstable for $\varkappa = 0$ [2]. But the stability of the equilibrium position of a system with periodic coefficients and that of the critical point of the mapping induced by the system over the period are equivalent; consequently, we obtain the following condition: if at the third-order resonance ($\alpha = 2\pi/3$) we have $\varkappa \neq 0$, then the critical point of the mapping is unstable. This condition expressed in terms of the expansion coefficients in the mapping (1.1.1) is equivalent to the validity of at least one of the following inequalities:

$$a_{20} - a_{02} - b_{11} \neq 0 \quad \text{or} \quad a_{11} - b_{02} - b_{20} \neq 0. \tag{1.1.10}$$

1.1.3. *The case $\alpha = \pi/2$ (fourth-order resonance and no resonance).* If $\alpha \neq 2\pi/3$, the whole quadratic part of the mapping (1.4) vanishes under the normalizing change of variables u, $v \to \xi$, η. Hence we shall assume that the mapping (1.1.1) originally had no quadratic terms.

Similarly to the previous case, on simplifying the cubic part of the mapping, we find that certain third-order terms ($c_{03}\eta_0^3$ and $d_{30}\xi_0^3$) do not vanish at $\alpha = \pi/2$ when fourth-order resonance occurs, and certain other terms ($c_{21}\xi_0^2\eta_0$ and $d_{12}\xi_0\eta_0^2$) never vanish at all. The normal mapping form up to third-order terms (inclusive) is:

$$\begin{aligned} \xi_1 &= e^{i\alpha}\xi_0 + c_{21}\xi_0^2\eta_0 + c_{03}\eta_0^3 + \dots , \\ \eta_1 &= e^{-i\alpha}\xi_0 + d_{12}\xi_0\eta_0^2 + d_{30}\xi_0^3 + \dots . \end{aligned} \tag{1.1.11}$$

As in the previous case, we shall try to find a Hamiltonian system inducing a mapping of the form (1.1.11) over a period. Transforming again to the variables φ and r according to (1.1.8), we obtain the Hamiltonian of this system in the form

$$H = \frac{\alpha}{2\pi} r + r^2\left[M - K\sin\left(4\varphi - \frac{2\alpha}{\pi}t\right) - L\cos\left(4\varphi - \frac{2\alpha}{\pi}t\right)\right] + O(r^{5/2}), \tag{1.1.12}$$

where the quantities M, K, and L can be expressed in terms of the coefficients of the original mapping (1.1.1).

If there is no resonance ($L = K = 0$), then, according to [3], the equilibrium position $r = 0$ of the system with the Hamiltonian (1.1.12) is stable whenever $M \neq 0$. But if there is the resonance ($L^2 + K^2 \neq 0$), then according to [2] the equilibrium position $r = 0$ is unstable whenever $|M| > \sqrt{K^2 + L^2}$ and stable whenever $|M| < \sqrt{K^2 + L^2}$.

We shall express the above conditions in terms of coefficients in (1.1.1).

(1) The case of no resonance ($\alpha \neq \pi/2$, $\alpha \neq 2\pi/3$). If

$$3a_{30} + b_{21} + a_{12} + 3b_{03} \neq 0, \tag{1.1.13}$$

then the critical point of the mapping (1.1.1) is stable. From the relation between mapping coefficients (1.1.1), inequality (1.1.13) is readily shown to be equivalent to the following one:

$$3b_{30} - a_{21} + b_{12} - 3a_{03} \neq 0.$$

(2) The case of resonance ($\alpha = \pi/2$). If the inequality

$$(a_{30} + b_{03})(a_{30} + 2a_{12} + b_{03}) - 2(a_{03} - b_{30})^2 > 0 \tag{1.1.14}$$

holds, the critical point is stable and if

$$(a_{30} + b_{03})(a_{30} + 2a_{12} + b_{03}) - 2(a_{03} - b_{30})^2 < 0, \tag{1.1.15}$$

then the critical point is unstable.

1.2. The case $|\operatorname{tr} A| = 2$. In this case the mapping can be transformed by a linear change of variables to a form where the linear part is given by one of the following matrices, depending on the rank of the matrix $A - \rho E$:

(1) $\begin{Vmatrix} 1 & 1 \\ 0 & 1 \end{Vmatrix}$ for $rK(A - \rho E) = 1$ and $\rho_1 = \rho_2 = 1$,

(2) $\begin{Vmatrix} -1 & 1 \\ 0 & -1 \end{Vmatrix}$ for $rK(A - \rho E) = 1$ and $\rho_1 = \rho_2 = -1$,

(3) $\begin{Vmatrix} 1 & 0 \\ 0 & 1 \end{Vmatrix}$ for $rK(A - \rho E) = 0$ and $\rho_1 = \rho_2 = 1$,

(4) $\begin{Vmatrix} -1 & 0 \\ 0 & -1 \end{Vmatrix}$ for $rK(A - \rho E) = 0$ and $\rho_1 = \rho_2 = -1$.

Below we shall assume that the appropriate change of variables has already been made.

1.2.1. *The case* $\rho_1 = \rho_2 = 1$ *and* $rK(A - \rho E) = 1$. The mapping is of the form (1.2.1)

$$q_1 = q_0 + p_0 + a_{20}q_0^2 + a_{11}q_0p_0 + a_{02}p_0^2 + a_{30}q_0^3 + a_{21}q_0^2p_0 + a_{12}q_0p_0^2 + a_{03}p_0^3,$$
$$p_1 = p_0 + b_{20}q_0^2 + b_{11}q_0p_0 + b_{02}p_0^2 + b_{30}q_0^3 + b_{21}q_0^2p_0 + b_{12}q_0p_0^2 + b_{03}p_0^3.$$

The area preservation condition (1.1.2) provides the following relation for the coefficients:

$$2a_{20} + b_{11} - 2b_{20} = 0, \qquad a_{11} + 2b_{02} - b_{11} = 0,$$
$$3a_{30} + b_{21} + 2a_{20}b_{11} - 3b_{30} - 2a_{11}b_{20} = 0,$$
$$a_{12} + 3b_{03} + 2a_{11}b_{02} - b_{12} - 2a_{02}b_{11} = 0,$$
$$a_{21} + b_{12} - b_{21} - 2a_{02}b_{20} + 2a_{20}b_{02} = 0.$$

The mapping (1.2.1) is specified by some T-periodic Hamiltonian system. Assume that $T = 1$. Then the linear part of the mapping (2.1) agrees with the quadratic Hamiltonian $H = p^2/2$. After being normalized up to third-order terms (inclusive), the T-periodic Hamiltonian with such a quadratic part can be written [4, 5] in the form

$$H = \frac{1}{2}\eta^2 + \frac{\gamma}{3}\xi^3 + \dots, \tag{1.2.2}$$

where $\gamma =$ constant, and the dots stand for terms in ξ and η of degree higher than 3, with T-periodic coefficients.

The following assertion [4, 5] is valid: if $\gamma \neq 0$, the solution $\xi = 0$, $\eta = 0$ is unstable.

The mapping induced by the Hamiltonian (1.2.2) during the period is of the form

$$\begin{aligned} \xi_1 &= \xi_0 + \eta_0 - \gamma\left(\tfrac{1}{2}\xi_0^2 + \tfrac{1}{3}\xi_0\eta_0 + \tfrac{1}{12}\eta_0^2\right) + \dots, \\ \eta_1 &= \eta_0 - \gamma\left(\xi_0^2 + \xi_0\eta_0 + \tfrac{1}{3}\eta_0^2\right) + \dots. \end{aligned} \tag{1.2.3}$$

On obtaining the normal form of the mapping (1.2.1) and comparing it with the mapping (1.2.3), we find that $\gamma = -b_{20}$.

Therefore, if

$$b_{20} \neq 0, \tag{1.2.4}$$

then the critical mapping point (1.2.1) is unstable.

Now let $b_{20} = 0$. By a suitable canonical change of variables $q, p \to \xi, \eta$, the entire quadratic part in the mapping (1.2.1) can be eliminated, and the third-order terms can be simplified. The mapping normalized up to third-order terms (inclusive) is induced by a T-periodic Hamiltonian of the following normal form:

$$H = \tfrac{1}{2}\eta^2 + \tfrac{1}{4}\mu\xi^4 + \dots, \tag{1.2.5}$$

where

$$\mu = -\left(\tfrac{1}{2}b_{11}^2 + b_{30}\right). \tag{1.2.6}$$

For systems with the Hamilton function (1.2.5), the following results [4, 5] can be mentioned: for $\mu > 0$ the trivial solution $\xi = 0$, $\eta = 0$ is stable, and for $\mu < 0$ it is unstable. From (1.2.6) we obtain the following condition: if

$$\tfrac{1}{2}b_{11}^2 + b_{30} < 0, \tag{1.2.7}$$

then the critical point of the mapping (1.2.1) is stable; if

$$\tfrac{1}{2}b_{11}^2 + b_{30} > 0, \tag{1.2.8}$$

then the critical point is unstable.

1.2.2. *The case $\rho_1 = \rho_2 = -1$ and $rK(A - \rho E) = 1$.* The mapping is of the form

$$\begin{aligned} q_1 &= -q_0 + p_0 + a_{20}q_0^2 + a_{11}q_0p_0 + a_{02}p_0^2 \\ &\quad + a_{30}q_0^3 + a_{21}q_0^2p_0 + a_{12}q_0p_0^2 + a_{03}p_0^3 + \dots, \\ p_1 &= -p_0 + b_{20}q_0^2 + b_{11}q_0p_0 + b_{02}p_0^2 \\ &\quad + b_{30}q_0^3 + b_{21}q_0^2p_0 + b_{12}q_0p_0^2 + b_{03}p_0^3 + \dots. \end{aligned} \tag{1.2.9}$$

The area preservation condition (1.1.2) yields the following relations for the coefficients (1.2.9):

$$\begin{gathered} 2a_{20} + b_{11} + 2b_{20} = 0, \qquad a_{11} + 2b_{02} + b_{11} = 0, \\ 3a_{30} + b_{21} + 3b_{30} + 2a_{11}b_{20} = 0, \\ a_{12} + 3b_{03} + b_{12} + 2a_{02}b_{11} = 0, \\ a_{21} + b_{12} + b_{21} - 2a_{20}b_{02} + 2a_{02}b_{20} = 0. \end{gathered}$$

It can be readily seen that there is no quadratic Hamiltonian with constant coefficients which would induce the linear part of the mapping (1.2.9). However, the case under consideration can be investigated along lines similar to the previous one. To demonstrate this, we shall first raise the mapping to the second power. The matrix of the linear part of the mapping induced over two periods is of the form

$$\begin{Vmatrix} 1 & -2 \\ 0 & 1 \end{Vmatrix}.$$

By the canonical change of variables $q, p \to \tilde{q}, \tilde{p}$ of the form $q = \tilde{q}$, $p = -\tilde{p}/2$, we obtain a mapping whose linear part contains the matrix

$$\begin{Vmatrix} 1 & 1 \\ 0 & 1 \end{Vmatrix}$$

which is the same as in the previous case $\rho_1 = \rho_2 = 1$. Note that in the mapping relation for the variable $\tilde{p}$, the coefficient $\tilde{b}_{20}$ of $\tilde{q}_0^2$ is zero, the coefficient $\tilde{b}_{11}$ equals $-2b_{20}$, and the coefficient $\tilde{b}_{30}$ of $\tilde{q}_0^3$ equals $4(b_{30} - b_{20}^2)$.

The stability conditions for the critical point of the mapping (1.2.9) can now be written in the form of inequalities (1.2.7) and (1.2.8), where the quantities b_{ij} must be replaced by $\tilde{b}_{ij}$. Thus we obtain the following conditions: if

$$b_{30} - \tfrac{1}{2}b_{20}^2 < 0\,, \tag{1.2.10}$$

then the critical point of the mapping (1.2.9) is stable; if

$$b_{30} - \tfrac{1}{2}b_{20}^2 > 0\,, \tag{1.2.11}$$

then the critical point is unstable.

1.2.3. *The case $\rho_1 = \rho_2 = 1$ and $rK(A - \rho E) = 0$.* The mapping is of the form

$$\begin{aligned} q_1 &= q_0 + a_{20}q_0^2 + a_{11}q_0p_0 + a_{02}p_0^2 \\ &\qquad + a_{30}q_0^3 + a_{21}q_0^2p_0 + a_{12}q_0p_0^2 + a_{03}p_0^3 + \dots\,, \\ p_1 &= p_0 + b_{20}q_0^2 + b_{11}q_0p_0 + b_{02}p_0^2 \\ &\qquad + b_{30}q_0^3 + b_{21}q_0^2p_0 + b_{12}q_0p_0^2 + b_{03}p_0^3 + \dots\,. \end{aligned} \tag{1.2.12}$$

The area preservation condition (1.1.2) yields the following relations for the coefficients (1.2.12):

$$\begin{gathered} 2a_{20} + b_{11} = 0\,, \qquad a_{11} + 2b_{02} = 0\,, \\ 3a_{30} + b_{21} + 2a_{20}b_{11} - 2a_{11}b_{20} = 0\,, \\ a_{12} + 3b_{03} + 2a_{11}b_{02} - 2a_{02}b_{11} = 0\,, \\ 2a_{21} + 2b_{12} - a_{11}b_{11} - 2a_{02}b_{20} = 0. \end{gathered}$$

The T-periodic Hamiltonian that induces this mapping is defined in the variables r, φ by

$$a = \sqrt{2r}\sin\varphi\,, \qquad p = \sqrt{2r}\cos\varphi$$

can be written in the form

$$H = r\sqrt{r}\,(a\sin\varphi + b\cos\varphi + c\sin 3\varphi + d\cos 3\varphi) + O(r^2)\,,$$

where

$$a = \frac{\sqrt{2}}{2}(-b_{20} - b_{02})\,, \qquad b = \frac{\sqrt{2}}{2}(a_{20} + a_{02})\,,$$
$$c = \frac{\sqrt{2}}{2}\left(\frac{1}{3}b_{20} - b_{02}\right), \qquad d = \frac{\sqrt{2}}{2}\left(-a_{20} + \frac{1}{3}a_{02}\right).$$

From the results of [5] we deduce that the critical point is unstable when at least one of the coefficients of the quadratic part of the mapping (1.2.12) is nonzero.

Assume now that there is no quadratic part in (1.2.12). Then the Hamiltonian specifying the mapping may be written in the variables r, φ as follows

$$H = r^2\Phi(\varphi),$$

$$\Phi(\varphi) = (a_{30} - b_{03})\sin 2\varphi - \tfrac{1}{2}(a_{30} + b_{03})\sin 4\varphi + \tfrac{1}{2}(a_{03} + b_{30})\cos 2\varphi$$

$$(1.2.13) \qquad -\tfrac{1}{4}(a_{21} - \tfrac{1}{2}a_{03} + \tfrac{1}{2}b_{30})\cos 4\varphi + \tfrac{1}{4}(\tfrac{3}{2}a_{03} + a_{21} - \tfrac{3}{2}b_{30}).$$

Using [5] again, we can obtain the following conditions: if the function $\Phi(\varphi)$ has no zeros in the interval $[0, 2\pi]$, then the critical point of the mapping (1.2.12) is stable; but if there exists a value $\varphi^* \in [0, 2\pi]$ such that $\Phi(\varphi^*) = 0$ and $\Phi'(\varphi^*) \neq 0$, then the critical point is unstable.

1.2.4. *The case $\rho_1 = \rho_2 = -1$ and $rK(A - \rho E) = 0$.* The mapping is of the form

$$(1.2.14) \qquad \begin{aligned} q_1 &= -q_0 + a_{20}q_0^2 + a_{11}q_0p_0 + a_{02}p_0^2 \\ &\quad + a_{30}q_0^3 + a_{21}q_0^2p_0 + a_{12}q_0p_0^2 + a_{03}p_0^3 + \dots, \\ p_1 &= -p_0 + b_{20}q_0^2 + b_{11}q_0p_0 + b_{02}p_0^2 \\ &\quad + b_{30}q_0^3 + b_{21}q_0^2p_0 + b_{12}q_0p_0^2 + b_{03}p_0^3 + \dots. \end{aligned}$$

The area preservation condition (1.1.2) yields the following relations for the coefficients (1.2.14):

$$\begin{aligned} 2a_{20} + b_{11} &= 0, \qquad a_{11} + 2b_{02} = 0, \\ 2a_{20}b_{11} - 3a_{30} - b_{21} - 2a_{11}b_{20} &= 0, \\ 2a_{11}b_{02} - a_{12} - 3b_{03} - 2a_{02}b_{11} &= 0, \\ 2a_{20}b_{02} - a_{21} - b_{12} - 2a_{02}b_{20} &= 0. \end{aligned}$$

We can show that the investigation of the stability of the critical point of the mapping (1.2.14) can be reduced to the case treated in 1.2.3. To demonstrate this, we shall first raise the mapping (1.2.14) to the second power and obtain the following mapping, induced over two periods, with no quadratic part:

$$\begin{aligned} q_2 &= q_0 + \tilde{a}_{30}q_0^3 + \tilde{a}_{21}q_0^2p_0 + \tilde{a}_{12}q_0p_0^2 + \tilde{a}_{03}p_0^3, \\ p_2 &= p_0 + \tilde{b}_{30}q_0^3 + \tilde{b}_{21}q_0^2p_0 + \tilde{b}_{12}q_0p_0^2 + \tilde{b}_{03}p_0^3. \end{aligned}$$

The coefficients $\tilde{a}_{ij}$ and $\tilde{b}_{ij}$ are expressed in terms of the coefficients a_{ij} and b_{ij} by the following relations:

$$(1.2.15) \qquad \begin{aligned} \tilde{a}_{30} &= -2a_{30} - 2a_{20}^2 + 2b_{02}b_{20}, & \tilde{b}_{30} &= -2b_{30}, \\ \tilde{a}_{21} &= -2a_{21} - 2a_{02}b_{20} - a_{11}a_{20}, & \tilde{b}_{21} &= -2b_{21} + 2b_{20}b_{02} - 2a_{20}^2, \\ \tilde{a}_{12} &= -2a_{12} + 2a_{02}a_{20} - 2b_{02}^2, & \tilde{b}_{12} &= -2b_{12} - 2a_{02}b_{20} - b_{11}b_{02}, \\ \tilde{a}_{03} &= -2a_{03}, & \tilde{b}_{03} &= -2b_{03} + 2a_{02}a_{20} - 2b_{02}^2. \end{aligned}$$

Thus the investigation of the stability of the critical point of the mapping (1.2.14) can be reduced to the study of the behavior of the function $\Phi(\varphi)$ defined by (1.2.13), with the quantities a_{ij} and b_{ij} replaced by $\tilde{a}_{ij}$ and $\tilde{b}_{ij}$ according to (1.2.15).

REMARK. In the above argument, area preserving pointwise mappings were associated with appropriate Hamiltonian systems without proving that such constructions are possible. In his paper [8], R. Douady proved that such a procedure is well defined in the class of infinitely differentiable periodic Hamiltonian systems.

§2. Stability of two-segment trajectories of Birkhoff billiards

2.0. Statement of the problem. Consider a point mass $m = 1$ moving inside a plane region bounded by curves specified by the equations $f_1(x) = l$ and $f_2(x) = a_2x^2 + a_3x^3 + a_4x^4 + \dots$. The impact of the point mass on the curves is taken to be absolutely elastic (the angle of incidence being equal to the angle of reflection); consequently the total mechanical energy

$$(*)\qquad E = \tfrac{1}{2}(\dot{x}^2 + \dot{y}^2)$$

is conserved in motion. Consider a particular motion of the point mass, namely the motion on the y axis. While in motion, the point mass alternately hits the curves $y = f_1(x)$ and $y = f_2(x)$. A trajectory of this type is called a *two-segment* trajectory. It is defined by

$$(**)\qquad x = 0\,, \qquad 0 \leqslant y \leqslant l\,.$$

Our concern will be with the orbital stability of the motion specified by $(**)$. The motion will be considered *stable* if for small initial values of $x(0)$ and $x'(0)$ the values of $x(t)$ and $x'(t)$ will remain small forever.

A linear stability problem for two-segment trajectory of the curvilinear Birkhoff billiards was investigated earlier [6, 7]. Our aim is to provide a rigorous solution of the nonlinear stability problem associated with the particular case of the billiards when one of its boundaries is rectilinear. In this we shall rely on the results of §1 concerning the stability of the critical point of an area preserving mapping.

For the sake of convenience, we introduce dimensionless coordinates x^* and y^* defined by $x^* = x/l$ and $y^* = y/l$. In this case the equations of the curves $y = f_1(x)$ and $y = f_2(x)$ (henceforth we are dropping the asterisk) are written as follows

$$f_1(x) = 1\,, \qquad f_2(x) = d_2x^2 + d_3x^3 + d_4x^4 + \dots\,,$$

where

$$d_i = \frac{f_2^{(i)}(0)}{i!}\, l^{i-1} = a_i l^{i-1}.$$

The motion under investigation $(**)$ is now described by the relations $x = 0$, $0 \leqslant y \leqslant 1$. We also choose a measurement scale of t such that the energy integral $(*)$ can be written in the form $\dot{x}^2 + \dot{y}^2 = 1$.

The stability and instability conditions of the two-segment trajectories considered above depend only on the type of the curve, and consequently they can be expressed in terms of the coefficients of the power series expansion of the function $y = f_2(x)$.

2.1. Construction of the mapping. To solve the stability problem, we shall construct a mapping describing the dynamical state of the point mass at the moment when it crosses the line $y = 1/2$. At the initial moment, let the moving point mass be on the line $y = 1/2$ with abscissa $x = x_0$, and let the direction of its motion make the angle θ_0 with the horizontal (see the Figure). Hence up to the moment of collision with the curve $y = f_2(x)$ at the point $P(\tilde{x}\,,\, f_2(\tilde{x}))$, the motion occurs along the line $y_{\mathrm{I}} = \frac{1}{2} - \tan\theta_0(x - x_0)$. Note that the abscissa $\tilde{x}$ of the point P can be found from the equation $\tilde{x} = x_0 + \cot\theta_0(\frac{1}{2} - f_2(\tilde{x}))$. Next, the point mass rebounds from the curve $y = f_2(x)$ and moves along the line

$$y_{\mathrm{II}} = f_2(\tilde{x}) + \tan(2\alpha + \theta_0)(x - \tilde{x}),$$

where α is the angle that the tangent to the curve $y = f_2(x)$ makes with the x axis at the point P, i.e., $\tan\alpha = f_2'(\tilde{x})$.

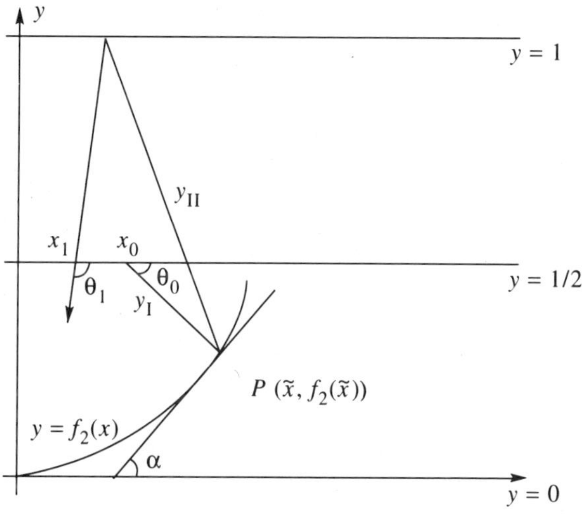

FIGURE

On the rebound from the line $y = 1$, the point mass moves back to the line $y = 1/2$ and crosses it at the point with abscissa x_1 satisfying the equation

$$x_1 = \tilde{x} + \cot(2\alpha + \theta_0)(\tfrac{3}{2} - f_2(\tilde{x})).$$

The angle θ_1 that the trajectory of the point mass makes with the line $y = 1/2$ can be determined from $\theta_1 = 2\alpha + \theta_0$.

In our case the canonically adjoint variables are the variables x and $\eta \equiv \dot{x} = \cos\theta$. Thus we have constructed the mapping

$$x_1 = x_1(x_0\,,\, \eta_0)\,, \qquad \eta_1 = \eta_1(x_0\,,\, \eta_0)\,.$$

Let us show that it is area preserving, i.e.,

$$\frac{\partial x_1}{\partial x_0}\frac{\partial \eta_1}{\partial \eta_0} - \frac{\partial x_1}{\partial \eta_0}\frac{\partial \eta_1}{\partial x_0} = 1\,. \tag{2.1.1}$$

Consider the functions

$$
\begin{aligned}
F &= x_0 + \cot\theta_0(\tfrac{1}{2} - f_2(\tilde{x})) = F(x_0, \theta_0, \tilde{x}), \\
G &= \cos(2\alpha + \theta_0) = G(\alpha, \theta_0), \\
H &= \tilde{x} + \cot(2\alpha + \theta_0)(\tfrac{3}{2} - f_2(\tilde{x})) = H(\tilde{x}, \alpha, \theta_0).
\end{aligned}
\tag{2.1.2}
$$

The following relations hold:

$$
\begin{gathered}
\tilde{x} = F(x_0, \theta_0, \tilde{x}), \quad \eta_1 = G(\alpha, \theta_0), \quad x_1 = H(\tilde{x}, \alpha, \theta_0), \\
\tan\alpha = f_2'(\tilde{x}), \quad \eta_0 = \cos\theta_0.
\end{gathered}
\tag{2.1.3}
$$

Taking the total differentials of the above functions and carrying out the necessary manipulations, we obtain

$$
\begin{aligned}
dx_1 &= \frac{1}{1 - F_{\tilde{x}}}\Big\{ F_{x_0}[H_{\tilde{x}} + \cos^2\alpha f_2'(\tilde{x}) H_\alpha]\, dx_0 \\
&\qquad - \frac{1}{\sin\theta_0}[F_{\theta_0} H_\alpha \cos^2\alpha f_2''(\tilde{x}) + H_{\theta_0} - H_{\theta_0} F_{\tilde{x}} + H_{\tilde{x}} F_{\theta_0}]\, d\eta_0 \Big\}, \\
d\eta_1 &= \frac{1}{1 - F_{\tilde{x}}}\Big\{ \cos^2\alpha f_2''(\tilde{x}) F_{x_0}\, dx_0 \\
&\qquad - \frac{1}{\sin\theta_0}[F_{\theta_0} G_\alpha \cos^2\alpha f_2''(\tilde{x}) G_\alpha + G_{\theta_0} - G_{\theta_0} F_{\tilde{x}}]\, d\eta_0 \Big\}.
\end{aligned}
\tag{2.1.4}
$$

These relations yield the necessary partial derivatives

$$
\frac{\partial x_1}{\partial x_0}, \quad \frac{\partial x_1}{\partial \eta_0}, \quad \frac{\partial \eta_1}{\partial x_0}, \quad \frac{\partial \eta_1}{\partial \eta_0}.
$$

Using the functions (2.1.2) and (2.1.3) and the expressions for the partial derivatives, we can prove the relation (2.1.1).

Using (2.1.2) and (2.1.3), we can obtain the mapping $x_0, \eta_0 \to x_1, \eta_1$ in the form

$$
\begin{aligned}
x_1 &= (1 - 6d_2)x_0 + (2 - 3d_2)\eta_0 - 9d_3 x_0^2 - 9d_3 x_0\eta_0 - \tfrac{9}{4} d_3 \eta_0^2 \\
&\quad - 4(-d_2^2 - 6d_2^3 + 3d_4)x_0^3 - 2(d_2 - 18d_2^2 + 18d_2^3 + 9d_4)x_0^2\eta_0 \\
&\quad + 2(-8d_2 + 33d_2^2 - 18d_2^3 - 9d_4)x_0\eta_0^2 \\
&\quad + (1 - 5d_2 + 8d_2^2 - 3d_2^3 - \tfrac{3}{2}d_4)\eta_0^3, \\
\eta_1 &= -4d_2 x_0 + (1 - 2d_2)\eta_0 - 6d_3 x_0^2 - 6d_3 x_0\eta_0 - \tfrac{3}{2} d_3\eta_0^2 \\
&\quad - 8(-2d_2^3 + d_4)x_0^3 - 4(d_2^2 - 6d_2^3 + 3d_4)x_0^2\eta_0 \\
&\quad + 2(d_2 - 2d_2^2 + 6d_2^3 - 3d_4)x_0\eta_0^2 - (d_2^2 - 2d_2^3 + d_4)\eta_0^3.
\end{aligned}
\tag{2.1.5}
$$

The two-segment trajectory considered above can be associated with the critical point $x_0 = 0$, $\eta_0 = 0$ of the mapping (2.1.5). Thus the investigation of stability of a two-segment periodic trajectory of the Birkhoff billiards amounts to solving the stability problem for the critical point of an area preserving mapping, which was treated in §1.

2.2. A linear problem. In matrix form, the mapping (2.1.5) can be written as follows

$$\begin{Vmatrix} x_1 \\ \eta_1 \end{Vmatrix} = A \begin{Vmatrix} x_0 \\ \eta_0 \end{Vmatrix} + \begin{Vmatrix} F^{(2)} \\ G^{(2)} \end{Vmatrix} + \begin{Vmatrix} F^{(3)} \\ G^{(3)} \end{Vmatrix} + \dots, \tag{2.2.1}$$

where $F^{(2)}$, $G^{(2)}$ and $F^{(3)}$, $G^{(3)}$ are homogeneous forms of the second and the third degree in x_0, η_0. The matrix A of the linear part of the mapping (2.2.1) is of the form

$$\begin{Vmatrix} 1-6d_2 & 2-3d_2 \\ -4d_2 & 1-2d_2 \end{Vmatrix}. \tag{2.2.2}$$

In our case, the necessary condition of linear stability $|\operatorname{tr} A| \leqslant 2$ can be reduced to the inequality

$$0 \leqslant d_2 \leqslant 1/2. \tag{2.2.3}$$

The coefficient d_2 has the simple geometrical meaning: $d_2 = l/2R$, where R is the curvature radius of the curve $y = f_2(x)$ at the point $x = 0$. Geometrically, inequality (2.2.3) has the following interpretation: if the two-segment trajectory is stable, then the length of the trajectory segment cannot be greater than the curvature radius of the curve $y = f_2(x)$ at the point $x = 0$.

For $d_2 = 0$ and $d_2 = 1/2$, we obtain the boundary of the linear stability region; the other values of d_2 inside the interval (2.2.3) correspond to the interior points of the region. If d_2 is outside the interval (2.2.3) $(l > R)$, then the two-segment trajectory is unstable regardless of the nonlinear terms of the mapping.

Now we consider a rigorous nonlinear stability problem for the case when the necessary condition (2.2.3) of stability holds in the linear approximation.

2.3. The case $0 < d_2 < 1/2$ ($|\operatorname{tr} A| < 2$).

2.3.1. *Linear normalization.* Within the linear stability region, i.e., for $0 < d_2 < 1/2$, there exists a canonical change of variables

$$\begin{Vmatrix} x \\ \eta \end{Vmatrix} = N \begin{Vmatrix} q \\ p \end{Vmatrix} \tag{2.3.1}$$

such that

$$N^{-1}AN = \begin{Vmatrix} \cos\alpha & \sin\alpha \\ -\sin\alpha & \cos\alpha \end{Vmatrix},$$

i.e., in the new variables the linear part of the mapping is a rotation by the angle α. It follows from §1 that the matrix N of the linear normalizing transformation is of the form

$$N = \begin{Vmatrix} 0 & -\dfrac{1}{2}\sqrt{\dfrac{5+3\cos\alpha}{\sin\alpha}} \\ 2\sqrt{\dfrac{\sin\alpha}{5+3\cos\alpha}} & \dfrac{\cos\alpha - 1}{\sqrt{\sin\alpha\,(5+3\cos\alpha)}} \end{Vmatrix},$$

the angle α being related to the coefficient d_2 by the formula $\cos\alpha = 1 - 4d_2$. In the new variables q, p, the mapping (2.2.1) has the form

$$\begin{Vmatrix} q_1 \\ p_1 \end{Vmatrix} = \begin{Vmatrix} \cos\alpha & \sin\alpha \\ -\sin\alpha & \cos\alpha \end{Vmatrix} \begin{Vmatrix} q_0 \\ p_0 \end{Vmatrix} + \begin{Vmatrix} Q^{(2)} \\ P^{(2)} \end{Vmatrix} + \begin{Vmatrix} Q^{(3)} \\ P^{(3)} \end{Vmatrix} + \dots,$$

where

$$\begin{Vmatrix} Q^{(2)} \\ P^{(2)} \end{Vmatrix} = N^{-1} \begin{Vmatrix} F^{(2)} \\ G^{(2)} \end{Vmatrix}, \qquad \begin{Vmatrix} Q^{(3)} \\ P^{(3)} \end{Vmatrix} = N^{-1} \begin{Vmatrix} F^{(3)} \\ G^{(3)} \end{Vmatrix}. \tag{2.3.2}$$

Here the variables x_0, η_0 in the forms $F^{(2)}$, $G^{(2)}$ and $F^{(3)}$, $G^{(3)}$ must be expressed in terms of q_0, p_0 according to (2.3.1).

2.3.2. *Third-order resonance.* According to 1.1.2, the third-order resonance occurs for $\alpha = 2\pi/3$. This value of α corresponds to $d_2 = 3/8$. The curvature radius R of the curve $y = f_2(x)$ at the point $x = 0$ equals $4l/3$.

The computations carried out according to (2.3.2) show that in the case of third-order resonance the coefficients of the quadratic forms

$$Q^{(2)} = a_{20}q_0^2 + a_{11}q_0p_0 + a_{02}p_0^2, \qquad P^{(2)} = b_{20}q_0^2 + b_{11}q_0p_0 + b_{02}p_0^2$$

are given by the following formulas:

$$a_{20} = \frac{6\sqrt[4]{3}}{7\sqrt{7}} d_3, \qquad a_{11} = -\frac{20\sqrt[4]{3}\sqrt{3}}{7\sqrt{7}} d_3, \qquad a_{02} = \frac{50\sqrt[4]{3}}{7\sqrt{7}} d_3,$$
$$b_{20} = \frac{18\sqrt[4]{3}}{7\sqrt{7}} d_3, \qquad b_{11} = -\frac{180\sqrt[4]{3}}{7\sqrt{7}} d_3, \qquad b_{02} = \frac{150\sqrt[4]{3}\sqrt{3}}{7\sqrt{7}} d_3.$$

In the case of third-order resonance, the stability condition (1.1.10) for the critical point of an area preserving mapping shows that for $d_3 \neq 0$ the two-segment trajectory is unstable.

2.3.3. *Fourth-order resonance.* Consider the fourth-order resonance which is associated with $\alpha = \pi/2$ and, consequently, with $d_2 = 1/4$ and $R = 2l$. For simplicity, we shall consider curves characterized by $d_3 = 0$. The computations carried out according to (2.3.2) in the case of $\alpha = \pi/2$ yield the values of the coefficients of the cubic forms

$$Q^{(3)} = a_{30}q_0^3 + a_{21}q_0^2p_0 + a_{12}q_0p_0^2 + a_{03}p_0^3,$$
$$P^{(3)} = b_{30}q_0^3 + b_{21}q_0^2p_0 + b_{12}q_0p_0^2 + b_{03}p_0^3,$$

which we shall use to find the quantity

$$(a_{30} + 2a_{12} + b_{03})(a_{30} + b_{03}) - 2(a_{03} - b_{30})^2.$$

This quantity enters the stability condition (1.1.15) for the critical point of an area preserving mapping in the case of fourth-order resonance. A small manipulation with the expression above gives a quadratic trinomial in d_4 of the form

$$p_2(d_4) = \frac{3\,584}{125} d_4^2 - \frac{1\,263}{500} d_4 + \frac{681}{40\,960}.$$

This trinomial has two real roots

$$d_4^{(1)} = \frac{1\,263 - \sqrt{1\,118\,469}}{28\,672} \approx 0.00716\,, \quad d_4^{(2)} = \frac{1\,263 + \sqrt{1\,118\,469}}{28\,672} \approx 0.08094\,.$$

It follows from (1.1.14) and (1.1.15) that the two-segment trajectory is unstable for $d_4^{(1)} < d_4 < d_4^{(2)}$, but it is stable for $d_4 < d_4^{(1)}$ or $d_4 > d_4^{(2)}$.

2.3.4. *The case of no resonance.* Now consider the case when $0 < d_2 < 1/2$, where $d_2 \neq 3/8$ and $d_2 \neq 1/4$, i.e., there is no resonance. To solve the stability problem, we must find whether inequality (1.1.13) holds in our case. We shall restrict our considerations to the case when the coefficient d_3 in the expansion of the function $y = f_2(x)$ is zero: $d_3 = 0$. Then the left-hand member of inequality (1.1.13) becomes a function of the two parameters d_2 and d_4. The zeros of this function were computed numerically, and the results are summarized in the table below. Thus, if the point with coordinates $(d_2\,,\,d_4)$ does not belong to this curve, the two-segment trajectory is stable; but if d_2 and d_4 are such that the point $(d_2\,,\,d_4)$ lies on the curve, then in order to solve the stability problem we must consider terms in the mapping (2.1.5) in x_0 and η_0 of order higher than 3.

TABLE

d_2	d_4	d_2	d_4	d_2	d_4
0.02	0.000055	0.18	0.025937	0.34	0.105751
0.04	0.000419	0.20	0.033520	0.36	0.117554
0.06	0.001343	0.22	0.041989	0.38	0.129465
0.08	0.003019	0.24	0.051254	0.40	0.141440
0.10	0.005585	0.26	0.061210	0.42	0.153466
0.12	0.009131	0.28	0.071748	0.44	0.165564
0.14	0.013705	0.30	0.082755	0.46	0.177794
0.16	0.019315	0.32	0.094123	0.48	0.190263

2.4. The case $d_2 = 0$ and $d_2 = 1/2$ ($|\operatorname{tr} A| = 2$).

2.4.1. *The case $d_2 = 0$.* For $d_2 = 0$, the matrix (2.2.2) is of the form

$$A = \begin{Vmatrix} 1 & 2 \\ 0 & 1 \end{Vmatrix}.$$

The eigenvalues ρ_1 and ρ_2 of the matrix A are equal to 1, and $rK(A - \rho E) = 1$. The change of variables $x = q$, $\eta = p/2$ reduces the matrix A to the form

$$\begin{Vmatrix} 1 & 1 \\ 0 & 1 \end{Vmatrix}.$$

The coefficient b_{20} of q_0^2 is equal to $-12d_3$, and the coefficient b_{30} of q_0^3 is equal to $-16d_4$.

According to (1.2.4), the two-segment trajectory is unstable whenever $d_3 \neq 0$. But if $d_3 = 0$, the mapping (2.1.5) has no quadratic part. Then it follows from (1.2.7) and (1.2.8) that for $d_4 > 0$ the two-segment trajectory is stable, and for $d_4 < 0$ it is unstable.

2.4.2. *The case $d_2 = 1/2$.* For $d_2 = 1/2$, the matrix (2.2.2) is of the form

$$A = \begin{Vmatrix} -2 & 1/2 \\ -2 & 0 \end{Vmatrix}.$$

The eigenvalues ρ_1 and ρ_2 of the matrix A are equal to -1, and $rK(A - \rho E) = 1$. The change of variables

$$\begin{Vmatrix} x \\ \eta \end{Vmatrix} = \begin{Vmatrix} \sqrt{2}/2 & -\sqrt{2}/2 \\ \sqrt{2} & 0 \end{Vmatrix} \begin{Vmatrix} q \\ p \end{Vmatrix},$$

reduces the matrix A to the form

$$\begin{Vmatrix} -1 & 1 \\ 0 & -1 \end{Vmatrix}.$$

The coefficient b_{20} of q_0^2 is equal to $12\sqrt{3}\, d_3$, and the coefficient b_{30} of q_0^3 is equal to $32d_4 - 7$. Then it follows from (1.2.10) and (1.2.11) that for

$$d_4 > \frac{7}{32} + \frac{9}{2} d_3^2$$

the two-segment trajectory is unstable, and for

$$d_4 < \frac{7}{32} + \frac{9}{2} d_3^2$$

it is stable.

References

1. C. L. Siegel, *Vorlesungen über Himmelsmechanik*, Springer-Verlag, Berlin and Heidelberg, 1956.
2. A. P. Markeev, *Libration points in celestial mechanics and astrodynamics*, "Nauka", Moscow, 1978. (Russian)
3. J. K. Moser, *Lectures on Hamiltonian systems*, Mem. Amer. Math. Soc., vol. 81, 1968.
4. G. A. Merman, *On the instability of the periodic solution of a canonical system with one degree of freedom in the case of essential resonance*, Problems of the Motion of Artificial Celestial Bodies, Akad. Nauk SSSR, Moscow, 1963. (Russian)
5. A. P. Ivanov and A. G. Sokolovskiĭ, *On the stability of a nonautonomous Hamiltonian system under a parametric resonance of essential type*, Prikl. Mat. Mekh. **44** (1980), no. 6, 963–970; English transl., J. Appl. Math. Mekh **44** (1980), no. 6, 687–691.
6. V. M. Babich and V. S. Buldyrev, *Asymptotic methods in the problems of the diffraction of short waves*, "Nauka", Moscow, 1972. (Russian)
7. V. V. Kozlov and D. V. Treshchev, *Billiards. A genetic introduction to the dynamics of systems with impacts*, Amer. Math. Soc., Providence, RI, 1991.
8. R. Douady, *Une démonstration directe de l'équivalance des théorèmes de tores invariants pour difféomorphismes et champs des vecteurs*, C. R. Acad. Sci Paris Sér. I Math. **295** (1982), 201–204.

Translated by N. K. KULMAN

Amer. Math. Soc. Transl.
(2) Vol. **168**, 1995

Hydrodynamics of Noncommutative Integration of Hamiltonian Systems

VALERY V. KOZLOV

§1. Hydrodynamics of Hamiltonian systems

First we recall some known results from hydrodynamics. Let v be a stationary velocity field of a homogeneous ideal fluid. If the flow is barotropic and the external forces are potential, then

$$v \times \operatorname{curl} v = \operatorname{grad} f. \tag{1.1}$$

Here f is the *Bernoulli integral*: it is constant on the flow lines (the integral curves of the field v) and on vortex curves (the integral curves of the curl field $w = \operatorname{curl} v$). Equation (1.1) is called the *Lamb equation*. If the flow is rotational (i.e., the fields v and w are linearly independent), then the *Bernoulli surfaces* $B_c = \{f = c\}$ are regular. The compact surfaces B_c are diffeomorphic to two-dimensional tori and the motion of fluid particles are conditionally-periodic on these tori (since the fields v and w are tangent to B_c and their commutator $[v, w]$ vanishes).

It is shown in [1] that one can develop a multi-dimensional hydrodynamical theory for Hamiltonian systems.

Let N be the configuration space of a mechanical system with local coordinates $(x_1, \dots, x_n) = x$, let $M^{2n} = T^*N$ be the phase space, $(y_1, \dots, y_n) = y$ be the canonical momenta conjugated with the coordinates x and $H(x, y)$ be the Hamiltonian function. In the reversible case,

$$H = \frac{1}{2}\sum_{i,j=1}^{n} g_{ij}(x)y_iy_j + V(x). \tag{1.2}$$

Let $\Sigma \subset M^{2n}$ be an invariant n-dimensional surface for the Hamiltonian equations

$$\dot{x} = \frac{\partial H}{\partial y}, \qquad \dot{y} = -\frac{\partial H}{\partial x}. \tag{1.3}$$

We suppose that Σ is projected one-to-one on the configuration space N. Thus it can be represented by the equations

$$y_j = u_i(x_1, \dots, x_n), \qquad 1 \leqslant i \leqslant n.$$

1991 *Mathematics Subject Classification*. Primary 58F05; Secondary 35Q20, 76A02.

We put (for brevity) $u = (u_1, \dots, u_n)$; u is a covector field on N.

Let us consider the function $f(x) = H(x, u(x))$, the vector field

$$v(x) = \left.\frac{\partial H}{\partial y}\right|_{y=u(x)},$$

the 1-form $\omega = u(x)\,dx$, and its differential $\Omega = d\omega$. These objects have a sufficiently clear meaning. For example, the field v is the projection of the Hamiltonian vector field (1.3) (which is tangent to the surface Σ) on the configuration space N. The projection $T^*N \to N$ is natural: each point (x, y) is mapped to x.

If the n-dimensional invariant surface Σ exists, then the study of the Hamiltonian system (1.3) is reduced to the analysis of the dynamical system

$$\dot{x} = v(x) \tag{1.4}$$

on N. This system turns out to possess many properties typical for stationary flows of an ideal fluid.

It is shown in [1] that f, v, ω, and Ω are connected by the following relations:

$$i_v\Omega = -df, \qquad L_v\Omega = 0. \tag{1.5}$$

Here i is the interior product ($i_v\Omega = \Omega(v, \cdot)$) and L_v is the differentiation along the vector field v. The second equation (1.5) is a consequence of the first one.

A vector w is called a *vortex* vector if $i_w\Omega = 0$. At each point $x \in N$ the vortex vectors form a linear space. We denote it by W_x. Let us suppose that $\dim W_x = m = \text{const}$. Then the family $\{W_x\}$ generates an m-dimensional distribution of tangent planes on the configuration space N. It is shown in the paper [2] that this distribution is always integrable. Consequently, every point of N lies on some m-dimensional integral manifold of the distribution $\{W_x\}$. It is natural to call them *vortex manifolds*. The second equation (1.5) implies the generalized *Thompson theorem*: the phase flow of system (1.4) transforms vortex manifolds into vortex manifolds (see [1]; a nonstationary version of this statement is obtained there too).

For Hamiltonian systems the following analog of the *Bernoulli theorem* holds: the function f is constant on flow lines (integral curves of the field v) and on vortex manifolds.

The first equation (1.5) can be written in a more usual coordinate form:

$$(\operatorname{curl} u)v = -\frac{\partial f}{\partial x}. \tag{1.6}$$

Here $\operatorname{curl} u$ denotes the skew-symmetric square matrix of order n with entries

$$\frac{\partial u_i}{\partial x_j} - \frac{\partial u_j}{\partial x_i}.$$

The form of equation (1.6) coincides with that of equation (1.1); therefore, we call it the *Lamb equation* too.

For the Hamiltonian (1.2), the vector equation (1.6) has the following explicit form:

$$\sum_{j,k=1}^{n} g_{jk}\left(\frac{\partial u_i}{\partial x_j}-\frac{\partial u_j}{\partial x_i}\right)u_k=-\frac{1}{2}\frac{\partial}{\partial x_i}\left(\sum_{j,k=1}^{n} g_{jk}u_j u_k\right)-\frac{\partial V}{\partial x_i}, \qquad 1\leqslant i\leqslant n.$$

As in hydrodynamics, we refer to the phase flow of system (1.4) (the stationary "stream" on N) as a *rotational* (*potential*) flow if $\Omega\neq 0$ (respectively $\Omega=0$). We use analogous terms for the characterization of the field v and the surface Σ. In symplectic topology the potential surfaces Σ are called *Lagrangian*.

For a potential flow we have $\operatorname{curl} u\equiv 0$. Consequently, locally, $\omega=dS(x)$ or, equivalently, $u=\partial S/\partial x$. In accordance with (1.6), we have $\partial f/\partial x\equiv 0$. Hence $f=h=\text{const}$. In this case system (1.6) can be replaced by the single Hamilton–Jacobi equation

$$H(x, \partial S/\partial x)=h. \tag{1.7}$$

The derivation of (1.7) from (1.6) follows exactly the derivation of the Lagrange–Cauchy integral from the hydrodynamical equations of a potential flow of ideal fluid. Unfortunately, the analogy between hydrodynamics and the theory of Hamiltonian systems is usually sketched very superficially in literature.

§2. An application to the problem on the geodesic lines on Lie groups with left-invariant metric

The nontrivial problem of the existence of stationary invariant surfaces Σ plays an essential role in the hydrodynamic theory of §1. Such surfaces actually exist, however, in a number of problems of Hamiltonian mechanics.

Our basic example is the problem of geodesic lines on an n-dimensional Lie group G with a Riemann metric invariant with respect to left shifts. The mechanical aspect of this problem is inertial motion. The simplest example is the Euler problem of the motion of a rigid body in three-dimensional space about a fixed point (here $G=SO(3)$).

Let $v_1, \dots, v_n$ be independent right-invariant vector fields on G. The corresponding phase flows are families of left shifts on G. Since the metric is invariant with respect to all left shifts, the equations of the geodesic lines have n independent Noether integrals

$$(y, v_k)=c_k, \qquad 1\leqslant k\leqslant n. \tag{2.1}$$

These relations define n-dimensional invariant surfaces Σ_c, which are projected one-to-one on the group G.

The analysis of the corresponding stationary flows on Lie groups is an important and instructive problem. It was considered in detail in the paper [2] for the group $SO(3)$. Let us describe briefly the results of this paper.

The Noether integrals (2.1) imply that the angular momentum $\vec{K}$ of a body is constant in the fixed space. Let $\vec{K}\neq 0$ (if $\vec{K}=0$, then the body is at rest). We set $\vec{\gamma}=\vec{K}/|\vec{K}|$, $|\vec{K}|=k$. We see that $\vec{K}=k\vec{\gamma}$. Using this fact, one can obtain the angular velocity of the body as a single-valued function of its position. Thus on the group $SO(3)$, the configuration space of a rigid body with fixed point, a dynamical system of the form (1.4) arises. The following propositions hold:

a) The vortex fields w on the group $SO(3)$ commuting with the field of velocities v generate the rotations of a rigid body with angular velocity $\omega = \mu\gamma$, $\mu = \text{const}$. In particular, the vortex fields are right-invariant and all the vortex curves are closed. The fibration of the group $SO(3)$ by the vortex curves coincides with the well-known Hopf fibration.

b) These fields are defined by the relations

$$i_w\Omega = 0\,, \qquad i_w\omega = \text{const}\,.$$

c) The Hamiltonian system on $T^*SO(3)$ with the Hamiltonian $H' = |\vec{K}|^2/2$ has the same invariant surfaces $\vec{K} = k\vec{\gamma}$. The projection of the corresponding Hamiltonian vector field on $SO(3)$ is a vortex field commuting with the field v.

d) The metric on $SO(3)$ defined by the kinetic energy of the body allows us to calculate $\operatorname{curl} v$. This field turns out to be a vortex field.

e) The phase flow of system (1.4) on $SO(3)$ conserves the bilaterally invariant Haar measure. Recall that on each Lie group there exists a unique (up to a constant multiplier) measure invariant with respect to all left (right) group shifts. It is called the *left-sided* (*right-sided*) *Haar measure*. For unimodular Lie groups these two measures coincide. All the compact groups are unimodular.

f) The "Bernoulli integral" f equals $(I^{-1}\gamma\,,\,\gamma)/2$, where I is the inertia operator of the rigid body. If the operator I is not spherical, then the flow on the group $SO(3)$ is vortex. The critical points of the function f are the orbits of the constant rotations of a body about the principal axes of the inertia ellipsoid (with a fixed value of angular momentum $\vec{K}$) and the critical values coincide with the values of the energy on these rotations. If c is not a critical value of the function f, then the "Bernoulli surface"

$$B_c = \{f = c\}$$

is the two-dimensional torus with a conditionally-periodic motion. The tori B_c are the natural projections of the two-dimensional Liouville tori from the phase space $T^*SO(3)$ to the group $SO(3)$.

Statements a)–f) constitute the vortex theory of the Euler top. Some of them admit generalizations to arbitrary Lie groups. For example, property e) holds for unimodular groups. In the nonunimodular case, the phase flow of system (1.4) preserves the right Haar measure. This result was obtained recently by the author and Yaroshchuk.

It is useful to note that the Hamiltonian H' is a Casimir function on the algebra $so(3)$ of the rotations group: this function commutes with all functions on $so(3)$ with respect to the corresponding Lie–Poisson bracket. It is evident that $H' \equiv \text{const}$ are the invariant surfaces.

§3. Three lemmas on Poisson brackets

This section is auxiliary. Let $f_1, \ldots, f_n$ be functions on the phase space M^{2n}, and

$$\frac{\partial(f_1, \ldots, f_n)}{\partial(y_1, \ldots, y_n)} \neq 0\,.$$

Then in accordance with the implicit function theorem the system of equations

$$(3.1)\qquad f_1(x, y) = c_1, \dots, f_n(x, y) = c_n$$

can be solved (at least locally) with respect to momenta y:

$$y_1 = u_1(x, c), \dots, y_n = u_n(x, c), \qquad c = (c_1, \dots, c_n).$$

Let us introduce the Poisson bracket matrix $A = (\{f_i, f_j\})$. The following simple lemma holds

LEMMA 1. *We have* $\operatorname{rank} A = \operatorname{rank}(\operatorname{curl} u)$.

As a consequence, we obtain the well-known statement which is usually used in the proof of the Liouville theorem on the complete integrability of Hamiltonian systems: if the functions $f_1, \dots, f_n$ are (pairwise) in involution, then the 1-form $u_1(x, c)\,dx_1 + \dots + u_n(x, c)\,dx_n$ is closed for all values of c.

For the proof we note that the functions

$$(3.2)\qquad F_k(x, c) = f_k(x, u(x, c)), \qquad 1 \leqslant k \leqslant n,$$

are identically equal to c_k. Consequently,

$$\frac{\partial F_k}{\partial x_i} = \frac{\partial f_k}{\partial x_i} + \sum \frac{\partial f_k}{\partial y_j}\frac{\partial u_j}{\partial x_i} = 0, \qquad \frac{\partial F_s}{\partial x_i} = \frac{\partial f_s}{\partial x_i} + \sum \frac{\partial f_s}{\partial y_j}\frac{\partial u_j}{\partial x_i} = 0.$$

We multiply the first equality by $\partial f_s/\partial y_i$ and the second one by $-\partial f_k/\partial y_i$ and sum them over i. As a result we obtain the following relations:

$$(3.3)\qquad \{f_k, f_s\} + \sum_{i,j} \frac{\partial f_k}{\partial y_j}\frac{\partial f_s}{\partial y_i}\left(\frac{\partial u_j}{\partial x_i} - \frac{\partial u_i}{\partial x_j}\right) = 0, \qquad 1 \leqslant k, s \leqslant n.$$

We set $B = \operatorname{curl} u$ and represent these relations in the form

$$\left(B\frac{\partial f_k}{\partial y}, \frac{\partial f_s}{\partial y}\right) = a_{ks},$$

where a_{ks} are the elements of the matrix A formed by the Poisson brackets of the functions $f_1, \dots, f_n$.

In accordance with our assumption the vectors

$$\frac{\partial f_1}{\partial y}, \dots, \frac{\partial f_n}{\partial y}$$

are linearly independent. Consequently, the matrices A and B are connected by some relation of the form $CBD = A$, where C and D are nondegenerate matrices. It is well known from linear algebra that in this case the ranks of the matrices A and B coincide. □

Let $\Phi(x, y)$ be one more function on M^{2n} commuting with all the functions $f_1, \dots, f_n$. The Hamiltonian vector field v_Φ generated by the Hamiltonian function Φ is then tangent to the n-dimensional surface Σ_c defined by equations (3.1). Let w be the projection of the field v_Φ to the tangent to the coordinate space $\{x\}$ plane. It is evident that

$$w = \left.\frac{\partial \Phi}{\partial y}\right|_{y=u(x, c)}$$

LEMMA 2. *If Φ is a function of $f_1, \dots, f_n$ then w is a vortex vector:*

$$(\operatorname{curl} u) w = 0. \tag{3.4}$$

PROOF. Since each function F_k (see (3.2)) is identically equal to c_k, the following relations hold:

$$\frac{\partial F_i}{\partial c_j} = \delta_{ij} = \sum_{s=1}^{n} \frac{\partial u_i}{\partial c_s} \frac{\partial f_s}{\partial y_j}. \tag{3.5}$$

Here δ_{ij} is the Kronecker symbol.

Let us multiply equalities (3.3) by the derivatives $\partial u_p/\partial c_k$, $\partial u_q/\partial c_s$ and then sum them over k and s. As the result we get:

$$\sum_{k,s} \frac{\partial u_p}{\partial c_k} \frac{\partial u_q}{\partial c_s} \{f_k, f_s\} + \frac{\partial u_p}{\partial x_q} - \frac{\partial u_q}{\partial x_p} = 0.$$

To verify equality (3.4), we multiply these relations by

$$\left. \frac{\partial \Phi}{\partial y_p} \right|_{y=u(x,c)}$$

and sum them over p from 1 to n. We are interested in the values of

$$\sum_{k,s} \sum_p \frac{\partial u_p}{\partial c_k} \left. \frac{\partial \Phi}{\partial y_p} \right|_{y=u} \frac{\partial u_q}{\partial c_s} \{f_k, f_s\}.$$

In accordance with our assumption, the function $\varphi(x, c) = \Phi(x, u(x, c))$ depends only on $c_1 = f_1, \dots, c_n = f_n$. Consequently,

$$\sum_p \left. \frac{\partial \Phi}{\partial y_p} \right|_{y=u} \frac{\partial u_p}{\partial c_k} = \frac{\partial \varphi}{\partial c_k}.$$

The sum

$$\sum_k \frac{\partial \varphi}{\partial c_k} \{f_k, f_s\}$$

is equal to the Poisson bracket $\{\Phi, f_s\}$, which vanishes by our assumption. The lemma is proved.

The collection of functions $f_1, \dots, f_n$ is called *closed* if all the Poisson brackets $\{f_i, f_j\}$ are functions of $f_1, \dots, f_n$. Let us find a simple condition for being closed by using properties of the collection of the covector fields $u(x, c)$.

The covectors

$$a_1 = \frac{\partial u}{\partial c_1}, \quad \dots, \quad a_n = \frac{\partial u}{\partial c_n}$$

are linearly independent. The dual collection of n linearly independent vectors $b_1, \dots, b_n$ such that $(a_i, b_j) = \delta_{ij}$, $1 \leqslant i, j \leqslant n$, corresponds uniquely to them. In accordance with (3.5), $b_j = \partial f_j/\partial y$, $1 \leqslant j \leqslant n$.

LEMMA 3. *The collection of functions $f_1, \dots, f_n$ is closed if and only if for all $i, j = 1, \dots, n$ the functions (Bb_i, b_j) do not depend on x.*

Here, as above, $B = \operatorname{curl} u$. Thus the verification of the fact that the collection is closed requires only differentiations and algebraic operations, including the operation of inversion of functions.

Lemma 3 is a simple consequence of relations (3.3) and (3.5).

§4. Noncommutative integration

Let us suppose that a Hamiltonian system with n degrees of freedom has a collection of m independent integrals $F_1, \dots, F_m$ closed with respect to the Poisson brackets

$$\{F_i, F_j\} = f_{ij}(F_1, \dots, F_n), \qquad 1 \leqslant i, j \leqslant n.$$

The matrix $(f_{i,j})$ is skew-symmetric, therefore its rank is even. Assume that $\operatorname{rank}(f_{i,j}) = 2k$.

The basic assumption of the *noncommutative theory* of Hamiltonian equations integration is as follows:

$$m = n + k. \tag{4.1}$$

For $k = 0$ we obtain the classical Liouville condition of complete integrability.

One can formulate a geometric consequence of condition (4.1): if the surface of the joint levels of the integrals $F_1, \dots, F_m$ is regular and compact, then each of its connected component is a $(2n - m) = (n - k)$-dimensional torus; moreover, the phase flow of the Hamiltonian system in question is a "winding" on these tori. An analytical consequence of (4.1) is the integrability in quadratures of Hamiltonian equations.

The theory of noncommutative integration of Hamiltonian systems was developed in the papers of N. N. Nekhoroshev, A. S. Mishchenko, A. T. Fomenko, Ya. V. Tatarinov, A. V. Strel′tsov, A. V. Brailov, and others. It is set forth in detail in the book [3]; the necessary references can be found there as well.

According to the classical Lie–Cartan theorem (see [4, p. 126]) there exist $m - 2k$ functions $\Phi_1, \dots, \Phi_{m-2k}$ of $F_1, \dots, F_m$, that commute with all of these m functions, and $2k$ functions $\Psi_1, \dots, \Psi_{2k}$ of $F_1, \dots, F_m$ can be chosen in such a way that their Poisson bracket matrix is nondegenerate. Let us consider the $2(n - k)$-dimensional invariant surface

$$M_\alpha = \{\Psi_1 = \alpha_1, \dots, \Psi_{2k} = \alpha_{2k}\}.$$

It is a symplectic manifold: the restriction of the symplectic structure (2-form) $dx \wedge dy$ on M_α is a closed nondegenerate 2-form. The restriction of $n - k$ functions $\Phi_1, \dots, \Phi_{m-2k}$ on M_α are independent and commute (with respect to the symplectic structure on M_α). Since, in accordance with (4.1), we have

$$m - 2k = n - k = \dim M_\alpha / 2,$$

we can apply the conventional Liouville theorem on involutive integrals. In this way one can, in fact, prove the theorem on the fibration of the original $2n$-dimensional phase space by $(n - k)$-dimensional tori with conditionally-periodic motions.

It is essential that the Lie–Cartan theorem cannot be used to prove the integrability quadratures, since this theorem does not give a constructive way to obtain the functions Φ and Ψ. A. V. Brailov proved the exact integrability of Hamiltonian equations with condition (4.1) by using other arguments.

Condition (4.1) has a clear meaning: one cannot add other independent functions to the functions $F_1, \dots, F_m$ keeping the rank of their Poisson bracket matrix fixed.

The basic examples of the application of noncommutative integration theory are connected with the problem of inertial motion on a Lie group with left-invariant kinetic energy (see §2). Let us discuss these examples in more detail.

Let G be a Lie group, g its algebra, $n = \dim G$. The phase space T^*G can be represented as the direct product $g^* \times G$, where g^* is the dual to g. Lie algebra g is the space of velocities $\omega = (\omega_1, \dots, \omega_n)$, and g^* is the space of momenta $k = (k_1, \dots, k_n)$. These objects are connected by the linear relations

$$k_s = \sum_{p=1}^{n} I_{sp}\omega_p, \tag{4.2}$$

where $I = (I_{sp})$ is the constant inertia tensor of the system.

Hamiltonian equations on T^*G include n dynamical *Euler–Poincaré equations*

$$\dot{k}_i = \sum c_{ip}^l k_l \omega_p, \qquad 1 \leqslant i \leqslant n, \tag{4.3}$$

and n geometric relations; their exact form is irrelevant here. In (4.3), the coefficients c_{ip}^l denote structure constants of the algebra g. System (4.3) can be considered either as a closed system on the algebra g (then relations (4.2) must be substituted for k) or as a system on the coalgebra g^* (then (4.2) must be solved with respect to ω). For the Lie algebra $so(3)$, equations (4.3) coincide with the famous Euler dynamical equations of a rotating top.

The Lie–Poisson bracket

$$\{H, F\} = \sum c_{is}^j k_j \frac{\partial H}{\partial k_i}\frac{\partial F}{\partial k_s}$$

is closely connected to equations (4.3). To each ordered pair of functions H, F on g^* there corresponds the bracket $\{H, F\}$, which is a function on g^* too. The Lie–Poisson bracket is degenerate: there exist nonconstant functions on g^* commuting with all basic functions $k_1, \dots, k_n$. These nonconstant functions are called the *Casimir functions*. They can also be characterized in the following way: they are exactly the functions of n Noether integrals that do not depend on the position of a system. For example, for the algebra $so(3)$ the square of the length of the angular momentum vector $k_1^2 + k_2^2 + k_3^2$ (cf. with §3) is a Casimir function.

Authors, who studied the problem of exact integration of Hamilton equations with left-invariant metric on Lie groups, considered, as a rule, only the Euler–Poincaré equations (4.3) (one can find a survey of the corresponding results in [3]). Let us suppose that equations (4.3) admit l independent integrals which are not Casimir functions; let $2p$ be the maximum rank of their Poisson bracket matrix and $2k$ the maximum rank of the Noether integrals Poisson bracket matrix. If

$$l = k + p, \tag{4.4}$$

then equations (4.3) are integrable in quadratures and almost the entire g^* is fibered by $(k - p)$-dimensional tori with conditionally periodic trajectories.

The derivation of condition (4.4) is based on the following arguments. The joint level of Casimir functions in the typical situation are $2k$-dimensional invariant symplectic manifolds with l independent integrals. Thus condition (4.4) is the noncommutative integrability condition (4.1).

It is easy to understand that condition (4.4) guarantees the integrability of a Hamiltonian system on the whole phase space T^*G. To see that, we rewrite (4.4) in the following way:

$$n + l = n + (k + p)\,.$$

This is the integrability condition (4.1). Actually, the system of Hamiltonian equations admits $n+l$ independent integrals (l integrals of the Euler–Poincaré equations and n Noether integrals). Moreover, the rank of their Poisson bracket matrix equals $2(k+p)$ (since the integrals of equations (4.3) commute with the Noether integrals). Hence almost all the phase space $T^*G = g^* \times G$ is foliated by $2n-l-n = (n-k-p)$-dimensional tori with conditionally periodic motions.

Concluding this section, we consider Hamiltonian systems from the hydrodynamical point of view. Let $(c_1, \dots, c_n) = c$ be constants of the Noether integrals (2.1). Consider the typical case when for the given value of c the rank of the Poisson bracket matrix of these integrals is maximal (equal to $2k$). It was noted in §2 that a dynamical system on G arises after projecting the n-dimensional invariant surface $\Sigma_c \subset T^*G$ on the group G.

LEMMA 4. *Functions of $k_1, \dots, k_n$ commute with the Noether integrals.*

PROOF. Let Φ be a function on g^*. It can be regarded as a function on $T^*G = g^* \times G$ invariant with respect to left shifts. Consequently, according to the Noether theorem, the Hamiltonian vector field v_Φ admits n Noether integrals. □

Let Φ be a function on g^*. In accordance with Lemma 4, the Hamiltonian vector field v_Φ is tangent to Σ_c. Thus the projection onto G of the restriction of v_Φ to Σ_c is well defined.

Let $w_1, \dots, w_{n-2k}$ $(v_1, \dots, v_l)$ be tangent fields on G that are the projections of the Hamiltonian fields generated by the Casimir functions (along the remaining integrals of the Euler–Poincaré equations). These fields are independent and we have $[w_i, v_j] = 0$ for all $i = 1, \dots, n-2k$, $j = 1, \dots, l$.

We denote by $F_1, \dots, F_l$ independent integrals of equations (4.3) which are not Casimir functions. Let $f_1, \dots, f_l$ be functions on the group G defined uniquely by the commutative diagram

$$\begin{array}{ccc} \Sigma_c & \xrightarrow{\text{pr}} & G \\ & {\scriptstyle F|_{\Sigma_c}} \searrow \quad \swarrow {\scriptstyle f} & \\ & \mathbb{R} & \end{array}$$

PROPOSITION. *The fields $w_1, \dots, w_{n-2k}$ are vortex and $f_1, \dots, f_l$ are Bernoulli integrals (they are constant on the flow lines and on the vortex manifolds).*

This proposition is a simple consequence of Lemma 2 (see §3).

In the typical situation, all the vortex vectors are linear combinations of the vectors $w_1, \dots, w_{n-2k}$. However, when the rank of the Poisson bracket matrix drops, new vortex vectors arise.

Regular "Bernoulli surfaces"

$$B_\alpha = \{f_1 = \alpha_1, \dots, f_l = \alpha_l\}$$

are $n - l = (n - k - p)$-dimensional tori with conditionally-periodic motions. They are projections onto the group G of Liouville tori foliating $2n$-dimensional phase space in accordance with the noncommutative integration theorem.

These observations extend conclusion c) (see §2) associated with the Euler top in three-dimensional space to the multi-dimensional case. It is easy to understand that the vortex fields $w_1, \dots, w_{n-2k}$ are right invariant and the vortex manifolds are closed (this is the analog of conclusion a)). We show that $i_{w_s}\omega = \text{const}$, $1 \leqslant s \leqslant n$. Indeed,

$$i_{w_s}\omega = \left(u, \left.\frac{\partial \Phi}{\partial y}\right|_{y=u}\right), \tag{4.5}$$

where Φ is the corresponding Casimir function. Since these functions are homogeneous polynomials of the Noether integrals, by the Euler theorem, the right-hand side of (4.5) is a constant (depends only on $c_1, \dots, c_n$). These arguments prove conclusion b). An analog of conclusion f) was formulated above. It would be interesting to obtain a natural multi-dimensional extension of property d).

§5. The vortex integration method

We shall follow the integration method of the equations for geodesic lines on Lie groups and set forth an integration method of Hamiltonian equations (1.3) given in the phase space $M^{2n} = T^*N$. Let $H(x, y)$ be a Hamiltonian function. We rewrite the Lamb equation (1.6) in a more explicit form:

$$(\operatorname{curl} u)\left(\left.\frac{\partial H}{\partial y}\right|_{y=u}\right) = -\frac{\partial H(x, u)}{\partial x}. \tag{5.1}$$

Definition 1. A solution $u(x, c_1, \dots, c_n)$ of system (5.1) is called *complete* if

$$\det \frac{\partial u}{\partial c} \neq 0.$$

Definition 2. A solution $u(x, c_1, \dots, c_n)$ is called *closed* if the conditions of Lemma 3 (see §3) hold.

Before the formulation of our result, we need the following remark. To each function $\Phi(x, y)$ defined on M^{2n} there corresponds the equation

$$(\operatorname{curl} u)\left(\left.\frac{\partial \Phi}{\partial y}\right|_{y=u}\right) = -\frac{\partial \Phi(x, u)}{\partial x} \tag{5.2}$$

of the form (1.5). Every solution $u(x)$ of (5.2) forms an n-dimensional surface $y = u(x)$ which is tangent to the Hamiltonian vector field v_Φ.

THEOREM. *Suppose a complete and closed n-parametric solution $u(x, c)$ of equation (5.1) is found and*

$$\operatorname{rank}\left(\frac{\partial u_i}{\partial x_j} - \frac{\partial u_j}{\partial x_i}\right) = 2k .$$

Moreover, let a closed collection of l integrals $\Phi_1, \dots, \Phi_l$ of Hamiltonian equations (1.3) exist and also suppose that

1) $\operatorname{rank}(\{\Phi_i, \Phi_j\}) = 2p$,

2) *the functions $u(x, c)$ satisfy all equations obtained from (5.2), by replacing the function Φ with the integrals $\Phi_1, \dots, \Phi_l$.*

3) *$\Phi_1(x, u(x, c)), \dots, \Phi_k(x, u(x, c))$ are independent as functions of x.*

If $l = k + p$, then the Hamiltonian equations (1.3) can be integrated in quadratures.

This result contains as a special case the Jacobi theorem on the complete integrals. Indeed, let

$$u = \frac{\partial S(x, c)}{\partial x} .$$

Then $\operatorname{curl} u \equiv 0$, consequently $k = 0$ and the function $S(x, c)$ satisfies the Hamilton–Jacobi equation. It remains to set $l = p = 0$.

One of the most interesting cases is contained in the following

COROLLARY. *Suppose a complete and closed family of solutions $u(x, c)$ of equation (5.1) is found and*

1) $\operatorname{rank}(\operatorname{curl} u) = 2$,

2) $d_x H(x, u(x, c)) \neq 0$.

Then the Hamiltonian equations with the Hamiltonian H are integrable in quadratures.

In this case $l = 1$, $p = 0$, and the Hamiltonian function is an additional integral. This statement is more effective for $n = 3$ because $k \leqslant 1$ in this case.

PROOF OF THE THEOREM. By the completeness of the solution, the equations

$$y_1 = u_1(x, c_1, \dots, c_n), \dots, y_n = u_n(x, c_1, \dots, c_n)$$

are locally solvable with respect to $c_1, \dots, c_n$:

$$c_1 = F_1(x, y), \dots, c_n = F_n(x, y) .$$

The functions $F_1, \dots, F_n$ form a collection of integrals of equations (1.3). Moreover,

1) the Poisson brackets $\{F_i, F_j\}$ depend only on $F_1, \dots, F_n$ (Lemma 3),

2) $\operatorname{rank}(\{f_i, F_j\}) = 2k$ (Lemma 1).

By the nondegeneracy condition of the solution $u(x, c)$ and assumption 3 of the theorem, the functions

$$F_1, \dots, F_n, \Phi_1, \dots, \Phi_l \tag{5.3}$$

are independent. Let $v_1, \dots, v_l$ be the Hamiltonian vector fields corresponding to the Hamiltonians $\Phi_1, \dots, \Phi_l$. Due to assumption 2 of the theorem, the fields

v_i are tangent to the invariant surfaces $\Sigma_c = \{x, y : y = u(x, c)\}$ for all values of c. Since $F_1, \ldots, F_n$ are independent, these functions are integrals of the fields $v_1, \ldots, v_l$. Consequently, all the Poisson brackets $\{F_i, \Phi_j\}$ vanish. This in turn implies that the rank of the Poisson bracket matrix of integrals (5.3) equals $k + p$. Hence (due to the assumption), $n + l = n + (k + p)$ and condition (4.1) of the noncommutative theory holds. This guarantees the exact integrability in quadratures of the Hamiltonian equations. The theorem is proved.

The method for finding a complete integral based on the separation of variables is known for the Hamilton–Jacobi equation. We know of nothing similar for the more general equation (5.1).

In conclusion, we consider again the rotation of the Euler top, which is described by the Euler dynamical equations

$$\dot{k} = k \times \omega, \tag{5.4}$$

and the geometrical relations for the fixed unit vectors α, β, γ:

$$\dot{\alpha} = \alpha \times \omega, \quad \dot{\beta} = \beta \times \omega, \quad \dot{\gamma} = \gamma \times \omega. \tag{5.5}$$

Here k is the angular momentum of the rigid body; it is equal to $I\omega$, where I is the constant inertia operator. It is clear that equations (5.4)–(5.5) are equivalent to the Hamiltonian equations of the Euler problem.

We see three-dimensional invariant surfaces that are projected one-to-one onto the group $SO(3)$. This means that the momentum k is a function of α, β, γ. From (5.4), we obtain the partial equation in vector form

$$\frac{\partial k}{\partial \alpha}(\alpha \times \omega) + \frac{\partial k}{\partial \beta}(\beta \times \omega) + \frac{\partial k}{\partial \gamma}(\gamma \times \omega) = k \times \omega. \tag{5.6}$$

Here instead of ω one must set $I^{-1}k$. Obviously, this equation is equivalent to equation (5.1) presented in noncanonical variables. The passage from (5.1) to (5.6) is completely similar to the passage from the Lagrangian (Hamiltonian) equations to the Euler–Poincaré (Chetaev) equations on Lie algebras.

It is evident that the function

$$k = c_1\alpha + c_2\beta + c_3\gamma \tag{5.7}$$

is one of the complete solutions of (5.6), where c_1, c_2, c_3 are arbitrary constants. This is a consequence of the invariance of the vector k in the fixed space. The fact that the solution (5.7) is closed is evident.

If $c \neq 0$ and the inertial tensor is not spherical, then assumptions 1 and 2 of the corollary hold.

References

1. V. V. Kozlov, *Hydrodynamics of Hamiltonian systems*, Vestnik Moskov. Univ. Ser. I Mat. Mekh. **1983**, no. 6, 10–22; English transl. in Moscow Univ. Math. Bull. **38** (1983).
2. ———, *Vortex top theory*, Vestnik Moskov. Univ. Ser. I Mat. Mekh. **1990**, no. 4, 56–62; English transl. in Moscow Univ. Math. Bull. **45** (1990).
3. A. T. Fomenko, *Integrability and nonintegrability in geometry and mechanics*, Kluwer, Dordrecht, 1988.
4. E. Cartan, *Leçons sur les invariants intègraux*, Hermann, Paris, 1922.

Translated by D. TRESHCHEV

DEPARTMENT OF MATHEMATICS, MOSCOW STATE UNIVERSITY, MOSCOW 119899, RUSSIA

Amer. Math. Soc. Transl.
(2) Vol. **168**, 1995

Problemata Nova, ad Quorum Solutionem Mathematici Invitantur

VALERY V. KOZLOV

We would like to draw the attention of the reader to some mathematical problems of classical mechanics. They came out in connection with investigations at the Theoretical Mechanics Department at Moscow State University. It is worth mentioning that, beginning with Newton, classical mechanics was always a source of new mathematical problems. Recall the Kepler equation

$$u - e\sin u = \zeta\,,$$

relating the eccentric anomaly u of the orbit to the mean anomaly ζ, which is a linear function of time. Lagrange, when he was solving the Kepler equation, was one of the first to use Fourier series. He obtained the following expression

$$u = \zeta + 2\sum_{m=1}^{\infty} \frac{J_m(me)}{m} \sin m\zeta\,.$$

Here $J_m(z)$ is the Bessel function of mth order, which was first introduced precisely in this problem. Looking for a representation of the solution of the Kepler equation as a power series in eccentricity, Lagrange came to the general theorem on local inversion of holomorphic functions (it is known now as the Burman–Lagrange theorem). According to Lagrange,

$$u = \sum_{m=0}^{\infty} c_m(\zeta)\,\frac{e^m}{m!}\,, \qquad c_0 = \zeta\,,\ c_m = \frac{d^{m-1}}{d\zeta^{m-1}} \sin^m \zeta \quad (m \geqslant 1)\,.$$

Last, but not least, let us mention that the main motivation that led Cauchy to his discoveries in complex analysis was to determine rigorously the region of convergence for the Lagrange power series (it is convergent for $e \leqslant 0.6627434\dots$). Of course, the problems presented below do not pretend to be of such global importance. They are intended in the first place for young mathematicians who would like to try their skills in this interesting field.

We shall start our discussion with some problems of stability theory.

1991 *Mathematics Subject Classification*. Primary 70–02, 73–02.

1. According to the famous *Lagrange–Dirichlet theorem*, a position of equilibrium is stable if the potential energy has a strict local minimum in this position. Unfortunately, the inverse theorem is not true. This is shown by a simple counterexample suggested by Painlevé (to find an example is a good problem for those who are not acquainted with this result). However, in the Painlevé example, the potential energy is of only finite smoothness. Later Wintner provided a counterexample with the infinitely smooth potential.

In 1892 Lyapunov formulated the inversion problem of the Lagrange–Dirichlet theorem in the analytic case. In spite of serious efforts of many mathematicians and experts in mechanics, a complete positive solution of the Lyapunov problem was obtained only hundred years later by Palamodov (see his article in this volume). To prove instability, he constructed a suitable Lyapunov function.

Merging Palamodov's theorem with the result of [1], it is possible to prove that if $x = (x_1, \dots, x_n) = 0$ is an *isolated* position of equilibrium that is not a point of local minimum for the potential energy, then the equations of motion have a solution $x(t)$ such that $x(t) \to 0$ as $t \to -\infty$.

Incidentally, since the dynamics equations are reversible, $x(-t)$ also is an asymptotic solution: it tends to the equilibrium $x = 0$ as $t \to \infty$.

a. *Does the existence of an asymptotic solution hold in general (without the assumption that the critical point $x = 0$ of the potential energy is isolated)?*

Earlier the problem of inversion of the Lagrange–Dirichlet theorem was treated by the author (in collaboration with Palamodov), using the first Lyapunov method. This method is based on constructing asymptotic solutions of dynamic equations in the form of a certain series (see the paper [2], where references to preceding publications can be found).

Let $V = V_2 + V_3 + \dots$ be the Maclaurin series for the potential energy; V_k is a homogeneous form in x of degree $k \geqslant 2$. If $x = 0$ is not a minimum for V_2, then, as was proved by Lyapunov, the equations of motion have a nontrivial asymptotic solution of the form

$$\sum_{m=1}^{\infty} x_m(t) e^{-m\lambda t}, \qquad \lambda = \text{const} > 0, \tag{1}$$

where $x_m(t)$ is a polynomial in t with constant coefficients. In the analytic case this series is convergent for all sufficiently large values of t. By reversibility, the equilibrium $x = 0$ is unstable.

The more general case

$$V = V_2 + V_m + V_{m+1} + \dots, \qquad m \geqslant 3,$$

is considered in [2]. Here the form V_2 is nonnegative and the dimension of the plane $\Pi = \{x : V_2(x) = 0\}$ is positive. Let W_m be the restriction of the form V_m to the plane Π. It turns out that if $x = 0$ is not a minimum for W_m, the system has an asymptotic solution in the form of a series

$$\sum_{m=1}^{\infty} \frac{x_m(\ln t)}{t^{m\mu}}, \qquad \mu = \text{const} > 0, \tag{2}$$

where $x_m(\cdot)$ are polynomials with constant coefficients.

b. *Is it always true that if a critical point of the potential energy is not a minimum, then the equations of motion have an asymptotic solution of the form* (1) *or* (2), *or series with multiple logarithms are needed?*

If the system is analytic and $V_2 \equiv 0$, the series (2) is convergent for sufficiently large t (Kozlov, Palamodov; 1982). However, if $V_2 \not\equiv 0$, the series (2) are, as a rule, divergent. Here is a simple model example:

$$(3) \qquad \ddot{x} = -\frac{\partial V}{\partial x}, \quad \ddot{y} = \dot{x}^2 - \frac{\partial V}{\partial y}, \qquad V = -6x^3 + \frac{y^3}{2}.$$

The presence of the term $\dot{x}^2$ means that the kinetic energy is noneuclidean. Equations (3) have a formal solution

$$(4) \qquad x = \frac{1}{2t^2}, \quad y = \frac{1}{t^6}\sum_{n=0}^{\infty} \frac{a_{2n}}{t^{2n}}, \qquad a_{2n} = \frac{(-1)^n(2n+5)!}{120}.$$

The radius of convergence of the series for y is zero.

However, according to the remarkable theorem by Kuznetsov [3], even in the case when the series (2) is divergent, the equations of motion have a solution such that the series (2) is its asymptotic representation. Incidentally, the work [3] was stimulated by investigations on the inversion of the Lagrange–Dirichlet theorem. As an illustration, we shall give the exact asymptotic solution of the system (3) corresponding to the formal series (4):

$$(5) \qquad x(t) = \frac{1}{2t^2}, \qquad y(t) = -\sin t \int_t^{\infty} \frac{\cos s}{s^6}\, ds + \cos t \int_t^{\infty} \frac{\sin s}{s^6}\, ds.$$

Performing successive integration by parts, it is possible to obtain the series (4) from these formulas. The function $y(\cdot)$ in (5) can be regarded as a sum (in a generalized sense) of the divergent series (4).

c. *Is it always possible to represent asymptotic solutions corresponding to divergent series* (2) *in an integral form like* (5) *with finite "kernels"?*

c′. *Is it true that in the analytic case to every formal solution in the form of a series there corresponds a unique "real" solution such that this series is its asymptotic representation?*

2. Now let us turn to problems of another nature, connected with the existence of *tensor conservation laws*.

Let M be a two-dimensional closed analytic surface that serves as the configuration space for a mechanical system with two degrees of freedom. We shall study only inertial motion and assume that the kinetic energy H (Riemannian metric on M) is an analytic function on TM, quadratic in velocity. According to the Maupertuis principle, all trajectories of the system are geodesics. Hence the corresponding dynamical system on the invariant three-dimensional surface $H = 1$ is often called a geodesic flow.

In 1979 the author proved the following theorem: if the genus of the surface M is greater than one, the geodesic flow does not admits nonconstant analytic integrals.

This result allows us to speak about purely topological obstacles to the integrability of reversible systems.

a. *What multidimensional surfaces M admit geodesic flows with a complete set of independent analytic integrals?*

Significant progress in this problem was achieved by Taĭmanov [4]. He proved, in particular, that the Betti numbers of a connected n-dimensional configuration space with a complete set of first integrals satisfy the inequalities

$$b_k \leqslant \binom{n}{k}, \qquad 0 \leqslant k \leqslant n.$$

Unfortunately, this elegant result does not solve the whole problem.

In the two-dimensional case, geodesic flows with first integrals that are independent of H may exist only on the sphere S^2 and on the torus T^2 (we consider the orientable case). The case of the two-dimensional torus is the most interesting, since it is possible to introduce global isotermic angle coordinates q_1, q_2 on the torus, so that the Hamiltonian function takes the simple form:

$$H = \Lambda(q_1, q_2)(p_1^2 + p_2^2)/2. \tag{6}$$

Here p_s is canonical momentum dual to q_s.

Every analytic integral can be expanded in a Maclaurin series in powers of momenta. It is evident that every homogeneous part of the series is also an integral. Hence, following Birkhoff, we should study systems admitting first integrals that are polynomial in momentums. Birkhoff showed that the existence of linear integrals is related to hidden cyclic coordinates, and quadratic integrals are related to separable variables. For the case of the torus, global versions of these results are given in [5].

Bolotin suggested the following conjecture.

b. *If the system on the torus with the Hamiltonian* (6) *has an independent polynomial integral of degree k, then there exists an integral independent of H that is linear or quadratic in momentum.*

Is this true?

From the point of view of the Maupertuis principle, the problem of existence of polynomial integrals for Hamiltonian systems with the Hamiltonian function $H = (p_1^2 + p_2^2)/2 + V(q_1, q_2)$, is closely related to the previous problem. Here V is an analytic function on the torus $\mathbb{T}^2$. In [6] this problem is treated for systems on the n-dimensional torus with a Hamiltonian of a similar form. The potential energy is assumed to be a *trigonometric polynomial* on $\mathbb{T}^n$. It is proved that if the Hamiltonian equations admit n independent polynomial integrals, then there exist n independent integrals in involution of first or second degree. From the technical point of view, the proof is rather complicated. It is based on the classical approach of perturbation theory.

c. *Is this statement true for Hamiltonian systems with general analytic periodic potentials?*

For the case of integrals of third or fourth degree in momentum, the positive answer was obtained by Bialy [7].

The same questions can be formulated for reversible systems with the two-dimensional sphere as the configuration space. However, in this case the situation is completely different. There exist metrics on the two-dimensional sphere such that the corresponding geodesic flow admits homogeneous integrals of third and fourth degree that cannot be reduced to linear or quadratic integrals. The reader is challenged to find an example. This is an interesting and enlightening problem. Let us consider the Hamiltonian function on T^*S^n $(n \geqslant 1)$ of the form $T + V$, where T is the standard Riemannian metric on the standard n-dimensional sphere and V is an analytic function on S^n.

d. *Is it possible for this system to possess a complete "irreducible" set of integrals of arbitrary high order in momentum*?

Probably, as a first step, the case when S^n is the standard sphere in $\mathbb{R}^{n+1}$ should be considered. Note that in the integrable Neumann problem (V is quadratic on $\mathbb{R}^n$), all integrals are quadratic in velocity.

3. Integrals are the simplest tensor invariants (type $(0, 0)$). The next class (in simplicity) of tensor invariants are invariants of type $(1, 0)$, i.e., symmetry fields. A *symmetry field* is a vector field u commuting with the field v that defines the dynamical system.

For Hamiltonian systems the problem of symmetry fields includes the problem of integrals. The reason is that to any function F on the phase space there corresponds a Hamiltonian vector field v_F. If F is an integral, then $u = v_F$ is a symmetry field.

In the paper [8] it is proved that a reversible analytic system with a surface of genus greater than one as the configuration space admits no nontrivial symmetry fields: any symmetry field is of the form $u = \lambda v_H$, where λ is an analytic function of the Hamiltonian H. It follows that there are no analytic integrals independent of H (cf. §3). The field λv_H is Hamiltonian: the Hamiltonian function is

$$\int \lambda(H)\, dH.$$

a. *Is it true that if there exists a nontrivial symmetry field, then Hamiltonian equations admit an integral independent of H*?

Since a reversible system is homogeneous, we should look for symmetry fields of the form

$$L_u = Q_1 \frac{\partial}{\partial q_1} + Q_2 \frac{\partial}{\partial q_2} + P_1 \frac{\partial}{\partial p_1} + P_2 \frac{\partial}{\partial p_2},$$

where Q_s (P_s) are homogeneous polynomials in p_1, p_2 of degree $k-1$ (respectively k). It is natural to say that a field is homogeneous of degree k if its Lie operator is of the above form (if F is a homogeneous polynomial in impulses of degree k, then the field v_F is homogeneous of degree k).

Recently, Denisova obtained a positive answer to problem **a** in the case $k \leqslant 2$.

b. *Investigate problem* **a**, *replacing the two-dimensional torus with the two-dimensional sphere.*

c. *Find conditions for the existence of nontrivial symmetry fields for nonreversible systems (with gyroscopic forces) with two degrees of freedom.*

Topological restrictions for the existence of first integrals of nonreversible systems were obtained by Bolotin [9].

4. Every Hamiltonian system with two degrees of freedom has the tensor invariants H, ω, $\omega^2 = \omega \wedge \omega$. Here H is the Hamiltonian function and $\omega = dp_1 \wedge dq_1 + dp_2 \wedge dq_2$ is the symplectic form. The invariance of the form ω^2 is equivalent to the Liouville theorem on the conservation of phase volume. The tensor fields (8) are of the type $(0, 0)$, $(0, 2)$, and $(0, 4)$ respectively.

Describe all analytic tensor invariants of geodesic flows on surfaces with negative Euler characteristics.

It is possible that there are no other nontrivial invariants.

5. In applications we often encounter differential equations with quadratic right-hand side:

$$\dot{x}_i = v_i(x), \quad 1 \leqslant i \leqslant n, \qquad v(\lambda x) = \lambda^2 v(x). \tag{9}$$

The most important example is the *Euler–Poincaré equations* on Lie algebras:

$$\dot{m}_k = \sum c^j_{ki} m_j \omega_i, \qquad m_s = \sum I_{sp}\omega_p. \tag{10}$$

Here $\omega = (\omega_1, \dots, \omega_n)$ is the velocity of the the system, $m = (m_1, \dots, m_n)$ is the momentum, c^j_{ki} are the structure constants of the algebra, and $\|I_{sp}\|$ is the constant inertial tensor. If we substitute the expressions for m_i in terms of ω_i into equations (10), we obtain a system of equations on the Lie algebra. The inverse transformation yields a dynamical system on the dual space. In both cases we get equations of type (9).

Kovalevskaya and Lyapunov introduced a method yielding necessary conditions for solutions to be univalent or meromorphic on the complex time plane. We shall describe it briefly for systems of type (9). The first step is to find a solution of the form

$$x = c/t. \tag{11}$$

The complex vector c satisfies the algebraic equation $v(c) = -c$. Let us write out the first variation equation for the solution (11):

$$\dot{\xi} = t^{-1}A\xi, \qquad A = (\partial v/\partial x)(c).$$

This is a Fuchsian system. It has particular solutions of the form $\xi = \varphi t^{\rho-1}$, where ρ is an eigenvalue and φ an eigenvector of the matrix $K = A + E$. This matrix is called *Kovalevskaya matrix*, and its eigenvalues are *Kovalevskaya exponents*. It can be proved that if the general solution of system (9) is represented by univalent (respectively meromorphic) functions of complex time, then the Kovalevskaya exponents are integers (respectively nonnegative integers except one, which is equal to -1).

Ioshida proved in 1983 that if $f(x)$ is a homogeneous integral of the system (9) and $df(c) \neq 0$, then $\rho = m$ is a Kovalevskaya exponent. This result establishes a remarkable connection between the univalence property of the general solution and existence of nonconstant integrals. An extension of Ioshida's theorem to the case of tensor invariants with homogeneous components appears in [10].

a. *Apply the Kovalevskaya–Lyapunov method to the Euler–Poincaré equations.*

Conditions for the general solution to be univalent or meromorphic will include restrictions on the structure of the Lie algebra and on the inertia tensor. For certain Lie algebras (for example, $so(4)$) this method was applied in the work of Adler and van Moerbeke (see, for example, [11]). It is possible to consider other simplified versions of problem **a**. For example, one can try to find all algebras such that solutions of the Euler–Poincaré equations are univalent for any choice of the inertia tensor (one of these algebras is $so(3)$).

b. *Apply the Kovalevskaya–Lyapunov method to the Euler–Poincaré–Suslov equations.*

The EPS-equations appear, for example, in the article by Fedorov and the author in this volume. They also are of the form (9).

Now let us discuss some variational problems. The first belongs to the realm of Morse theory.

6. Let $M = N \times S^1$ be a Riemannian manifold diffeomorphic to the Cartesian product of a compact manifold N and the circle S^1. We look for closed geodesics that are homotopic to the curves $n \times S$, where n is a point in N.

Find a lower bound for the number of these geodesics in terms of topological invariants of the surface N.

Probably this lower bound is at least $\operatorname{cat}(N)$. The *category* of the manifold N is defined as the smallest number of closed subsets of N retractable to a point, that cover N. For example, $\operatorname{cat}\mathbb{T}^k = k+1$. The category of N provides a lower bound for the number of critical points of a smooth function on N.

An example from dynamics is provided by the problem of periodic oscillations of an n-link pendulum. The configuration space M is the n-dimensional torus $\mathbb{T}^n$. According to the Maupertuis principle, for fixed energy h greater then the maximum of the potential energy (when all the rods are pulled up), trajectories of the pendulum are geodesics in M. In this case, for a given h there exist at least n different periodic motions of energy h in any homotopy class. This means that each link makes the prescribed number of rotations in a period.

7. Let M be the configuration space, T the kinetic energy, $V\colon M \to \mathbb{R}$ the potential energy. The equations of motion have the energy integral $T + V = h$. Since $T \geqslant 0$, trajectories with energy h are contained in the *region of possible motion*

$$\Sigma_h = \{x \in M : V(x) \leqslant h\}.$$

We assume that Σ is compact, its boundary $\partial\Sigma$ is nonempty, and $\partial\Sigma$ contains no positions of equilibrium. The latter is equivalent to the assumption that h is not a critical value of the function V.

According to the Maupertuis principle, trajectories of energy h contained in the interior of Σ are geodesics in the Jacobi metric $(h - V)T$. This metric is degenerate on the boundary $\partial\Sigma$.

There are two types of periodic orbits with energy h: *rotations* and *librations*. Trajectories of rotations have no points on the boundary. For the trajectory of a libration there are exactly two such points. The velocity periodically become zero (as for librations of the ordinary pendulum).

Bolotin proved that if these assumptions are satisfied, there exists at least one libration [12]. Much earlier (1948) this result was proved by the well-known topologist Herbert Seifert for the case when Σ is diffeomorphic to the ball. He stated the following conjecture: *if Σ is diffeomorphic to the n-dimensional ball, then there exist at least n different librations.* This conjecture is not proved yet. It is easy to give an example where there exist exactly n librations (try!).

Let $\Sigma = I \times N$, where I is the segment $[0, 1]$ and N is a closed manifold.

The problem is to find a lower estimate for the number of librations in terms of topological invariants of the surface N. Is it true that in this case there are at least two librations?

For a start, one may consider the simplest case of the two-dimensional annulus, when $N = S^1$. In [13] the number of librations was estimated for the not simply connected case by the rank of the fundamental group of the space $\Sigma/\partial\Sigma$.

8. In a paper written when he was still a student, Chaplygin (1890) studied the motion of a heavy plate in a boundless ideal fluid. He obtained the elegant equation

$$\ddot{x} = t^2 \sin x . \tag{12}$$

The coordinate x is the double rotation angle of the plate and t is a parameter proportional to physical time. Later we shall discuss some properties of generic solutions of equation (12). Here we consider special doubly asymptotic solutions $x(t)$ that tend to the unstable equilibrium $x = 0 \mod 2\pi$ as $t \to \pm\infty$.

It is relatively easy to prove the existence of a doubly asymptotic solution $x(t)$ that performs exactly one full rotation:

$$\lim_{t \to +\infty} x(t) - \lim_{t \to -\infty} x(t) = 2\pi . \tag{13}$$

This result was obtained in [14] by means of the Hamilton variational principle. Bolotin showed that it is possible to modify the proof in [14] and to prove the existence of a doubly asymptotic solution such that the difference (13) is equal to $2\pi n$ with arbitrary integer n.

It is possible to generalize equation (12). Let M be a compact configuration space, T the kinetic energy, and $p(t)V$ the potential energy. Here V is a smooth function on M that has a strict nondegenerate maximum at $a \in M$ and $p(\cdot)$ is a nonnegative function of time. Suppose that $p(t)$ is monotone for $|t| >$ const and tends to infinity as $t \to \pm\infty$. In the local coordinates $x_1, \dots, x_n$ the motion is described by the Lagrange equations

$$\frac{d}{dt}\frac{\partial T}{\partial \dot{x}_i} - \frac{\partial T}{\partial x_i} = -p(t)\frac{\partial V}{\partial x_i}, \qquad 1 \leqslant i \leqslant n .$$

It is clear that $x = a$ is an unstable equilibrium.

The question is, *does there exist a nontrivial solution $x(t)$, such that $x(t) \to a$ as $t \to \pm\infty$?* Probably the number of these solutions is always infinite.

Note that for the ordinary pendulum (when there is no multiplier t^2 in (13)), there exist only two different doubly asymptotic trajectories (they correspond to rotation numbers $n = \pm 1$).

9. Now let us discuss some questions connected with the existence of periodic orbits for nonreversible systems. Again let M be the configuration space, ω a closed 2-form on M, and V the potential energy. In the nonreversible case, the equations of motion are:

$$\frac{d}{dt}\frac{\partial T}{\partial \dot{x}} - \frac{\partial T}{\partial x} = \omega(\dot{x}\,,\,\cdot\,) - \frac{\partial V}{\partial x}\,. \tag{14}$$

It is natural to interpret the term $\omega(\dot{x}\,,\,\cdot\,)$ as an additional force acting on the mechanical system. Since it does not prevent the conservation of full energy $T + V = h$, it is usually called a *gyroscopic force*. Hence the 2-form ω is said to be the form of gyroscopic forces. The nature of these forces can be quite diverse. For example, they appear when we use a rotating frame of reference, after the reduction of systems with symmetries, or when we study the motion of a charged particle in a magnetic field. The assumption that the form ω is closed implies that locally solutions of equation (14) are extremals of the variational problem with the Lagrangian $L = T - \varphi - V$, $d\varphi = \omega$.

Suppose that M is compact and $h > \max_M V$. Then the velocity of the system is never zero.

The question is: *does there exist a periodic motion with fixed energy* $h > \max V$?

S. P. Novikov formulated sufficient conditions. For example, it is sufficient to assume that M is simply connected and $H^2(M) \neq 0$ (see the survey [15], where there are references to preceding publications). The case of two-dimensional sphere $M = S^2$, important for applications, is included here.

However, full and rigorous proofs of these results are unknown as of now. In the recent paper [16], Taĭmanov proved the existence of a closed orbit on the two-dimensional sphere under the additional assumption that the isoenergetic action functional

$$F^* = \int_{\partial\Pi} \sqrt{2(h-V)T} - \iint_{\Pi} \omega$$

assumes a negative value on some two-dimensional surface Π (with boundary), imbedded into M. The orientation of Π is induced by the orientation of M. This assumption is satisfied if the 2-form ω changes sign somewhere on the sphere. An analogous result is announced in [16] for all two-dimensional closed surfaces M. Earlier in [17] the opposite case was considered, when the form of gyroscopic forces does not vanish anywhere on the two-dimensional torus. This case is one of the most interesting for physical applications. Using the generalization of the last geometric theorem of Poincaré suggested by Arnold, the existence of three different closed orbits for the inertial motion of a particle on a flat torus (the curvature of the Riemannian metric is zero), was proved [17].

a. *It is interesting to extend this method to the case of an arbitrary metric on the two-dimensional torus.*

Suppose now that $h < \max_M V$ and h is a regular value of the potential energy.

b. *Does a periodic trajectory with a given energy h always exist*?

This problem seems more difficult then the problem considered by Novikov.

10. Once again consider the motion on a compact manifold M under the action of a potential force field. Let a, b be different points of M.

The question is, *does there exist an orbit with energy h, joining these points*?

If $h > \max_M V$, the positive answer is well known (it follows from the Maupertuis principle and the Hopf–Rinov theorem in Riemannian geometry). If $h < \max_M V$, for arbitrary points in the connected region of possible motion

$$\Sigma_h = \{x \in M : V(x) \leqslant h\},$$

the answer is negative (give an example!).

Suppose that V has no critical points on the boundary of Σ_h. Then it is easy to show that any two points $a, b \in \Sigma$ can be joined by a piecewise smooth broken trajectory (each link is a trajectory of energy h). In other words, it is possible to get from the point a to the point b by applying a finite number of isoenergetic impulses. Hence our problem is connected with control theory.

Let $k(a, b)$ be the least possible number of impulses needed for getting from the position a to the position b. By reversibility, $k(a, b) = k(b, a)$. It turns out that $K = \max_{a, b \in \Sigma} k(a, b)$ is finite. This is an important geometric characteristic of the system. Of course, it depends on the energy h.

We have the *maximum principle*:

$$K = \max_{a, b \in \partial\Sigma} k(a, b).$$

This a simple corollary of the *boundary hit theorem* (see [17]): for every point $a \in \Sigma$ there exists a motion $x(t)$, $0 \leqslant t \leqslant \tau$, such that $x(0) \in \partial\Sigma$ and $x(\tau) = a$.

a. *Obtain upper and lower estimates for K. Is it true that always $K \leqslant 2$?*

Let us make the problem more complicated by adding a linear nonintegrable constraint

$$(c(x), \dot{x}) = 0, \qquad c \neq 0. \tag{15}$$

The equations of motion are replaced by the nonholonomic equations with a multiplier λ:

$$\frac{d}{dt}\frac{\partial T}{\partial \dot{x}} - \frac{\partial T}{\partial x} = -\frac{\partial V}{\partial x} + \lambda c. \tag{16}$$

Equations (15), (16) form a complete system. Once again the energy $T + V$ is constant on every solution.

b. *Is the boundary hit theorem true in the case of a nonintegrable constraint* (15)?

For the nonholonomic problem it is also possible to introduce the number K. Once again it is finite. This fact is less evident than in the holonomic case. The proof uses the Rashevskiĭ–Chow theorem (every two points of a connected manifold M can be joined by an admissible curve satisfying equation (15)). The reader can try to prove that K is finite himself.

c. *Is the maximum principle true in the nonholonomic case?*

Note that the Hopf–Rinov theorem does not hold in the presence of a nonintegrable constraint (try to find an example). The reason is that solutions of the nonholonomic equations in general cannot be described as extremals of any smooth functional.

Now let us discuss some problems of celestial mechanics.

11. It is well known that equations of the famous restricted circular three-body problem in suitable units for length, time, and mass, are as follows

$$
\begin{aligned}
&\ddot{x} = 2\dot{y} + \partial W/\partial x\,, \qquad \ddot{y} = -2\dot{x} + \partial W/\partial y\,,\\
&W = (x^2 + y^2)/2 + (1-\mu)/\rho_1 + \mu/\rho_2\,,\\
&\rho_1^2 = (x+\mu)^2 + y^2\,, \qquad \rho_1^2 = (x+\mu)^2 + y^2.
\end{aligned} \tag{17}
$$

Here x, y are coordinates of the asteroid in the rotating coordinate system, and μ is the ratio of the masses of Jupiter and the Sun ($0 \leqslant \mu \leqslant 1/2$). Equations (17) have the Jacobi integral

$$(\dot{x}^2 + \dot{y}^2)/2 - W = h = \text{const}\,.$$

For a fixed value of h, motion takes place in the region of admissible motions

$$\Sigma_h = \{x\,, y : -W \leqslant h\}\,,$$

called the *Hill region* in celestial mechanics. Its geometry is well known from analytic and numerical investigations. Since equations (17) are nonreversible, the boundary hit theorem is no longer true (see §11). Define the set K_h as the closure of the union of trajectories that start at the boundary of the Hill region.

a. *It is interesting to study numerically the structure of the set K_h for different values of h.*

The boundary of K_h includes envelopes for the family of trajectories starting on $\partial\Sigma_h$. For integrable nonreversible systems K usually differs from Σ. Since the three-body problem is nonintegrable, it is natural to ask the following question.

b. *Is it possible that for the restricted three-body problem we have $K_h = \Sigma_h$ for some h?*

12. Consider another variant of the spatial three-body problem, in which two points of equal masses move in the x, y plane along elliptic orbits that are symmetric with respect to z-axis while the third point of zero mass lies always on this axis. The motion of the third point is governed by the differential equation $\ddot{z} = -z[z^2 + r^2(t)]^{-3/2}$, where

$$r(t) = (1 + e\cos\varphi(t))^{-1}\,, \qquad \dot{\varphi} = (1 + e\cos\varphi)^2\,, \quad \varphi(0) = 0\,.$$

This problem was suggested by Kolmogorov in order to verify the Chazy conjecture on the existence of oscillating motions in the three-body problem. Alekseev [18] established the quasirandom character of oscillations for equation (18) if the amplitude is sufficiently large. In particular, there exist infinitely many long-periodic unstable motions.

If the eccentricity e of elliptic orbits of massive bodies is zero, equation (18) is autonomous and hence integrable. For negative values of energy the phase trajectories are closed curves. Thus it is possible to introduce the action-angle variables and for small e consider the Kolmogorov problem as a perturbation of a completely integrable system.

The perturbed system is nonintegrable and satisfies the assumptions of the well-known *Poincaré theorem*, i.e., the criterion for the existence of nondegenerate long-periodic orbits. For each sufficiently large integer n and small e equation (18) has a pair of nondegenerate periodic solutions. One of them is elliptic (the first variation equation is stable) and the other hyperbolic (unstable).

Poincaré periodic solutions depend on two parameters: continuous e and discrete n. The question arises about the behavior of these solutions when we increase e.

Is it true that if e increases up to the limit $\sim 1/n$, the multiplicators λ, λ^{-1} of the Poincaré periodic solution, starting from the point $\lambda = \lambda^{-1} = 1$ for $e = 0$, in the hyperbolic case move in the opposite directions along the real axes, and in the elliptic case revolve along the unit circle on the complex plane until the collision at the point $\lambda = \lambda^{-1} = -1$, and then move in the opposite directions along the negative real half-axes?

This conjecture is based on the result of Dovbysh [19] about the behavior of Poincaré periodic solutions near the split separatrices when the small parameter is increased. If the answer is positive, it will provide a connection between the Poincaré and Alekseev periodic solutions.

13. The potential of the gravitational interaction has two fundamental properties. On one hand, it is a harmonic function on the three-dimensional space (i.e., satisfies the Laplace equation), and on the other hand only this potential (and the potential of an elastic spring) generates the central field where all bounded orbits are closed (Bertrand theorem). It appears that in the more general situation of motion in a constant curvature space these properties remain true [20].

As a matter of convenience, we shall consider motion on the three-dimensional sphere with unit radius. Let a particle m of unit mass move in the force field with potential V depending only on the distance between the particle and some fixed point M. Let ϑ be the length of the arc of the great circle connecting m and M (ϑ is measured in radians). Then the function V depends only on ϑ. The Laplace equation is replaced by the Laplace–Beltrami equation:

$$\Delta V = \sin^{-2}\vartheta \frac{\partial}{\partial\vartheta}\left(\sin^2\vartheta \frac{\partial V}{\partial\vartheta}\right) = 0 .$$

Its solution is

$$V = -\gamma \tan^{-1}\vartheta + \alpha , \qquad \alpha , \gamma = \text{const} . \tag{19}$$

The constant α is irrelevant. The parameter γ plays the role of the gravitational constant.

The dual potential is

$$V = (k/2)\tan^2\vartheta , \qquad k = \text{const} . \tag{20}$$

This is the analog of the potential of an elastic string. It appears that all orbits in central fields with the potentials (19) and (20) are closed.

This discussion leads to the natural generalization of the n-body problem: n particles move in the three-dimensional constant curvature space and their interaction is governed by the potential (19). The two-body problem is especially interesting. Contrary to the plane case, it cannot be reduced to the generalized Kepler problem.

a. *Are the bounded orbits of the generalized two-body problem closed*?

b. *Is the analog of the Sundman theorem (on expansion of solutions in convergent series for all t) true for the generalized three-body problem?*

The main difficulty is to exclude triple collisions. In the flat case it is sufficient to assume that the angular momentum of gravitating particles with respect to their center of mass does not vanish.

Concluding the discussion of this topic, consider a Hamiltonian system with two degrees of freedom and the following Hamilton function:

$$H = \frac{\Lambda(q_1, q_2)}{2}(p_1^2 + p_2^2) - \frac{f(q_1, q_2)}{\sqrt{q_1^2 + q_2^2}}. \tag{21}$$

Here Λ and f are positive analytic functions. We already mentioned that locally the kinetic energy always can be transformed to the given form. Coordinates q_1, q_2 are isotermic coordinates. The potential in the Hamiltonian (21) is said to be a *Newtonian-type* potential.

c. *Find all functions Λ and f such that all bounded orbits of the system with the Hamiltonian function* (21) *are closed.*

This is a generalization of the Bertrand problem. *Are there any solutions except the constant curvature metric and the potential of type* (19)?

In conclusion let us formulate two separate problems.

14. Let us return to the nonlinear Chaplygin equation (12). In [14] it is proved that for almost all initial conditions the solutions tend to the stable equilibrium (to the point $x = \pi \mod 2\pi$) as $t \to \infty$. These solutions have the asymptotics

(22)
$$x(t) = \pi + 2\pi n_+ + \frac{a_+}{\sqrt{t}} \sin\frac{t^2}{2} + \frac{b_+}{\sqrt{t}} \cos\frac{t^2}{2} + O\left(\frac{1}{\sqrt{t}}\right), \qquad n_+ \in \mathbb{Z},\ a_+, b_+ \in \mathbb{R}.$$

a. *Is it true that the numbers n_+, a_+, b_+, define the solution of the Chaplygin equation?*

If we replace t by $-t$, the equation (12) does not change. Hence for almost all solutions, we have an asymptotic representation of type (22) as $t \to -\infty$ (the numbers n_+, a_+, b_+ must be replaced by n_-, a_-, b_-).

b. *Study the properties of the correspondence $S_n: (a_-, b_-) \to (a_+, b_+)$.*

This nonlinear scattering problem depends on the discrete parameter $n = n_+ - n_-$, which is the number of half-rotations of the falling plate while t changes from $-\infty$ to $+\infty$.

The map S is not determined for some solutions. Among them are the doubly asymptotic solutions considered in §9. The simplest is the solution $x_a(t)$ satisfying the condition

$$x_a(t) \to \begin{cases} 0 & \text{as } t \to -\infty, \\ 2\pi & \text{as } t \to +\infty. \end{cases}$$

Asymptotically, as $t \to -\infty$ and $t \to +\infty$, a certain point of the plate descends along vertical lines. The distance between these lines is given, up to a constant coefficient, by the integral

$$I = \int_{-\infty}^{+\infty} t \sin x_a(t)\, dt.$$

It is easy to see that the integral is convergent.

c. *Is it possible to express I in terms of known mathematical constants?*

This problem can be generalized to the case when the solution $x_a(t)$ is replaced by one of the infinite sequence of solutions considered in §9.

15. Let us recall that a dynamical system

$$\dot{x} = v(x) \tag{23}$$

is called *Hamiltonian* if there exist a closed nondegenerate 2-form ω and a function $H(x)$ such that $\omega(v, \cdot) = -dH$. In local *canonical* coordinates p, q such that $\omega = \sum dp \wedge dq$, equation (23) takes the usual form

$$\dot{p} = -\partial F/\partial q, \qquad \dot{q} = \partial F/\partial p. \tag{24}$$

Here F is the function H expressed in the coordinates p, q. The form ω is called the *symplectic structure*, and H is the Hamilton function.

It follows that in order to find out whether or not a given dynamical system is Hamiltonian, we need to search for two objects: the symplectic structure and the Hamiltonian. If the system does not have the form (24), this does not mean that it is not Hamiltonian: it might just be represented in noncanonical variables.

Let us give a simple example demonstrating a hidden Hamiltonian structure. Consider a linear system with constant coefficients

$$\dot{x} = Ax, \tag{25}$$

admitting a quadratic integral $f = (Bx, x)/2$. It turns out that if the matrices A and B are nondegenerate, the system (25) is Hamiltonian. The Hamiltonian function is the integral f. The reader can verify this by producing a suitable symplectic structure.

The problem of recognizing the Hamiltonian nature of a dynamical system is a difficult and, probably, unsolvable problem. It makes sense to consider it for dynamical systems from particular classes. One of the approaches is to consider systems that are close to completely integrable systems:

$$\dot{I}_i = \varepsilon F_i(I, \varphi) + \dots, \qquad \dot{\varphi}_j = \omega_j(I) + \varepsilon G(I, \varphi) + \dots, \tag{26}$$

$1 \leqslant i, j \leqslant n$. Here $I = (I_1, \dots, I_n)$ are the slow variables and $\varphi = (\varphi_1, \dots, \varphi_n)$ are the fast angle variables. The functions $F_i, G_j, \dots$ are 2π-periodic in $\varphi_1, \dots, \varphi_n$. Dots mean terms of order ε^2, where ε is a small parameter. Equations like (26) are often encountered in applications. Of course, it also makes sense to study systems with different numbers of fast and slow variables.

Let us consider the most important case when, for $\varepsilon = 0$, system (26) is nondegenerate:

$$\frac{\partial(\omega_1, \dots, \omega_n)}{\partial(I_1, \dots, I_n)} \neq 0.$$

When it is possible to find ω and H in the form of series in ε

$$\omega_\varepsilon = \omega_0 + \varepsilon\omega_1 + \dots, \qquad H_\varepsilon = H_0 + \varepsilon H_1 + \dots,$$

with 2π-periodic in φ coefficients, such that the Hamiltonian condition $\omega_\varepsilon(v_\varepsilon, \cdot) = -dH_\varepsilon$ *is satisfied*? Here v_ε is the vector field defined by system (26).

The idea of this problem goes back to Poincaré who was the first to consider the problem of existence of "univalued" (periodic in the angles φ) integrals for Hamiltonian equations of type (26), which are represented in the form of power series in ε. This problem is closely related to the so called *small denominators problem*, first encountered in celestial mechanics. Related problems of the existence of integral invariants and symmetry fields in the form of power series in ε for systems of type (26) were considered in [21] and [22]. In the symmetry fields problem, small denominators also play the central role.

References

1. V. V. Kozlov, *Conjecture on the existence of asymptotic motions in classical mechanics*, Funktsional. Anal. i Prilozhen. **16** (1992), no. 4, 72–73; English transl. in Functional Anal. Appl. **16** (1992), no. 4.
2. ———, *Asymptotic motions and the inversion problem of the Lagrange–Dirichlet theorem*, Prikl. Mat. Mekh. **50** (1986), no. 6, 928–937; English transl. in J. Appl. Math. Mech. **50** (1986).
3. A. N. Kuznetsov, *On the existence of solutions entering a critical point of an autonomous system possessing a formal solution*, Funktsional. Anal. i Prilozhen. **23** (1989), no. 4, 63–74; English transl. in Functional Anal. Appl. **23** (1989), no. 4.
4. I. A. Taĭmanov, *On topological properties of integrable geodesic flows*, Mat. Zametki **44** (1988), no. 2, 283–284; English transl. in Math. Notes **44** (1988).
5. V. V. Kozlov, *Integrable and nonintegrable Hamiltonian systems*, Soviet Sci. Rev. Sect. C: Math. Phys. Rev., vol. 8, Part 1, Harood Academic, Chur, 1989, pp. 1–81.
6. V. V. Kozlov and D. V. Treshchev, *The integrability of Hamiltonian systems with configuration torus*, Mat. Sb. **135** (1988), no. 1, 119–138; English transl. in Math. USSR-Sb. **63** (1989).
7. M. L. Bialy, *On polynomial in integrals first integrals of a mechanical system on the two-dimensional torus*, Funktsional. Anal. i Prilozhen. **21** (1987), no. 4, 64–65; English transl. in Functional Anal. Appl. **21** (1987), no. 4.
8. V. V. Kozlov, *On symmetry groups of geodesic flows on closed surfaces*, Mat. Zametki **48** (1990), no. 5, 62–67; English transl. in Math. Notes **48** (1990).
9. S. V. Bolotin, *On first integrals of systems with gyroscopic forces*, Vestnik Moskov. Univ. Ser. I Matem. Mekh. **1984**, no. 6, 75–82; English transl. in Moscow Univ. Math. Bull. **39** (1984).
10. V. V. Kozlov, *Tensor invariants of quasihomogeneous systems of differential equations and the asymptotic method of Kovalevskaya–Lyapunov*, Mat. Zametki **51** (1992), no. 2, 46–52; English transl. in Math. Notes **51** (1992).
11. M. Adler and P. van Moerbeke, *Kowalewski's asymptotic method, Kac-Moody Lie algebras and regularization*, Comm. Math. Phys. **83** (1982), 83–106.
12. S. V. Bolotin, *Libration motions of natural dynamical systems*, Vestnik Moskov. Univ. Ser. I Matem. Mekh. **1978**, no. 6, 72–77; English transl. in Moscow Univ. Math. Bull. **32** (1978).
13. S. V. Bolotin and V. V. Kozlov, *Libration in systems with many degrees of freedom*, Prikl. Mat. Mekh. **42** (1978), no. 2, 245–250; English transl. in J. Appl. Math. Mech. **42** (1978).
14. V. V. Kozlov, *On the fall of a heavy rigid body in an ideal fluid*, Mekh. Tverd. Tela **5** (1989), 10–17. (Russian)
15. S. P. Novikov, *Hamiltonian formalism and a multivalued analog of Morse theory*, Uspekhi Mat. Nauk **37** (1982), no. 5, 3–49; English transl. in Russian Math. Surveys **37** (1982).
16. I. A. Taĭmanov, *Nonselfintersecting closed extremals of multivalued or non everywhere positive functionals*, Izv. Akad. Nauk SSSR Ser. Mat. **55** (1991), no. 2, 367–383; English transl. in Math. USSR-Izv. **38** (1992).
17. V. V. Kozlov, *Calculus of variations in the large and classical mechanics*, Uspekhi Mat. Nauk **40** (1985), no. 2, 33–60; English transl. in Russian Math. Surveys **40** (1985).
18. V. M. Alekseev, *Quasirandom dynamical systems*. I, Mat. Sb. **76** (1968), no. 1, 72–134 II, Mat. Sb. **77** (1968), no. 4, 545–601; III, Mat. Sb. **78** (1969), no. 1, 3–50; English transl. in Math. USSR-Sb **5** (1968); **6** (1968); **7** (1969).

19. S. A. Dovbysh, *Splitting of separatrices and creating of isolated periodic solutions in Hamiltonian systems with one and a half degrees of freedom*, Uspekhi Mat. Nauk **44** (1989), no. 2, 229–230; English transl. in Russian Math. Surveys **44** (1989).
20. V. V. Kozlov and A. Kharin, *Kepler's problem in constant curvature spaces*, Celestial Mech. (1992).
21. V. V. Kozlov, *On the existence of integral invariants for smooth dynamical systems*, Prikl. Mat. Mekh. **51** (1987), no. 4, 538–545; English transl. in J. Appl. Math. Mech. **51** (1987).
22. V. V. Kozlov, *On symmetry groups of dynamical systems*, Prikl. Mat. Mekh. **52** (1988), no. 4, 531–541; English transl. in J. Appl. Math. Mech. **52** (1988).

Translated by S. BOLOTIN

DEPARTMENT OF MATHEMATICS, MOSCOW STATE UNIVERSITY, MOSCOW 119899, RUSSIA

Other Titles in This Series

(*Continued from the front of this publication*)

129 **S. Ya. Khavinson,** Two Papers on Extremal Problems in Complex Analysis
128 **I. K. Zhuk et al.,** Thirteen Papers in Algebra and Number Theory
127 **P. L. Shabalin et al.,** Eleven Papers in Analysis
126 **S. A. Akhmedov et al.,** Eleven Papers on Differential Equations
125 **D. V. Anosov et al.,** Seven Papers in Applied Mathematics
124 **B. P. Allakhverdiev et al.,** Fifteen Papers on Functional Analysis
123 **V. G. Maz′ya et al.,** Elliptic Boundary Value Problems
122 **N. U. Arakelyan et al.,** Ten Papers on Complex Analysis
121 **V. D. Mazurov, Yu. I. Merzlyakov, and V. A. Churkin, Editors,** The Kourovka Notebook: Unsolved Problems in Group Theory
120 **M. G. Kreĭn and V. A. Jakubovič,** Four Papers on Ordinary Differential Equations
119 **V. A. Dem′janenko et al.,** Twelve Papers in Algebra
118 **Ju. V. Egorov et al.,** Sixteen Papers on Differential Equations
117 **S. V. Bočkarev et al.,** Eight Lectures Delivered at the International Congress of Mathematicians in Helsinki, 1978
116 **A. G. Kušnirenko, A. B. Katok, and V. M. Alekseev,** Three Papers on Dynamical Systems
115 **I. S. Belov et al.,** Twelve Papers in Analysis
114 **M. Š. Birman and M. Z. Solomjak,** Quantitative Analysis in Sobolev Imbedding Theorems and Applications to Spectral Theory
113 **A. F. Lavrik et al.,** Twelve Papers in Logic and Algebra
112 **D. A. Gudkov and G. A. Utkin,** Nine Papers on Hilbert's 16th Problem
111 **V. M. Adamjan et al.,** Nine Papers on Analysis
110 **M. S. Budjanu et al.,** Nine Papers on Analysis
109 **D. V. Anosov et al.,** Twenty Lectures Delivered at the International Congress of Mathematicians in Vancouver, 1974
108 **Ja. L. Geronimus and Gábor Szegő,** Two Papers on Special Functions
107 **A. P. Mišina and L. A. Skornjakov,** Abelian Groups and Modules
106 **M. Ja. Antonovskiĭ, V. G. Boltjanskiĭ, and T. A. Sarymsakov,** Topological Semifields and Their Applications to General Topology
105 **R. A. Aleksandrjan et al.,** Partial Differential Equations, Proceedings of a Symposium Dedicated to Academician S. L. Sobolev
104 **L. V. Ahlfors et al.,** Some Problems on Mathematics and Mechanics, On the Occasion of the Seventieth Birthday of Academician M. A. Lavrent′ev
103 **M. S. Brodskiĭ et al.,** Nine Papers in Analysis
102 **M. S. Budjanu et al.,** Ten Papers in Analysis
101 **B. M. Levitan, V. A. Marčenko, and B. L. Roždestvenskiĭ,** Six Papers in Analysis
100 **G. S. Ceĭtin et al.,** Fourteen Papers on Logic, Geometry, Topology and Algebra
99 **G. S. Ceĭtin et al.,** Five Papers on Logic and Foundations
98 **G. S. Ceĭtin et al.,** Five Papers on Logic and Foundations
97 **B. M. Budak et al.,** Eleven Papers on Logic, Algebra, Analysis and Topology
96 **N. D. Filippov et al.,** Ten Papers on Algebra and Functional Analysis
95 **V. M. Adamjan et al.,** Eleven Papers in Analysis
94 **V. A. Baranskiĭ et al.,** Sixteen Papers on Logic and Algebra
93 **Ju. M. Berezanskiĭ et al.,** Nine Papers on Functional Analysis
92 **A. M. Ančikov et al.,** Seventeen Papers on Topology and Differential Geometry
91 **L. I. Barklon et al.,** Eighteen Papers on Analysis and Quantum Mechanics
90 **Z. S. Agranovič et al.,** Thirteen Papers on Functional Analysis
89 **V. M. Alekseev et al.,** Thirteen Papers on Differential Equations

(See the AMS catalog for earlier titles)